T0327358

A Guide to Virology for Engineers and Applied Scientists

A Guide to Virology for Engineers and Applied Scientists

Epidemiology, Emergency Management, and Optimization

Megan M. Reynolds and Louis Theodore

Registered Office
John Wiley & Sons, Inc., 111 River Street, Hoboken, NJ 07030, USA

Editorial Office
111 River Street, Hoboken, NJ 07030, USA

For details of our global editorial offices, customer services, and more information about Wiley products visit us at www.wiley.com.

Wiley also publishes its books in a variety of electronic formats and by print-on-demand. Some content that appears in standard print versions of this book may not be available in other formats.

Library of Congress Cataloging-in-Publication Data applied for:
Hardback ISBN: 9781119853138

Cover Design: Wiley
Cover Image: © Yuichiro Chino/Getty Images

Set in 9.5/12.5pt STIXTwoText by Straive, Chennai, India

TO
My father, Dr. Joseph P. Reynolds (MMR)

AND

My friends and colleagues working in the field of virology (LT)

"When an epidemic of physical disease starts to spread, the community approves and joins in a quarantine of the patients in order to protect the health of the community against the spread of the disease…"

Franklin Delano Roosevelt (1882–1945)

Contents

Preface

As its title implies, this book offers a guide to virology which provides information on viruses from an engineer's and applied scientist's perspective. Concise and easy to use, this guide brings together a wealth of general information on viruses in one compact book. It additionally offers practical and technical information plus calculation details.

The guide has been written not only for students but also for those in technical roles, such as engineers and applied scientists who work in public health, pharmaceuticals, or other health-related fields. It is a tool that may be used whenever and wherever information about viruses is likely to be sought.

In the wake of the COVID-19 pandemic, it has become evident that knowledge of virology is no longer critical only to doctors or epidemiologists; there is an urgent need for cooperation among varied disciplines to address the current pandemic and prepare for the next one. The authors feel that no one source currently covers all of the information on viruses in the manner presented in this book. It is hoped that this book will serve to fill the growing need for concise and digestible information – both academically and professionally – in these fields.

The guide is divided into three parts. Part I provides an overview of the science of viruses, including what they are, how they work, and how illnesses can be prevented and treated. Parts II and III provide information on practical/technical considerations and on calculation details, respectively. In addition, a number of illustrative examples are included for each chapter.

Reasonable care has been taken to ensure the accuracy of the information contained in this book. However, the authors and the publisher cannot be held responsible for erroneous omissions in the information presented or for any consequences arising from the use of the information published in the book. Accordingly, reference to original sources is encouraged. Reporting of any errors or omissions is solicited in order to ensure that appropriate changes may be made in future editions.

The authors wish to thank the following individuals for their important contributions to the work: Marybeth R. Radics, Mary K. Theodore, Ann Marie Flynn, and Matthew C. Ogwu.

Finally, the authors also wish to acknowledge the contributing authors: Sarah Forster (Overview of Molecular Biology), Emma Parente (Safety Protocols and Personal Protection Equipment), Vishal Bhatty (Engineering Principles and Fundamentals), Paul DiGaetano, Jr. (Ethical Considerations in Virology), and Julian Theodore (Introduction to Mathematical Methods).

August 2022

Megan M. Reynolds
Merano, Italy
Louis Theodore
East Williston, NY

About the Authors

Megan M. Reynolds, BS, MS, MBA, is a freelance medical writer and editor with a particular focus on infectious diseases. With degrees in chemical engineering, international business, and medicine, she worked in the pharmaceutical field in various capacities for more than a dozen years in both the United States and Europe. Her experience encompasses manufacturing, sales, and marketing from managing production scale-up for the launch of new drug manufacturing lines to spearheading an education initiative for healthcare providers at a large New York public hospital aimed at increasing vaccine utilization. She also studied methods for minimizing bacterial resistance due to the overuse of antibiotics. Previous publications include textbook chapter contributions, a case report on the successful treatment of a patient with a rare, highly resistant infection, as well as a medical narrative on treating patients in severe pain. Her recent research interests have focused on addressing the high rate of hospital-acquired infections leading to sepsis and on reducing vaccine hesitancy towards measles, mumps, and rubella vaccine (MMR). Raised in New York City, Ms. Reynolds is multilingual and has lived and worked in several countries, including Italy, Spain, and Germany, and has studied in the United States, Mexico, and Grenada. She is currently based in northern Italy and enjoys living in the Alps while pursuing her passions in rock climbing, yoga, and skiing.

Born and raised in Hell's Kitchen, Louis Theodore received the degrees of MChE and EngScD from the New York University and a BChE from the Cooper Union. For over 50 years, Dr. Theodore was a chemical engineering professor, as well as graduate program director, researcher, professional innovator, and communicator in the engineering field. He has authored numerous texts and reference books, nearly 200 technical papers, and is section editor to the last four editions of Perry's Chemical Engineers' Handbook. He has served as a consultant to the US EPA, DOE and DOJ, and Theodore Tutorials. Dr. Theodore is a member of Phi Lambda Upsilon, Sigma Xi, Tau Beta Pi, American Chemical Society, American Society of Engineering Education, Royal Hellenic Society, and a fellow of the International

Air & Waste Management Association (AWMA). In addition to providing invited testimony to a Presidential (Ford) Crime Commission Hearing, Dr. Theodore was honored at Madison Square Garden in 2008 for his contributions to basketball and the youth of America. His current technical interests include risk management, desalination, and pandemic modeling.

Part I

Introduction to Viruses

Merriam-Webster defines *Introduction* as "something that introduces, such as,

- a part of a book or treatise preliminary to the main portion,
- a preliminary treatise or course of study" (Merriam-Webster 2022)

Indeed, that is exactly what this Part I of the book is all about. The chapters contain material that one might view as a prerequisite for the technical considerations and engineering calculations that are addressed in Parts II and III, respectively.

It is no secret that viruses are responsible for a host of diseases that can include something as simple as the so-called "common" cold to those that are more serious and fatal, i.e., COVID-19, West Nile, AIDS, Ebola, etc. The technical community began to realize that viruses, in general, were responsible for a range of diseases at the turn of the 20th century. The variation in disease severity occurs because various viruses attack different tissues and organs. In addition, one of the problems with virus detection has been the extremely small size of many of viruses, i.e., both the SARS-CoV-2 and polio viruses are in the 0.01-0.1-micron size range.

There are six chapters covering these issues in Part I. The chapter numbers and accompanying titles are listed below:

Chapter 1: Overview of Molecular Biology
Chapter 2: Basics of Virology
Chapter 3: Pandemics, Epidemics, and Outbreaks
Chapter 4: Virus Prevention, Diagnosis, and Treatment
Chapter 5: Safety Protocols and Personal Protection Equipment
Chapter 6: Epidemiology and Virus Transmission

A Guide to Virology for Engineers and Applied Scientists: Epidemiology, Emergency Management, and Optimization, First Edition. Megan M. Reynolds and Louis Theodore.
© 2023 John Wiley & Sons, Inc. Published 2023 by John Wiley & Sons, Inc.

1

Overview of Molecular Biology

Contributing Author: Sarah Forster

After much deliberation, the authors have decided to include a preliminary chapter concerned with molecular biology and the immune system. This decision was based on the fact that the book was written for engineers and applied scientists who may not have a background in biology. Why the inclusion? The authors felt that these topics, for those interested, could provide the readers with a better understanding of how cells function under normal circumstances, and thus better comprehend how viruses take over and use these mechanisms against the body.

Biology, as the science of life, involves the general study of living forms. *Molecular* biology, which includes *biophysics* and *biochemistry*, has made fundamental contributions to modern biology. Thus, more information is now available about the structure and function of nucleic acids – the base of DNA and proteins, and the key molecules of all living matter. *Cellular* biology is closely related to molecular biology (the title of this chapter). It primarily deals with the functions of the cell – the basic structural unit of life – which studies its components and their

A Guide to Virology for Engineers and Applied Scientists: Epidemiology, Emergency Management, and Optimization, First Edition. Megan M. Reynolds and Louis Theodore.

interactions. The life functions of multicellular organisms are governed by the activities and interactions of their cellular components. The study of organisms includes not only their growth and development but also how they function.

When a virus infects a host, it utilizes the genetic code of the invaded cell to hijack the normal replication process in order to replicate numerous copies of itself. Thus, it is helpful to have some understanding as to how genetic coding works under normal circumstances, in order to fully comprehend the complex mechanism with which the virus commandeers a cell for its own purposes. While viruses are not themselves cellular, they do contain the same basic genetic materials as cells, either DNA or RNA.

This chapter attempts to provide the reader with some of the key terms that have become integral to the study of molecular biology. The chapter also endeavors to offer a framework of normal cell functions critical to the understanding of virology. Chapter 2: Basics of Virology, the next chapter, utilizes this framework to depict how viruses invade and hijack standard cell function. Hopefully, the importance of the definitions and explanations in this earlier section will become clear. In the same vein, those already familiar with molecular biology may wish to skip this chapter in favor of Chapter 2.

1.1 CELL BASICS

This section describes the basic concepts of *eukaryotic* cells, which compose all multicellular organisms, such as humans, animals, and plants. Alternatively, single-celled microorganisms such as bacteria are referred to as *prokaryotes*. Different viruses infect different types of cells. As discussed above, Chapter 2 further examines how viruses invade and infect human cells.

Within each eukaryotic cell is a highly complex system of *organelles* – the tiny cellular structures that perform specific functions within the cell. These structures keep the cell running, much like organs do in the human body. Several key organelles are shown below in Figure 1.1.

(Louten 2016)

The following subsections below highlight a few organelles that play a crucial role during virus invasion leading to infection. These include the cytoplasm, ribosomes, and nucleus.

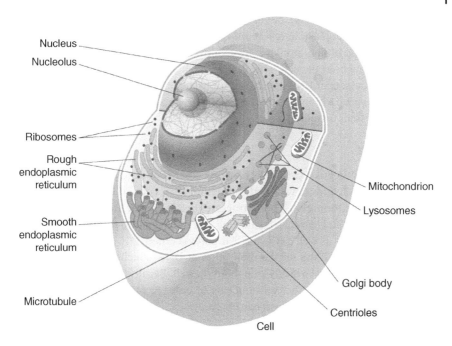

Nucleus

Nucleolus

Ribosomes

Rough endoplasmic reticulum

Smooth endoplasmic reticulum

Microtubule

Cell

Mitochondrion

Lysosomes

Golgi body

Centrioles

Figure 1.1 Illustration of organelles within a cell. *Source*: National Cancer Institute/U.S. Department of Health and Human Services/Public Domain.

1.1.1 Cytoplasm

The cytoplasm is one of the most important organelles within the cell membrane since it holds the other organelles together in its gel-like composition and allows for numerous processes to occur within the cell through the suspension of organelles and cellular molecules. Cytoplasm also allows for the occurrence of biochemical reactions within the cell, such as the replication of RNA viruses and protein synthesis (Denison 2008). The replication of RNA viruses occurs here in the cytoplasm as a majority of the enzymes used to replicate RNA are virally encoded.

1.1.2 Ribosomes

Ribosomes found in the cytosol play an important role in the manufacture of proteins within the cell. These ribosomes are located not only attached to the *rough endoplasmic reticulum (rER)*, but also floating within the cytosol. The ribosomes attached to the endoplasmic reticulum have the ability to create proteins. Once transferred to the lumen, proteins are modified to be utilized by the remaining organelles throughout the cell. This is all possible due to the binding of the ribosomes to the messenger RNA (mRNA) prior to the production of proteins (Louten 2016). Viruses have the ability to overtake the production of these proteins by the ribosomes for their own use.

1.1.3 NUCLEUS

The nucleus within eukaryotic cells contains organelles necessary for the regulation of cellular activities as well as the structures that contain the cell's DNA and other hereditary information. These structures inside the nucleus are comprised of chromosomes, the nuclear matrix, nucleoli, the nucleoplasm, the outer and inner nuclear membranes as well as the nuclear pores (Louten 2016).

The nucleus also allows for the replication of DNA which is then transcribed into messenger RNA to be used throughout the cell. Because of this, viruses must be able to have access to the cell's nucleus in order to replicate their DNA and attack other cells (Geer and Messersmith 2002).

1.2 CELL REPLICATION

Cell replication is a detailed process involving the copying of DNA to make new cells. DNA contains the genetic code that is present in every cell in the human body. DNA and RNA are both made up of nucleic acids, which are described below. They are critical to the process of replication and survival, not only for the cell but also for the invading virus (Denison 2008).

1.2.1 NUCLEIC ACIDS

This subsection will review the structure and function of DNA and RNA, which, as previously mentioned, are both made up of various nucleotides. Nucleotides are the basic building blocks for all living organisms, and are a crucial component in all cells. Nucleotides are comprised of three basic components:

- A five-carbon sugar molecule (deoxyribose for DNA or ribose for RNA)
- A phosphate group containing phosphorus and oxygen
- A *nitrogenous base*, a ringed molecule of nitrogen, oxygen, and hydrogen

There are four variations of nitrogenous bases, and together they form the basic building blocks for all living organisms: adenine (A), guanine (G), cytosine (C), and thymine (T). (Note: Uracil (U) replaces thymine in RNA) Together, these four different nucleotides combine to form polynucleotide base pairs. In DNA, adenine

Figure 1.2 DNA polynucleotide base pairs with sugar--phosphate backbone https://medlineplus.gov/genetics/understanding/basics/dna/

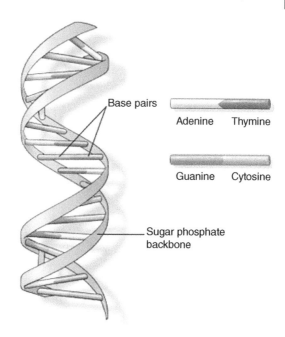

Base pairs

Adenine   Thymine

Guanine   Cytosine

Sugar phosphate backbone

always pairs with thymine, while cytosine pairs with guanine. The pairs are bound together by hydrogen bonds. These base pairs form the coding sequences within the DNA double helix, as shown in Figure 1.2. (Seladi-Schulman 2019; NIH 2010).

The double helix of DNA is structured as two complementary polynucleotide strands, with the leading strand running from the 5′ to 3′ carbon and the lagging strand running from the 3′ to 5′ carbon, as shown in the middle of Figure 1.3, below. The base pairs are located within the resulting double helix. The code, which is the order of nucleotides, determines which amino acids will be produced, and therefore, which proteins. Each amino acid is encoded by the order of three nucleotides (Geer and Messersmith 2002).

1.2.2 DNA REPLICATION

The cell cycle consists of four stages including gap 1(G_1), synthesis (S Phase), gap 2(G_2), and mitosis. Within these four stages, each cell has the ability to grow and divide while also replicating its DNA. The process of DNA replication occurs when the cell creates a direct copy of its chromosomes either during synthesis or during the s-phase of the cell cycle.

As depicted in Figure 1.3, the DNA molecule is untwined during replication, and the two DNA strands are separated from one another through the presence of *cellular enzymes* within the cell. *DNA polymerase* is one of the main

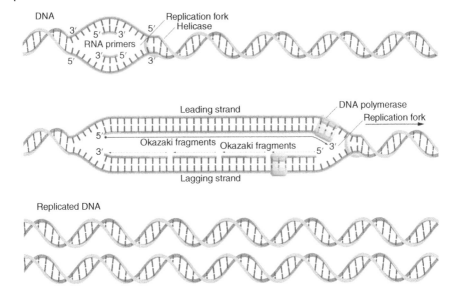

Figure 1.3 DNA Replication https://www.genome.gov/genetics-glossary/DNA-Replication

enzymes utilized in DNA replication due to its ability to place the complementary nucleotides of the new DNA strand in the 5′ to 3′ direction. DNA polymerase also adds in nucleotides based upon the complementary base pair rules as discussed in the previous section and is highly accurate, so there is a very low rate of misplaced nucleotides. This is shown in Figure 1.3. (Geer and Messersmith 2002).

Working alongside DNA polymerase is an enzyme known as *RNA polymerase* that synthesizes RNA. Similar to DNA polymerase, this enzyme uses a DNA template to produce a section of RNA that adds nucleotides in a 5′ to 3′ direction while also using the complementary base pair rule. RNA polymerase is known to have a lower rate of fidelity as compared to DNA polymerase (Louten 2016). This fact is highly relevant to virus replication, since an RNA virus tends to have more replicating errors—and as a result, more mutations than a DNA virus, as will be discussed in Chapter 2.

Another enzyme involved with DNA replication is known as *primase*, which is an enzyme that allows DNA polymerase to bind to a formerly single strand of DNA. Primase has the ability to form a double-stranded segment that allows for the binding of DNA polymerase through laying a complementary fragment of RNA on top of the single strand of DNA. (Geer and Messersmith 2002).

Because DNA replication is performed in the cell's nucleus, viruses must gain entry before taking advantage of DNA polymerase and other enzymes to replicate their own genomes and divide further (Louten 2016).

1.2.3 RNA Structure and Role

DNA and RNA are known to have both positive and negative strands. The positive strand of DNA is found within a single-stranded DNA virus and is referred to as any strand that has the same base sequence as a negative DNA strand. Meanwhile, a negative strand of DNA will have a base sequence complementary to that of the positive strand. The positive strand of RNA has the same polarity as viral mRNA while also containing codon sequences, "trinucleotide sequences of DNA or RNA that correspond to a specific amino acid…" [that may be translated to viral proteins] (genome.gov). Negative RNA strands are noncoding and must be copied by RNA polymerase in order to produce mRNA that is translatable (King et al. 2014). This background information is important for understanding key characteristics of various categories of viruses, (e.g., (+) or (-) sense DNA viruses).

DNA is able to use mRNA, transfer RNA (tRNA), and ribosomal RNA (rRNA) to replicate itself and repair mistakes during the process of replication. Messenger RNA is a single stranded temporary copy of a DNA molecule that is to be translated by the ribosome. Ribosomal RNA is known to help translate the information available in mRNA and change it into a protein to be used by the cell. The mechanism is illustrated below in Figure 1.4 below. The process of DNA replication and the transcription of DNA into mRNA occurs in the nucleus of the cell while mRNA is translated in the cell's cytosol by ribosomes, leading to the creation of a protein. RNA acts as a copy of a DNA molecules' hereditary genetic information. Ribosomes have the ability to create proteins through the use of amino acids using a sequence of nucleotides present in the RNA (Louten 2016).

As depicted in Figure 1.4, *transcription* refers to the creation of an RNA model from an original DNA sequence, while translation occurs when RNA goes through the process of being made up of nucleotides to becoming made up of proteins consisting of amino acids during the process of protein *translation*. These processes will be described in detail in the next subsection.

Nucleic acids have the ability to code for proteins through a series of three nucleotides that code for a specific amino acid. Because proteins are made up of amino acid chains, the nucleic acids within each amino acid directly impact each protein. mRNA of a previously known arrangement can be used to determine the amino acid sequence by directing the synthesis of a selected protein. In this manner, the genetic code could be decoded by comparing the original order of the mRNA with the synthesized proteins' amino acid sequence (Smith 2008).

1.2.4 Protein Synthesis

Protein synthesis is the process of creating protein strands through the use of ribosomes, mRNA, tRNA, and amino acids – mainly through transcription and translation, as shown in Figure 1.4.

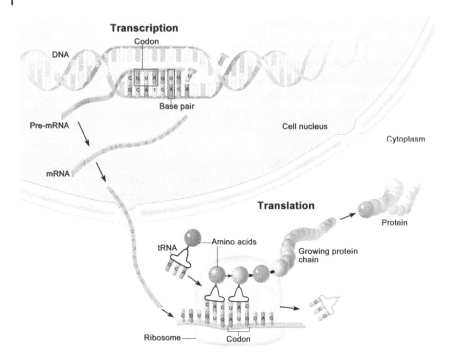

Figure 1.4   Protein Synthesis: Transcription and Translation https://nci-media.cancer
.gov/pdq/media/images/761782.jpg

Transcription occurs when a section of DNA is copied into mRNA. The template
strand of the two strands of DNA is known to act as a template for transcription.
RNA polymerase II is the enzyme that can synthesize DNA from that template,
which is recruited to the complex by transcription factors binding to the promoter
sequence. RNA polymerase II then reads the template strand of DNA using complementary base pairing rules specifically in the 3′ to 5′ direction (Louten 2016).

Translation is a three-part process consisting of initiation, elongation, and termination. Each of these three parts, discussed below, allows for the assembly of a
protein through amino acids from the ribosome (Geer and Messersmith 2002).

- *Initiation*
  This consists of a ribosome attached to a 5′ cap of mRNA transcript which then
  scans until a start codon is recognized. Corresponding tRNA along with a ribosomal subunit join the complex and elongation begins.
- *Elongation*
  During elongation, tRNA delivers amino acids to form a growing chain for each
  additional codon. Once the tRNA delivers its amino acid to the mRNA, it is
  then released from the ribosome and recharged by enzymes within the cell to
  be reused.

- *Termination*
This third and final stage of translation occurs when the moving ribosome encounters a stop codon, causing the ribosome to leave the mRNA when the protein is released as a result.

1.3 CELLULAR TRANSPORT

Cellular transport refers to the movement of resources or supplies through the plasma membrane of the cell. There are two general categories of transport: passive and active. Passive transport does not require any energy, while active transport does require it.

1.3.1 PLASMA MEMBRANE

The plasma membrane is the main layer of separation between a cell and its surrounding environment. Because of this, the plasma membrane is the first layer of contact a virus encounters while infecting a cell. The plasma membrane is most commonly represented by the fluid mosaic model, which portrays the integral proteins as being suspended in the lipid bilayer. The lipid bilayer is composed of amphipathic phospholipid molecules that contain both a hydrophobic tail and a hydrophilic head portion. The fatty tails of the phospholipids face together when placed in an aqueous solution, while the heads will be positioned on the outside. This way, the bilayer is able to form a barrier between the inside of the cell and the extracellular environment, creating a plasma membrane sufficient for cellular activities.

Integral membrane proteins (IMPs) are located in the lipid bilayer of the cell membrane. These integral proteins are necessary for a variety of extracellular functions, and, among other functions, can act as receptors or as cellular adhesion molecules (CAMs), which are used to adhere neighboring cells to each other. In addition to integral proteins are peripheral membrane proteins located on the surface of the plasma membrane associated with intracellular activities (Louten 2016).

Integral proteins also help with transporting substances, including those such as ions and other small molecules from one side of the cell to the other. However, some of these substances may be too large to fit between the channels and carrier proteins situated in the lipid bilayer and must be exported through processes such as exocytosis – a form of active transport out of a cell.

1.3.2 CELL SIGNALING

Also known as transduction, cell signaling is a vast network where the cell has the ability to communicate with other cells by releasing hormones and with other signaling molecules. Transduction would not be possible without the signal-transduction pathway, which allows for signals to be transmitted throughout the cell and results in a cellular response (Nair et al. 2019).

In order to communicate successfully with other cells, messages are first transferred from the *ligand* (a molecule that binds to a receptor) to the appropriate receptor, then decoded through a series of reactions of second messengers otherwise known as ions, kinases, and other small molecules. Afterward, the same message travels to the nucleus from the cell membrane, where several processes occur, including gene expressions, subsequent translations, as well as protein targeting. These processes target both the cell membrane and other organelles, all from the original message communicated. An intermediate is formed by a combination of intracellular signaling. This process is initiated by the response brought to the cell's ligands.

The process of cell signaling has the ability to control various multicellular activities including cell growth, differentiation, and other cell-specific functionalities. Due to this variability, signaling can be used in multiple areas, including endocrine, paracrine, juxtracrine, autocrine, and in neuronal neurotransmission. In addition to the various signals, the chemical makeup of the ligands is also differentiated in the inclusion of smaller molecules such as lipids, nucleic acids, and proteins, among others (Nair et al. 2019).

1.4 IMMUNE DEFENSE

The human body has various ways to defend itself against assault from infectious diseases and toxins. The immune system has the ability to determine the difference between "self" and "foreign" elements, and to mount an effective immune defense by acting against all "foreign" invaders. There are different protective mechanisms at the body's disposal, consisting of *innate* and *adaptive immunity*, both of which are explained below.

1.4.1 INNATE IMMUNITY

The innate immune system acts against all nonspecific invaders such as antigens and other harmful microorganisms rather than just attacking any one specific infectious microparasite. This nonspecific protective mechanism consists of three separate components: *physical barriers*, the *inflammatory response*, and *phagocytosis*.

Physical (and chemical) barriers that act as the body's primary line of defense are surrounded by epithelia. Epithelial surfaces have the ability to separate the body from pathogens present in the external environment and consist of the skin, respiratory and urogenital tracts, as well as the mucus membranes. When these barriers are penetrated by invaders and pathogens enter the body, an immune response occurs to rapidly eliminate the infection.

The inflammatory response occurs at the infected area of tissue injury caused by an influx of foreign invaders or from other bodily trauma. An influx of specialized cells to the injured tissue results in an inflammatory reaction leading to a release of chemicals. These specialized phagocytic cells, known as macrophages, have the ability to isolate and identify invasive cells while releasing active molecules. The surrounding capillaries then dilate as a result of the released inflammatory cytokines, which then increases blood flow, eventually causing fluid leakage that accounts for the side effects of inflammation. These effects may include swelling, redness, and pain.

Phagocytosis, the last step in innate immunity, is used to engulf and digest cells as a form of receptor-mediated endocytosis. This phase is oftentimes referred to as the cleansing or "healing" phase as it is where the inflammatory response is diminished, and the side effects of inflammation are reduced. Phagocytosis plays a role in protecting the body against viruses as macrophages and other specialized phagocytic cells help to digest bacteria and dead cells in an effort to defend against pathogens and other harmful microorganisms. In some instances, however, certain viruses are able to gain entry and take over a cell through phagocytosis.

1.4.2 Adaptive Immunity

Adaptive immunity is a specific protective mechanism that has two main components and five important characteristics. As compared to innate immunity, adaptive immunity allows the body to recognize and identify specific antigens and pathogens while also retaining their genetic information for future attacks. Because this immunity can recognize specific antigens, it activates lymphocytes (a type of white blood cell) with a plan of attack unique to that antigen. As a result, the humeral and the cell-mediated immune responses—the two main components of adaptive immunity—are put into place. Both these responses are discussed below.

1.4.2.1 Humoral Immunity

The *humoral* immune response mainly corresponds with the function of B cells and their production of immunoglobulins, or antigen-specific antibodies. Immunoglobulins are a type of antigen receptors found on B cells, and each immunoglobulin is only able to identify a single antibody. Because of this, the immune system contains a variety of antigens in order to create a successful defense against the wide array of pathogens and other foreign invaders the immune system encounters on a daily basis.

B cells have the ability to produce more cells with their particular antibodies when they identify and collide with their specific, corresponding antigens. Several of the new B cells created from this collision will transform into plasma cells, which are known to produce a high quantity of antibodies. The main function of these antibodies within the process of adaptive immunity is to bind with extracellular pathogens present in the blood or other bodily fluids.

1.4.2.2 Cellular Immunity

In addition to the humoral aspect of adaptive immunity is the *cell-mediated* immune response. Rather than relying on the function of B cells, cellular adaptive immunity focuses more on the work of T cells within the body in response to foreign invaders and potentially harmful substances. Much like B cells, T cells also have the ability to recognize antigens; however, their receptors are able to identify a large number of antigens through their similarities to immunoglobulins. B cells are only able to recognize one specific antigen.

Unlike B cells, T cells are only able to identify and attack an antigen or other harmful cell after it has already been processed by particular "antigen-presenting" cells such as macrophages. After the cell is processed, antigens on the cell's surface bind to the T-cell receptor, which then signals the T-cell to grow and divide while also going through cellular differentiation. The activated T cells are now able to travel to the site where the antigen entered the body and release *cytokines* – small proteins involved in the inflammatory process – while working together with other T cells, B cells, and phagocytic cells to rid the body of infectious invaders. Cytokines or "chemical messengers" are generated by a variety of cells and can have many different functions including initiating cellular activities and processes.

The above two immune systems work together to successfully eliminate and protect the body from foreign material and pathogenic invaders. Both adaptive and innate immunity are able to initiate the body's immune response, prompting the release and activation of *lymphocytes* – the inflammatory response and phagocytosis among other methods to eliminate the attacking antigens and to prepare the body for future invasions of hazardous microorganisms.

1.5 APPLICATIONS

The following four illustrative examples are intended to complement the above material and to provide a better understanding of the aspects discussed within the chapter.

Illustrative Example 1.1 Cytokines
Examine the role of cytokines in inflammation.

Solution

There are many types of cytokines produced by the body under stress or injury. The common inflammatory response induces heat, redness, swelling, and pain, as well as impaired function. The majority of cytokines are formed by activated macrophages and T cells. Some are proinflammatory, such as IL-1β, IL-6, and TNF-α, while others are activated in order to counteract the inflammatory process. Anti-inflammatory cytokines include IL-4 and IL-10, among others.

Illustrative Example 1.2 RNA

Describe the functions of mRNA, tRNA, and rRNA.

Solution

While there are other types of RNA, the main three are:

- Messenger RNA (mRNA), which is transcribed directly from DNA, goes on to produce proteins with the help of tRNA.
- Transfer RNA (tRNA) transcribes mRNA into proteins.
- Ribosomal RNA (rRNA) forms ribosomes, and are, therefore, critical for the synthesis of proteins.

Illustrative Example 1.3 LYMPHOCYTES

Explain the role of lymphocytes in adaptive immunity.

Solution

The lymphocytes are a part of the lymphoid system and are composed of B and T cells. The adaptive immune system is activated from these lymphocytes, and pathogens and other foreign invaders are disposed of before entering the bloodstream. The activation of large numbers of lymphocytes can occur as a result of antigen recognition, and the activated lymphocytes will have specified roles against particular antigens. This specific immunity can be seen in the humoral and cell-mediated immune responses against pathogens.

Illustrative Example 1.4 TRANSLATION AND TRANSCRIPTION

Differentiate between translation and transcription in protein synthesis.

Solution

Translation accounts for the assembly of a protein through amino acids from the ribosome, and is composed of three parts, including initiation, elongation, and termination. Through the use of these three processes, translation is able to assemble a protein made up of an amino acid sequence from RNA in the cell's ribosome. Meanwhile, transcription in protein synthesis takes place when mRNA is produced through the copying of a DNA section.

1.6 CHAPTER SUMMARY

- A highly complex system of organelles is present within each eukaryotic cell.
- One of the most important organelles is the cytoplasm, which has the ability to secure the organelles together, while also being the area many cellular processes occur by suspending other organelles and molecules in its characteristic gel-like composition.
- When DNA replication occurs, the cell is able to create a direct replicate copy of itself and its chromosomes during synthesis of the cell cycle.
- Protein synthesis utilizes mRNA and tRNA in addition to amino acids and ribosomes when creating protein strands through transcription and translation.
- The main layer of separation between cells and the outside extracellular environment is the plasma membrane, which also acts as the first layer of contact encountered by a virus or foreign invader when infecting a cell.
- Transduction, or cell signaling, has the ability to facilitate multiple processes such as the communication between cells.
- The immune system acts against all foreign invaders in different protective mechanisms consisting of innate and active immunity.

1.7 PROBLEMS

1 How many antigens are B cells able to recognize?

2 What are immunoglobulins? Cytokines?

3 When identifying nonspecific foreign invaders, which immunity is used?

4 How many nucleotides is each amino acid encoded by?

REFERENCES

Denison, M.R. (2008). Seeking membranes: positive-strand RNA virus replication complexes: E270. *PLoS Biology* 6 (10): e270. https://doi.org/10.1371/journal.pbio .0060270.

Geer, R.C. and Messersmith, D.J. (2002). Introduction to molecular biology information resources. National Center for Biotechnology Information. https:// www.ncbi.nlm.nih.gov/Class/MLACourse/ (accessed 6 January 2022).

King, R.C., Mulligan, P.K., and Stansfield, W.D. (2014). *A Dictionary of Genetics*. Oxford University Press https://www.oxfordreference.com/view/10.1093/oi/authority.20110803100332484.

Louten, J. (2016). Chapter 3 - features of host cells: cellular and molecular biology review. In: *Essential Human Virology*, 31–48. Academic Press https://doi.org/10.1016/B978-0-12-800947-5.00002-8.

Nair, A., Chauhan, P., Saha, B., and Kubatzky, K.F. (2019). Conceptual evolution of cell signaling. *International Journal of Molecular Sciences* 20 (13): 3292. https://doi.org/10.3390/ijms20133292.

National Cancer Institute. Surveillance, Epidemiology and End Results (SEER) Program, SEER Training (n.d.). *Cell Structure*. US National Institutes of Health https://training.seer.cancer.gov/anatomy/cells_tissues_membranes/cells/structure.html (accessed 6 March 2022).

National Institute of General Medical Sciences & Office of Communications and Public Liaison (2010). NIH publication no.10 662. https://www.nigms.nih.gov/education/Booklets/the-new-genetics/Documents/Booklet-The-New-Genetics.pdf (accessed 8 July 2021).

Seladi-Schulman, J., 2019. What is DNA? *Structure, Function, Pictures & Facts*. Healthline. https://www.healthline.com/health/what-is-dna (accessed 24 June 2021).

Smith, A.P. (2008). *Nucleic Acids to Amino Acids: DNA Specifies Protein*. Nature Education https://www.nature.com/scitable/topicpage/nucleic-acids-to-amino-acids-dna-specifies-935/.

2

Basics of Virology

Viruses that cause death and disease, such as Smallpox, Ebola, and HIV, are well known and have played a significant role in human history. At the time of this writing, the world is contending with just such a virus in SARS-CoV-2, the virus that causes COVID-19. The aim of this chapter is to explain what exactly defines a virus and to demonstrate the unique place viruses hold in the universe.

The chapter will also review classification, definitions, and the viral characteristics that allow them to invade cells and self-replicate. As was discussed, viruses are composed of nucleic acids, either DNA (deoxyribonucleic acid) or RNA (ribonucleic acid). Some background information about these basic genetic materials is vital to understanding how viruses behave. The reader is referred to Chapter 1: Overview of Molecular Biology, for a review of what nucleic acids are, how they function and maintain cells, and how they facilitate cellular reproduction.

2.1 VIRAL BASICS AND TERMINOLOGY

Viruses are complex biochemical entities often regarded as nonliving and are therefore classified separately from the nomenclature used to describe all living

A Guide to Virology for Engineers and Applied Scientists: Epidemiology, Emergency Management, and Optimization, First Edition. Megan M. Reynolds and Louis Theodore.

organisms. All viruses contain genetic material in the form of either DNA or RNA, but never both. Although they contain the same genetic material that exists in cells, viruses are not themselves cellular. Importantly, they can only replicate inside living cells, or host cells, which means that they must invade and commandeer a host cell to make it work for them. Unlike cells, they do not replicate in a binary fashion but instead rapidly multiply and self-assemble within the invaded cell and can release thousands of copies at the same time, often by *lysing*—thus destroying—the cell. The host cell can be a single-celled bacteria or a cell within a *eukaryote*—a multi-cellular organism—such as a plant, animal, or human (Taylor 2014, pp. 23–24).

A virus's nucleic acid core is its *genome*—genetic material—which is either DNA or RNA. This nucleic acid can be single- or double-stranded. It is possible for the genome to be *monopartite,* with all genes based on one nucleic acid molecule, or *segmented*, and distributed among multiple nucleic acid molecules (Burrell and Howard 2017, p. 30). The core is surrounded by a *capsid*, which is a protein barrier layer that protects the genetic material while also allowing the virus to attach to the host cell. Together, the genome and capsid are referred to as the *neocapsid*. In addition, some viruses also contain an outer *envelope*, an extra outer layer comprised of a lipid bilayer and surface proteins, which further aid in attaching to specific cells. The envelope is often originally derived from the membrane of a host cell, which was acquired upon the virus's exit. Envelopes perform various functions, including the facilitation of binding to the receptors of the target cell, and membrane fusion. In addition, since the envelope is cell-derived, it may serve to help the virus to evade the host's immune system while it searches for more cells to invade (Baron 1996; Burrell et al. 2017, pp. 36).

Virions are defined as complete virus particles with full infectivity. As noted, virus particles are only active once inside a host cell. This inactive state allows for protection from the external environment and facilitates the discovery of new host cells. The virion contains the neocapsid along with any enzymes needed for replication and any proteins on its surface that aid in the attachment process, which will be explained below.

There are various ways to classify viruses, and the most basic grouping is centered simply on which genetic material they contain, either DNA or RNA. While cells have double-stranded DNA and single-stranded RNA, viruses can contain either double-stranded or single-stranded versions of either DNA or RNA. This categorization helps to predict some basic traits within groups, such as higher degree of mutations in RNA viruses due to the lack of proofreading within the cell.

2.2 VIRAL LIFE CYCLE

Since virions only become active once they find a cell to invade, the virus life cycle is generally considered to begin with attachment to the host cell and to end with the destruction of that cell. When the host cell is destroyed, new virion copies are released into the extracellular environment. These virions are then free to continue the cycle by attacking nearby cells (Taylor 2014). The stages of viral replication are listed here and further explained in the subsections below (Louten 2016),

- Attachment (Cell Connection)
- Penetration (Cell Entry)
- Uncoating
- Replication
- Assembly
- Maturation
- Release

These stages are also depicted in Figures 2.1 and 2.2:

2.2.1 ATTACHMENT (CONNECTION)

Prior to infection, the virion must enter the cell. The first stage is attachment, when the virus uses the host's own surface proteins against it. In the cell's normal

Figure 2.1 Cell Connection and Entry. (Source: https://www.genome.gov/about-genomics/fact-sheets/Genomics-and-Virology.)

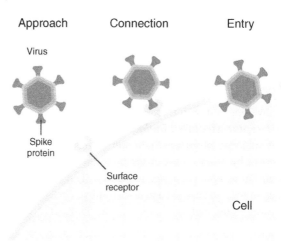

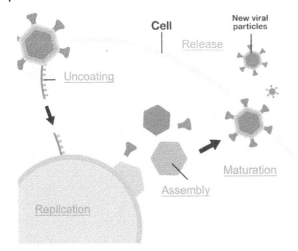

Figure 2.2 Virus Replication and Release. (Source: Adapted from: https://www.genome.gov/about-genomics/fact-sheets/Genomics-and-Virology.)

functioning, these proteins serve to transport ions and molecules across the phospholipid membrane for use within the cell. The surface proteins often serve as receptors for inbound proteins (Louten 2016). The viral surface proteins, or anti-receptors, mimic those specific proteins the cell would recognize, giving the virus access to the receptor sites for entry (Roizman 1996).

2.2.2 PENETRATION (ENTRY)

Upon successful attachment, the virus must enter the cell by crossing through the cell plasma membrane. This could occur in a few ways, including:

- **Entry by Direct Fusion**
  This method is available only to enveloped viruses, where the envelope fuses directly with the membrane, leaving the virus within the cell wall without its envelope, but with its capsid intact.
- **Receptor-Mediated Endocytosis**
  A majority of viruses, whether enveloped or not, rely on the cell to initiate entry by endocytosis. After bonding with the receptor on the cell surface, a virion–receptor complex is formed. The cell instantly responds, engulfing the complex in a vesicle coated with substances that can withstand the difficult trip through the membrane of the cell. (Ryu 2017).

2.2.3 UNCOATING

In the uncoating stage, the capsid—and envelope if it remains—is stripped away so that the virus genome can be exposed once it reaches its destination. Some viruses remain in the cell's *cytoplasm* and replicate from there (Louten 2016).

For the many viruses that replicate inside the cell nucleus, they must first cross through pores in the *nuclear envelope* prior to any gene expression. Some viruses are small enough to cross with their capsids intact. Larger viruses are usually able to attach to the surface of the nuclear envelope and then inject their genome into the nucleus (Ryu 2017).

2.2.4 REPLICATION

All viruses rely, to some extent, on their host for the ability to replicate, but the dependence varies widely. For all viruses, the main obstacle to replicating is that they do not carry *ribosomes*. This means that the virus must instead produce a readable *messenger RNA (mRNA)* code for the host to translate via *ribosomal proteins* and *RNA (rRNA)*, and then synthesize those into the necessary viral proteins during replication. The details of how viruses arrive at that point is specific to individual viral families and classes and will be further discussed in Section 2.3. For further explanation on the various types of RNA, the reader is referred to Chapter 1: Overview of Molecular Biology (Ryu 2017; Li 2019).

2.2.5 ASSEMBLY

Once the correct viral proteins have been produced and transported to the target location, these newly replicated viral proteins come together to form the initial stages of a virion. Two often-simultaneous phases of assembly take place. One is *genome packaging*, and the other is capsid assemblage (Ryu 2017; Louten 2016).

2.2.6 MATURATION AND RELEASE

Maturation is the final stage before the virus becomes infectious. The virus undergoes changes to its capsid structure, often making itself more chemically and physically stable and capable of withstanding temperature fluctuations, for example.

These last two phases of replication are virus-specific, and maturation can occur both intracellularly and extracellularly. Naked viruses (i.e., those without an envelope) mature within the host cell, usually destroying the host in the process of exiting. This eruption causes a flood of virus particles, both mature and immature, with the mature, virulent copies seeking out more cell hosts.

In contrast, enveloped viruses emerge by way of a process called *budding*, in which the virus particle instructs the host cell membrane to begin wrapping itself around the virus. This process eventually allows the virus to exit the cell surrounded by a segment of the cell plasma membrane, now the viral envelope. Some enveloped viruses are highly *cytolytic* which means they cause cell death upon release (similar to naked viruses), while others leave the cell intact. Regardless, the affected cell will be targeted by the host immune system (Roizman 1996).

2.3 VIRUS STRUCTURE AND CLASSIFICATION

When discussing viral structure, it is important to differentiate between the definition of a virus and that of a virion. The structure of a virion, fully assembled and infectious, is what is described here. In contrast, the virus, once it is intracellular, disassembles to reveal its genome and takes over the cell infrastructure. At that point, there is no physical structure to attribute to it. The terms have often been used incorrectly, but it has been accepted in recent years that the term virion is used to designate an entire virus particle, inert and intact, while searching for the next cell to invade (van Regenmortel and Mahy 2004).

Since viruses replicate by copying the same strand (or few strands) many times over, they rely on a repeating architecture when assembling their capsids around the newly formed genomes. This capsid architecture refers to a triangulation number (T) as the number of structural units required to form one face of the capsid. The most common virus structures are helical and icosahedral. The capsid structure is not dependent on what type of genome is inside, and can be formed on any of the virus classes to be discussed in this section. In addition, the capsid geometry is also independent of the presence or absence of an outer envelope (Louten 2016). Examples of various viral shapes are depicted in Figure 2.3, below:

Examples of viruses

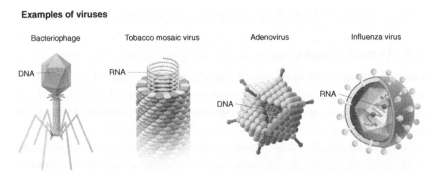

Figure 2.3 Examples of Virus Structures. (Source: https://www.genome.gov/genetics-glossary/Virus.)

The widely accepted Baltimore Classification, in use since the 1970s, breaks viruses down into seven classes, based on their genome and on how each uses the host cell to replicate, as noted in Section 2.2. The individual classes are listed below, and their mechanism of replication will be examined later in this section (Louten 2016).

DNA Viruses
- Class I: double-stranded DNA (dsDNA)
- Class II: single-stranded DNA (ssDNA)

RNA Viruses
- Class III: double-stranded RNA (dsRNA)
- Class IV: single-stranded positive sense RNA (+ssRNA)
- Class V: single-stranded negative sense RNA (-ssRNA)

Reverse Transcription Viruses (Retroviruses)
- Class VI: double-stranded reverse transcription DNA viruses
- Class VII: single-stranded reverse transcription RNA viruses

2.3.1 DNA Viruses

Class I and II consist of viruses that contain DNA. Class I viruses have double-stranded DNA, which makes this group the closest in genome to host cells. As such, they can best exploit the cells' proteins and enzymes that are normally used for cell DNA replication. Viruses such as Herpesvirus and Adenovirus are in this category.

Class II are single-stranded DNA viruses, such as Parvovirus. Much of this class infects bacteria and includes some of the smallest known viruses. Their initial stage of replication is to produce a copy of their own DNA genome and joining it to become double-stranded. The replication process then follows the trajectory of Class I viruses.

DNA viruses usually replicate in the nucleus of the host cell, although pox viruses, such as smallpox, are a notable exception. DNA viruses can be enveloped or naked.

2.3.2 RNA Viruses

Classes II through IV are all RNA viruses, and most do not replicate in the nucleus. Since they contain RNA instead of DNA, they can replicate using their own genomic template. For this reason, most RNA viruses carry their own versions of transcription enzymes.

RNA viruses are categorized by their genomes, with Class III carrying double-stranded RNA, a phenomenon that does not occur in living organisms. Class III are all naked viruses with segmented genomes.

Class IV viruses have single-stranded, *positive-sense* RNA. *Sense* refers to how the strand of viral RNA becomes mRNA, which then goes on to make the necessary proteins for viral replication. In positive-sense RNA viruses, the viral strand can be used as a direct substitution for the host's mRNA. *Negative-sense* (Class V) indicates that the virus must first manufacture a complementary strand of RNA which will then be used by the cell ribosomes to produce more virus copies.

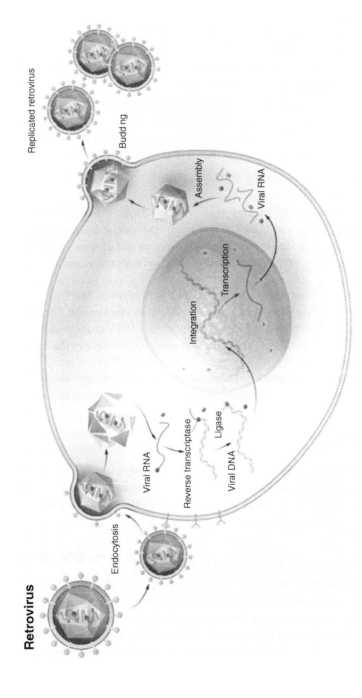

Figure 2.4 Replication of a Retrovirus. (Source: https://www.genome.gov/genetics-glossary/Retrovirus.)

2.3.3 Reverse Transcription Viruses (Retroviruses)

Retroviruses that reverse transcribe are rarer than other virus groups and have complex modes of replication. However, they are of significant importance medically because the HIV virus (and some types of cancer) falls within this category, specifically into Group VI. This group of RNA retroviruses makes use of an enzyme that allows them to transcribe a linear strand of complement dsDNA (cDNA) from their own ssRNA and fully incorporate it into the cell's own genome. In other words, the infected cell converts RNA into DNA which it then inserts into its own genetic material so that it becomes a permanent and integral part of the cell, and will be passed on to all subsequent cells. Any virus that reverse transcribes is referred to as a *retroid virus* (Louten 2016; Genome.gov 2022). See Figure 2.4 for a depiction of this process.

Class VII are dsDNA viruses that use reverse transcriptase to insert its genome into the host. This is done at the end of replication, rather than at the beginning as with Group VI. The only virus that infects humans in this class is Hepatitis B.

2.4 VIRUSES IN CONTEXT OF THE TREE OF LIFE

Central to any discussion of viruses is whether they are living or inert entities. The International Committee on Taxonomy of Viruses (ICTV), the leading authority on the naming and classification of viruses, has based their system and nomenclature on a hierarchy similar to, yet separate from, that of living organisms. This system recognizes the Baltimore Classification described in Section 2.3, but it also integrates other chemical and physical property similarities among the virions, such as structure and particle size (Louten 2016). It also makes clear, however, that viruses are classified as separate and distinct from living organisms. They are not included in the proverbial "Tree of Life."

The debate concerning which side of life viruses belong to has continued for decades, and will no doubt remain a subject of scientific discourse for many more. The arguments made on both sides of the issue are accurate and legitimate. Often the conversation diverges into the philosophical realm and, at its core, centers around the question of what it means to be alive.

Examining this discussion from various viewpoints is one of the most interesting and insightful exercises when learning what viruses are, how they are perceived, and how they have been studied. While the ICTV separates viruses from living species, citing them as nonliving entities, many scientists in recent years have rethought the role of viruses in evolution. Many are now studying viruses' ongoing role in aspects of the existence of life (Koonin and Starokadomskyy 2016).

In many ways, viruses exist in a gray area between living and nonliving because they possess qualities of both. When a virion enters a cell, it comes alive, at least metaphorically. The invaded cell is now a host, and the inert virion is now at work within the cell. In simplest terms, the virus sheds its coat and makes itself at home, revealing its own genetic code and using the host's replication apparatus to replicate itself.

The central argument against viruses being living organisms is that virions are not cells, and they cannot replicate without the aid of a host (van Regenmortel and Mahy 2004). Viruses are *obligate intracellular parasites* by this definition (Brown and Bhella 2016). It is important to note here that many viruses are not actually harmful to the host, and some can form mutualistic, as opposed to pathogenic, relationships with the affected organism (Dimijian 2000). While submicroscopic, viruses far outnumber all living things on the planet put together (Rice 2021).

Many microbiologists over the last century have also suggested that viruses are nonliving because they are noncellular, have only one type of nucleic acid and lack the ability to undergo binary fission or to produce power (Taylor 2014).

In contrast, the clearest argument for life is that they can be killed (Koonin and Starokadomskyy 2016). In the late nineteenth century, viruses were considered to be the tiniest and simplest of all living forms. Much like bacteria, which are categorically alive, viruses – such as rabies and foot-and-mouth disease – were known to cause illness, and at the time it was already understood that these diseases could be passed from one person to another. Recent discoveries have also shed light on the issue, with larger and more complex viruses having been found, and their genome sequenced. Scientist believe viruses have been instrumental in our own evolution, with fundamental changes to the eukaryotic genome over the course of millennia (Koonin and Starokadomskyy 2016). (Taylor 2014; Forterre 2010; Berman 2019).

Finally, viruses are highly adaptable to their environment and are subject to the laws of natural selection, in which viral variants best adapted to spreading throughout a population may prevail over other similar variants. A stark example of this occurred with the SARS-CoV-2 virus when several more infectious variants replaced the original strain of the virus, leading to a near complete substitution by these variants in only a few months.

2.5 VIRAL GENETICS

This section focuses on some of the methods in which viruses can evolve quickly, sometimes by encountering another, similar virus while both are attacking the same cell. More frequently, viruses evolve through various mechanisms of random mutations.

2.5.1 ANTIGENIC SHIFT

Antigenic shift, also known as reassortment, occurs when two different viruses coinfect the same cell. In this case, the viruses can exchange entire genomes. This transpires because viral copies of RNA will be replicated, but the reassembly will be random, and the result will be a new variant with potentially different traits such as virulence or transmissibility.

2.5.2 ANTIGENIC DRIFT

Less dramatic but far more common is antigenic drift. This refers to multiple minor mutations within the genome adding up over several replications that finally accumulate and change the surface proteins enough that they are no longer recognizable to the host immune system.

2.5.3 PHENOTYPIC MIXING

Phenotype refers to the physical properties displayed. Phenotyping mixing is the combination of surface antigens from two related viruses, thus conferring potentially improved mechanism of entry to specific cells for each virus.

2.5.4 COMPLEMENTATION

Complementation is the process by which one virus passes along traits to a similar virus so that it gains an additional function, such as increased infectivity. This occurs most frequently in larger DNA viruses when two highly mutated strains of the same virus infect the same cell.

2.6 APPLICATIONS

The following four illustrative examples represent various viruses and virus families that extensively and significantly effect humans. The same viruses examined here will also be presented in detail in the next several chapters focusing on impact of infection in the body, the role of vaccines and treatments, and the impact of outbreaks on societies.

Illustrative Example 2.1 INFLUENZA
Briefly describe how influenza is classified and what is significant about the envelope proteins specific to Influenza A.

Solution
Influenza viruses are a part of the *Orthomyxoviridae* family and are designated by types A through D. They are Class V from the Baltimore Classification system, so all have negative-sense, single-stranded, RNA genomes. Their genome is segmented, and they are enveloped viruses.

All influenza types can cause disease in mammals. Influenza A and B are much more common than the others and have the ability to cause more serious outbreaks. Influenza D has so far only known to infect cattle. Influenza A is further broken down to several subtypes which are designated by two of their key envelope proteins. There are eighteen different hemagglutinin (H1 through H18) and eleven neuraminidase proteins (N1 through N11), allowing for significant combinations. To date, however, only three H subtypes H1, H2, H3 have caused significant outbreaks in human populations.

Influenza is particularly known for its ability to rapidly evolve because it is highly adaptable and prone to both antigenic shift and antigenic drift. This is especially true for the hemagglutinin surface protein of Influenza A. Antigenic shift as discussed in Section 2.5 is the reassortment of two viral genomes within the same host cell. This reassortment allows for great variability and a rapid change to the characteristics of influenza virus subtypes.

Illustrative Example 2.2 Ebola
Briefly describe the structure of viruses from the Filoviridae family and what makes these viruses among the deadliest human pathogens.

Solution
Viruses from the Filoviridae family include Ebolavirus and Marburgvirus species. Ebola is a Baltimore Class V virus, a single-stranded negative sense RNA. These viruses are among the most lethal in existence and have caused significant deaths during several outbreaks in Africa since 2013.

Illustrative Example 2.3 Corona Viruses
Explain the similarities and differences among the species of Corona viruses. What is their structure?

Solution
Corona viruses are enveloped positive-sense single-stranded RNA viruses with a helical capsid symmetry, which make them Class IV viruses. The envelope is spherical with glycoproteins projecting out of the surface. This family includes viruses that cause the common cold, SARS-CoV (SARS), and Middle East Respiratory Syndrome, MERS-CoV (MERS).

Illustrative Example 2.4   HIV
Briefly describe Human Immunodeficiency Virus (HIV) and its mechanism of replication.

Solution
HIV is well-known as a retrovirus, for which, after over 20 years into an ongoing pandemic an effective cure remains frustratingly elusive. Officially called Human Immunodeficiency Virus, which causes acquired immunodeficiency syndrome (AIDS), HIV is a Class VI ssRNA reverse transcription viruses.

2.7   CHAPTER SUMMARY

- Viruses cause significant suffering and disease; they exist in a gray area on the edge of life which has long been a source of scientific dispute.
- Classification is based on their genomes which refer to whether they are DNA or RNA and how they replicate.
- Virus replication involves takeover of a host cell for its mechanisms; it involves entry into the cell, uncoating its capsid and revealing its genome to be replicated. Once the virus has produced enough copies, it will assemble, mature to its full infectivity, and exit the virus, which more often than not involves cell death.
- Several mechanisms exist that allows for viruses to evolve efficiently; these involve the swapping of entire genomes with a similar virus in the co-infected cell, which is called antigenic shift.
- Another method is antigenic drift, which refers to an accumulation of mutations over time that will ultimately confer new infectivity or other characteristics to the virus.

2.8   PROBLEMS

1   Which class of viruses in the Baltimore Classification system is most closely related to the human genome?

2   List three arguments each for and against the theory that viruses are alive.

3   List the seven steps in the viral lifecycle.

4   What is a virion?

REFERENCES

Baron, S. (ed.) (1996). Introduction to Virology. In: *Medical Microbiology*, 4e. Galveston (TX): University of Texas Medical Branch at Galveston Available from: https://www.ncbi.nlm.nih.gov/books/NBK8098/.

Berman, J.J. (2019). Viruses. *Taxonomic Guide to Infectious Diseases* 263–319: https://doi.org/10.1016/B978-0-12-817576-7.00007-9.

Brown, N. and Bhella, D. (2016). Are Viruses Alive? *Microbiology Society* 43 (2)· https://doi.org/https://microbiologysociety.org/publication/past-issues/what-is-life/article/are-viruses-alive-what-is-life.html.

Burrell, C. J., Howard, C. R., Chapter 3 - Virion structure and composition, *Fenner and White's Medical Virology (5)*, Academic Press, 2017, Pages 27–37, ISBN 9780123751560, https://doi.org/10.1016/B978-0-12-375156-0.00003-5.

Burrell, C.J., Howard, C.R., and Murphy, F.A. (2017). Virus replication. *Fenner and White's Medical Virology* 39–55: https://doi.org/10.1016/B978-0-12-375156-0.00004-7.

Dimijian, G.G. (2000). Evolving together: the biology of symbiosis, Part 1. *Proceedings (Baylor University. Medical Center)* 13 (3): 217–226.

Forterre, P. (2010). Defining life: the virus viewpoint. *Origins of Life and Evolution of the Biosphere: The Journal of the International Society for the Study of the Origin of Life* 40 (2): 151–160. https://doi.org/10.1007/s11084-010-9194-1.

Genome.gov. (2022). *Retrovirus*. Retrieved August 10, 2022, from https://www.genome.gov/genetics-glossary/Retrovirus.

Koonin, E.V. and Starokadomskyy, P. (2016). Are viruses alive? The replicator paradigm sheds decisive light on an old but misguided question. *Studies in History and Philosophy of Biological and Biomedical Sciences* 59: 125–134. https://doi.org/10.1016/j.shpsc.2016.02.016.

Li, S. (2019). Regulation of Ribosomal Proteins on Viral Infection. *Cells* 8 (5): 508. https://doi.org/10.3390/cells8050508.

Louten, J. (2016). *Essential human virology*. London, UK: Elsevier.

van Regenmortel, M.H. and Mahy, B.W. (2004). Emerging Issues in virus taxonomy. *Emerging Infectious Diseases* 10 (1): 8–13. https://doi.org/10.3201/eid1001.030279.

Rice, G. (2021) Are viruses alive? Microbial Life - Educational Resources. https://serc.carleton.edu/microbelife/yellowstone/viruslive.html (accessed 13 January 2021).

Roizman, B. (1996). Multiplication, Chapter 42. In: *Medical Microbiology*, 4e (ed. S. Baron). Galveston (TX): University of Texas Medical Branch at Galveston Available from: https://www.ncbi.nlm.nih.gov/books/NBK8181/.

Ryu, W.S. (2017). Virus life cycle. *Molecular Virology of Human Pathogenic Viruses* 31–45: https://doi.org/10.1016/B978-0-12-800838-6.00003-5.

Taylor, M.W. (2014). What is a virus? *Viruses and Man: A History of Interactions* 23–40: https://doi.org/10.1007/978-3-319-07758-1_2.

3

Pandemics, Epidemics, and Outbreaks

Thousands of viruses exist on earth, many already known and others yet to be discovered. For every living organism, there are many viruses that can infect it. These are cataloged by the International Committee on Taxonomy of Viruses (ICTV). Many viruses do not fit within the standard classification system, however, so the study of virology continues to evolve with new techniques and discoveries (Mahmoudabadi and Phillips 2018).

For humans, there are hundreds of viral species known to cause disease. This chapter will focus on key viruses that cause death and disease on a grand scale, triggering local outbreaks, regional epidemics, or global pandemics. Such emphasis has taken on particular importance since 2020 as SARS-CoV-2, the virus which causes COVID-19, continues to cause death, suffering, and economic and political devastation, as the world looks to an uncertain future.

A Guide to Virology for Engineers and Applied Scientists: Epidemiology, Emergency Management, and Optimization, First Edition. Megan M. Reynolds and Louis Theodore.
© 2023 John Wiley & Sons, Inc. Published 2023 by John Wiley & Sons, Inc.

3.1 HUMAN VIRAL DISEASES

Viruses are the smallest of all microbes, ranging on average from 0.02 to 0.30 micrometers, although some, such as the aptly named megavirus, can measure up to 1 micrometer (Kramer 2020). Unlike bacteria and fungi, most viruses cannot be identified by examination under a light microscope. Viral infections are among the most challenging to treat due to various factors. First, detection is more difficult because, unlike bacteria, a viral particle is inert until it enters a host cell, meaning it does not immediately arouse the suspicion of the immune system. Once inside, the virus takes over the host cell's reproduction mechanisms to produce numerous copies of itself. The entire replication process takes place intracellularly—completely within the host cell—so that by the time the infection is discovered, the *viral load* is often quite high (Roizman 1996).

While antibiotics are an effective treatment against bacterial illnesses, they are not useful in treating viral infections. Antibiotics work by targeting processes necessary for bacterial cell survival, such as *cell metabolism* or reproduction, or they attack part of the bacterial structure, such as the cell wall in order to neutralize the bacteria and clear the infection. These approaches are useful because, as previously mentioned, the bacteria are not hidden inside of the host cells. For viruses, none of those treatment options apply. Viruses are also highly *mutagenic* and frequent mutations in the viral *genome* change its characteristics and capabilities. This has the potential for increased contagion and severity of illness, as well as the ability to overcome vaccines or natural immunity. Because mutations are random, natural selection leads to the survival of more aggressive and sustainable variants (Bouvier and Palese 2008).

Antiviral drugs and vaccines must be specially tailored to each virus, and at times, each strain or subtype. For example, annual flu vaccines are targeted at specific strains based on forecasting models that predict the likely dominant strains for the next season. If a different strain becomes dominant that year, which does happen, the vaccine available will not be as effective (Becker et al. 2021).

The difficulties in treating and controlling viruses have opened the door to viral pandemics and outbreaks that have changed the course of history. Smallpox is highly contagious and *virulent (i.e.,* severe) and is believed to have been around for three thousand years, killing untold millions. Overall, 30% of those infected died. In the eleventh century, the vast movement of people during the crusades facilitated the spread of smallpox, and it was brought to the new world by European settlers in the seventeenth century. The practice of *variolation*— introducing minute amounts of infected material from a patient with a mild case of smallpox to a healthy individual in order to stimulate their immune system without making them seriously ill—was first documented in China and India as far back as the fifteenth century. The scientific breakthrough against smallpox that would eventually give the world modern vaccines began in 1796 with Edward Jenner, an English doctor who realized that exposure to the related, less virulent cowpox virus could protect people against later smallpox infection (CDC 2021a; WHO 2016).

Smallpox was also the first virus to be eradicated by vaccine through an immense, well-coordinated, and worldwide effort. The scourge that had lasted millennia was finally declared ended on 8 May 1980, by the 33rd World Health Assembly. Successes in mass vaccination programs that have eliminated or greatly reduced infections and deaths from polio, yellow fever, and measles have followed (WHO 2021c). Vaccine development, achievements, and challenges will be further discussed in Chapter 4: Prevention, Diagnosis, and Treatment.

Despite vaccine breakthroughs, however, several novel viruses have emerged within the last century that continue to threaten humanity. Of these, Ebola virus (and closely related Marburg virus), human immunodeficiency virus (HIV), influenza, and coronaviruses will be discussed in detail in this chapter.

3.2 EBOLA AND MARBURG VIRUSES

Among the deadliest *pathogens* in existence are viruses from the Filoviridae family, which cause hemorrhagic fever, a critical illness that effects multiple organ systems, causes high fevers, and damages the cardiovascular system and blood vessels, causing severe systemic bleeding. These viruses include several species of Ebola virus and the Marburg virus. The infections, which are clinically similar, have *case fatality rates* (CFRs)—the proportion of deaths to those who contract the illness—of 24–88% for Marburg and up to 90% in the case of Zaire Ebola Disease (WHO 2021a).

Marburg viral disease (MVD) was first discovered in its namesake city in Germany in 1967, when a laboratory incident caused a local outbreak that also led to outbreaks in Frankfurt and Belgrade, Serbia. The source of the outbreak was traced back to monkeys imported from Uganda. Outbreaks have since occurred in several countries on the African continent. Most cases of human-to-human transmission involve direct contact through bodily fluids such as blood, feces, saliva, vomit, or semen. Direct transmission to humans from animal hosts, such as bats or monkeys, have also caused outbreaks (WHO 2021d).

Ebola virus disease (EVD)—previously referred to as Ebola hemorrhagic fever—has a similar *pathogenicity* to Marburg and has caused several outbreaks since its discovery in Central Africa in 1976, including a protracted and widespread epidemic from 2014 to 2016. Like MVD, EVD can be spread to humans through contact with bats, who are thought to be natural hosts (i.e., are carriers of the virus), but do not become sick from it. Monkeys and other primates can also transmit the virus, but, like humans, can become ill from it. Ebola is transmitted through bodily fluids and can be spread by contact with contaminated surfaces and fabrics. This method of transmission leaves healthcare workers particularly susceptible to infection, and, without the strictest infection control precautions, has led to many deaths among first line workers. Ebola virus can survive for several days after death of the host, making the transport and burial of bodies extremely hazardous. CFRs for EVD range from 25 to 90%, with an overall average of 50% (WHO 2021a).

3.2.1 Symptoms

According to the World Health Organization (WHO), the *incubation period* from infection to symptom onset ranges from two days to three weeks. Ebola is not transmissible until symptoms begin, and then patients remain infectious as long as the virus remains in their blood. Recent studies have revealed that viral load in semen is high enough among survivors that sexual transmission of Ebola is possible potentially for years after initial infection (Thorson et al. 2021).

Sudden onset of symptoms is common and include (WHO 2021a):

- Fever
- Fatigue
- Muscle pain
- Headache
- Sore throat
- Loss of appetite

Initial symptoms are quickly followed by:

- Unrelenting high fever
- Unexplained hemorrhaging, bleeding, or bruising
- Abdominal pain
- Vomiting
- Diarrhea
- Signs of kidney and liver failure
- Rash

In fatal cases of Ebola, death usually occurs within eight to 10 days with severe blood loss and shock, a critical condition caused by a sudden and significant loss of blood flow within the body. Patients who survive Ebola can often experience long-lasting symptoms, including impacts to the brain and other parts of the central nervous system, and may be critically ill for several weeks. Survivors have measurable antibodies for up to 10 years after infection. It is believed that survivors may have some natural immunity that would prevent reinfection by the same strain of Ebola that sickened them (CDC 2021b).

3.2.2 Diagnosis

Since Ebola and Marburg are such deadly pathogens, early detection and accurate diagnosis are critical. As soon as they suspect a case, healthcare workers must use the strictest precautions possible to avoid direct contact with the virus in order to protect themselves, other patients, and the community from an outbreak. A complicating factor in the diagnosis of EVD and MVD is that the early symptoms can mimic many other illnesses, such as malaria, typhoid fever, shigellosis, meningitis, and other viral infections. It is also important to differentiate between Marburg and Ebola, since they are so clinically similar, and determine the strain of the virus. Finding the origin of any outbreak, no matter how small, and vigilance with contact tracing and quarantine are critical to isolating potential carriers and curbing transmission.

The following are the diagnostic tools recommended by the CDC for both Ebola and Marburg confirmation: (CDC 2021b)

- Antibody-capture enzyme-linked immunosorbent assay (ELISA)
- Antigen-capture detection tests
- Serum neutralization test
- Reverse transcriptase polymerase chain reaction (RT-PCR) assay
- Electron microscopy
- Virus isolation by cell culture

Any collected samples of EVD and MVD are highly lethal, Biohazard Level 4 (BSL-4) contaminants, and must be transported accordingly. For further details regarding safe handling of infectious materials, the reader is referred to Chapter 5: Safety Protocols and Personal Protective Equipment. In addition, details on virus diagnostic methods are available in Chapter 4: Prevention, Diagnosis, and Treatment (Bayot and King 2021).

3.2.3 Prevention and Treatment

The keys to preventing outbreaks of hemorrhagic diseases such as Ebola and Marburg are vigilance, early detection, healthcare facility training and preparedness, as well as community support and involvement. To limit spread by sexual transmission, it is suggested that male survivors use safe sex practices for at least one year from onset of symptoms or until they have twice tested negative for the virus.

Treatment options for Ebola and Marburg are limited. However, early interventions with fluids and treatment of fever and other symptoms increase chances of survival. In 2020, two *monoclonal antibody treatments* were approved in the United States by the Food and Drug Administration (FDA), specifically for the Zaire Ebola virus strain, which is the most pathogenic. A vaccine was also approved in late 2019 by the FDA as a single shot dose for those at high risk to exposure (WHO 2022d).

3.3 HUMAN IMMUNODEFICIENCY DISEASE (HIV)

Acquired immunodeficiency syndrome (AIDS) is caused by HIV. HIV is a *retrovirus* that attacks and destroys the immune system, specifically *CD4+ T-cells*, which play a critical role in the immune system's ability to recognize invasion. When HIV infects and destroys these cells, the body cannot mount an effective defense against infections from bacteria, fungus, parasites, or other viruses (Battistini Garcia and Guzman 2021).

Epidemiological studies have traced the origin of HIV to simian immunodeficiency virus (SIV) among chimpanzees. It is believed that the virus may have crossed over to humans as early as the late seventeenth century in Central Africa among hunters who came into contact with the blood of the animals, and that the virus then slowly spread (and continued to mutate) across Africa and then globally. HIV arrived in the United States as early as the mid-1970s. AIDS, first recognized in 1981 primarily in the gay communities of New York and San Francisco, reached epidemic proportions in the early 1980s in the United States and around the world. Apparently healthy young men were suddenly struck by infections that normally plague immunosuppressed patients – with diseases such as Pneumocystis pneumonia, cytomegalovirus, mucosal candida infections, and the malignant Karposi sarcoma, a previously rare cancer that causes lesions on the skin, mucosa, lymph nodes, and internal organs (CDC 2021c).

In 1982, the year the term "AIDS" was coined, HIV was recognized as a global emergency. Blood was identified as a key method of transmission as the CDC identified the first case of HIV from a blood transfusion, and transmission through drug use and from exposure to healthcare settings were also reported. In 1983, all major routes of transmission were identified, and the scientific community was informed by the CDC that HIV could not be spread through casual contact with an infected person (CDC 2021c; CDC 2022b).

By 1992, AIDS had become the number one killer of men aged 25–44 in the United States. While the first antiretroviral therapy was introduced in 1987, the standard of treatment referred to as HAART—highly active antiretroviral therapy, which was introduced in 1996—changed the landscape of the epidemic. The year 1997 marked the first year of significant decline in AIDS deaths, with a 47% drop in mortality from that of the prior year. Since then, great effort has been made to improve public awareness of safe sex practices and disposable needle use, as well to protect international blood supplies (CDC 2021c).

Globally, case rates have exploded over the last thirty years, with more than 37 million people currently living with HIV. Africa is the current epicenter of the global HIV epidemic, with nearly two-thirds of cases worldwide. Although the survival rate for those living with HIV has greatly improved since the advent of antiretrovirals, an estimated 36 million people have died from AIDS (WHO 2022d).

3.3.1 HIV SYMPTOMS

3.3.1.1 Stage 1: Acute Infection

According to the CDC, the acute phase of virus may be asymptomatic or may be flu-like and occur between two and four weeks following infection. Symptoms, if any, may include the following (CDC 2021c):

- Fatigue
- Fever
- Chills
- Muscle aches
- Night sweats
- Sore throat
- Rash
- Swollen lymph nodes
- Mouth ulcers

During the acute phase of infection with HIV, the viral load is very high, as is the possibility of infecting others through sexual contact, shared needles, blood donation, and from pregnant mother to fetus (CDC 2021c).

3.3.1.2 Stage 2: Chronic HIV Infection (Latent Phase)

The latent phase of HIV infection is asymptomatic and can last as long as a decade or more, even without treatment. In some patients, the course of the disease progresses more quickly. With proper oral antiretroviral treatment, taken exactly as prescribed for the rest of their lives, patients may be able to manage their HIV disease as a chronic illness and remain in the latent phase indefinitely.

For those with access to advanced medicines and healthcare, HIV has transformed from a death sentence to a manageable, life-long chronic disease. But despite years of research, there is still no cure and no vaccine available.

3.3.1.3 Stage 3: Acquired Immunodeficiency Syndrome (AIDS)

The final stage of HIV is AIDS, which is the full expression of the disease, and is the stage at which the patient is severely ill. Medically, AIDS is defined as an HIV diagnosis and a CD4+ T-cell count below 200 cells/mm^3 (In a healthy person, a normal CD4+ count is 500–1000 cells/mm^3.) At this stage, the patient is highly infectious with very high viral loads. AIDS is the most severe phase of HIV infection, with typical survival of approximately three years without treatment. Most deaths occur due to opportunistic infections, such as those mentioned previously (Battistini Garcia and Guzman 2021).

3.3.2 DIAGNOSIS

In recent years, HIV diagnosis has advanced significantly, and it is now possible to diagnose infections much earlier than in the past. The eclipse period, which indicates the time since exposure in which it is too early to test for infection, has shortened from six months to less than one month.

Current HIV tests include (CDC 2021c):

- Nucleic acid tests (NATs)
- Antigen/antibody combination tests
- Antibody tests

NATs detect virus *ribonucleic acid (RNA)*, and of all available testing, can be used at the earliest time since exposure. Results are usually accurate within 10 days to one month. Antibody tests measure both the level of *immunoglobulin M (IgM)* and *immunoglobulin G (IgG)*, antibodies which detect recent and past exposure to the virus. The antigen test measures a portion of the viral capsid – the *p24 antigen* – that can detect the presence of the HIV virus before the body begins to produce any antibodies. The combination antigen/antibody tests have a window period of 18–45 days. The standalone antibody test may not be accurate until three months after infection (NIAID 2020).

3.3.3 HIV PREVENTION AND TREATMENT

While antiretroviral therapies and diagnostic tools have improved greatly since the early days of the HIV epidemic, a cure or vaccine has been elusive despite decades of research and international investment. This failure attests to the unique challenges of retroviruses, which utilize *reverse transcriptase* to insert their own viral code directly into the host cell DNA. More detail on the different types of viruses and their mode of infection was discussed earlier in Chapter 2.

In the absence of a vaccine, HIV prevention through condom use, disposable needle use, comprehensive infection precautions in healthcare settings, and regular screenings are all highly effective. When exposure is suspected, *post-exposure prophylaxis (PEP)*, the immediate use of a combination of antiretroviral medications as a precaution is recommended within the first 72 hours after contact. PEP is a standard procedure in healthcare settings whenever a healthcare worker has potentially come into contact with blood or fluids from an individual of unknown HIV status. It is also recommended for the public after possible exposure from unprotected sexual encounter or drug use (WHO 2022d).

Medical prophylaxis – taking regular antiretroviral medications – for those at high risk of contracting HIV is considered safe and is highly effective. Pre-exposure prophylaxis (PrEP), first approved in the United States in 2012, is an exceptionally successful combination of antiretroviral medications. When taken as prescribed, PrEP can reduce HIV risk from sex by 99% and risk from IV drug use by at least 74%. In general, medications approved for PrEP are well-tolerated and do not cause major side effects, with adverse reactions mainly gastrointestinal in nature, such as nausea and diarrhea. Longer-term side effects related to renal or liver toxicity are usually short-lived (Tetteh et al. 2017; CDC 2021c). Antiretroviral therapies are discussed further in Chapter 4: Prevention, Diagnosis, and Treatment.

3.4 INFLUENZA

Influenza viruses, which are a part of the *Orthomyxoviridae* virus family, are designated types A through D. Influenza A and B viruses circulate and cause seasonal epidemics ranging from mild to severe. In addition, influenza A can cause pandemics when new strains emerge in humans, usually originating from within the animal population. These are referred to as zoonotic viruses and are the most likely to cause severe disease due to a lack of natural protection in the human population. Seasonal influenza refers to the annual epidemics of influenza that emerge around the globe during the coldest months. According to the WHO, there are three to five million influenza cases yearly, causing between 290,000 and 650,000 deaths worldwide (WHO 2018a).

Influenza A, which is the only type that has so far caused pandemics, is further broken down into several subtypes that are designated by two of their key surface proteins, hemagglutinin (HA) and neuraminidase (NA).

While seasonal influenza epidemics occur every year, the transmissibility and severity of the circulating virus varies depending on several factors, including mutations and natural immunity of the population from prior exposure, if any. Influenza A has the capability to rapidly evolve since it is highly adaptable and prone to both *antigenic shift* and *antigenic drift*. This topic was discussed in more detail earlier in Chapter 2.

3.4.1 INFLUENZA SYMPTOMS

Influenza can cause mild to severe illness, with a sudden onset of symptoms following an incubation period of one to four days. Pre-symptomatic transmission – passing of the virus to another person even before there are symptoms – is possible one day prior to symptom onset, which increases the chance of spread of an already highly contagious pathogen such as influenza.

Symptoms include (WHO 2018a):

- Fever and chills
- Fatigue
- Headache
- Vomiting and diarrhea (more common in children)
- Sore throat
- Runny/stuffy nose
- Muscle or body aches
- Dry cough

Seasonal influenza usually subsides in healthy people within one to two weeks without medical intervention, with a cough possibly persisting. However, influenza can cause severe illness or death, especially in people at high risk, such as the immunocompromised, elderly, or those in lower-income countries without easy access to healthcare (WHO 2018a).

3.4.2 INFLUENZA DIAGNOSIS

According to the CDC, several diagnostic tools to detect influenza A or B are currently available. Approved tests are shown in Table 3.1, below. (CDC 2020)

3.4.3 INFLUENZA PREVENTION AND TREATMENT

Each year, the CDC analyzes the global trends of influenza subtypes circulating in the population. It then makes recommendations, based on epidemiological modeling, as to the most likely landscape of the next seasonal epidemic,

Table 3.1 Influenza A and B Diagnostic Testing Techniques and Timing of Results.

Test Type	Result Timing
Rapid Influenza Diagnostic Tests (antigen detection)	15 minutes
Rapid Molecular Assay (viral RNA or nucleic acid detection)	15–30 minutes
Immunofluorescence, Fluorescent Antibody Staining (antigen detection)	1–4 hours
RT-PCR7 and other molecular assays (viral RNA or nucleic acid detection)	1–8 hours, varies by test
Rapid Cell Culture (yields live virus)	1–3 days
Viral Tissue Cell Culture (yields live virus)	3–10 days

commonly referred to as "flu season." Influenza vaccines targeting those subtypes are synthesized by various manufacturers as approved by the FDA. The influenza vaccines are either trivalent or quadrivalent, signifying that they are designed to target either three or four different strains of the virus, typically two type A and one or two type B strains. The CDC recommends that everyone six months and older receive the influenza vaccine annually (CDC 2021e).

There are several antiviral treatment options for patients who become infected with the flu. However, for the treatment to be effective, it should be started within 48 hours of the patient developing symptoms. Antivirals can shorten the length and severity of the illness and reduce the risk of complications (CDC 2021e).

3.4.4 INFLUENZA PANDEMICS

Influenza pandemics occur when a novel strain of influenza A emerges, to which populations have no natural immunity and which has transformed enough from previous subtypes so that existing vaccines are inadequate. Usually, the source is a currently circulating zoonotic virus, such as avian (bird) or swine flu, that then jumps from animal to human. For this influenza strain to reach pandemic proportions, the virus would have to adapt in its ability to infect humans, such that sustained human-to-human transmission is possible.

Zoonotic influenza viruses are carefully tracked by the WHO, the CDC, and many other organizations and governments worldwide. Surveillance is crucial to preventing such outbreaks from spiraling out of control. Several avian viruses are of great concern due to their potential to mutate, such as Influenza A(H5) and A(H7), which can cause severe disease. Although rare cases of patients contracting the disease directly from another patient have been reported, there has not been sustained human-to-human transmission (WHO 2018b).

The H1N1 influenza pandemic of 1918 was by far the worst pandemic in the twentieth century, with an estimated two-thirds of the world's then-population inflicted, and at least 50 million deaths. Unlike other influenza outbreaks, previously young, healthy adults were most at risk of dying. This particular H1N1 virus has been vigorously studied, yet its origins are still unclear (CDC 2018).

Other notable pandemics include:

- 1957–1958 Pandemic (H2N2 virus) of avian origin
- 1968 Pandemic (H3N2 virus) of avian origin
- 2009 Pandemic (H1N1)pdm09 new strain not previously identified

Further details on pandemic preparedness are available in Part II, Chapter 9: Emergency Planning and Response.

3.5 CORONAVIRUSES

Most coronaviruses, which were first identified in the 1960s, are zoonotic, infecting animals including bats, cats, pigs, and camels. There are seven coronaviruses known to have spilled over to infect humans, of which four are responsible for the common cold. The other three – SARS-CoV, MERS-CoV, and SARS-CoV-2 – are considerably more pathogenic and capable of causing epidemics and pandemics to varying degree (NIAID 2022).

SARS-CoV was first detected in early 2003 (WHO 2022d) in China when an outbreak arose of what is now called severe acute respiratory syndrome. Within a few months, it had spread to multiple countries, primarily through travel. Before the virus disappeared in 2004, eight thousand people had become ill. As the first novel virus to emerge in the twenty-first century, SARS-CoV garnered much international attention from the scientific community. Research for an effective SARS vaccine has been ongoing (WHO 2022d; NIAID 2022).

Middle East respiratory syndrome (MERS) emerged as a highly virulent infection in 2012 in Saudi Arabia. MERS coronavirus (MERS-CoV) was identified as the culprit, and the second new deadly coronavirus to emerge since 2003, although this one was transmitted to humans by ill camels. MERS-CoV continues to cause outbreaks sporadically, and most infections can be traced back directly to contact with camels. In the healthcare setting, some cases of human-to-human transmission have been documented, but fortunately this has not been a significant route of infection to-date. The mortality rate is very high, estimated at 30–35%. The MERS virus' poor ability to spread efficiently among humans has prevented larger epidemics (WHO 2019).

In December 2019 in China, the third novel coronavirus of the twenty-first century emerged and was identified in January 2020 as SARS-CoV-2. Within the first three months of discovery, more than one million people around the world were infected, and more than 50,000 had died. On 11 March 2020, the WHO declared a global pandemic of Coronavirus Disease 2019. Within six months of the virus' discovery, worldwide cases surpassed 10 million with more than 500,000 deaths (WHO 2020). By the end of 2021, the death toll had surpassed 5 million (WHO 2021e). At the time of this writing, the world is nearing the two-year mark of the COVID-19 pandemic. No corner of the world has been unaffected, and global deaths have exceeded 6 million as of July 2022 (WHO 2022b). It would not be an overstatement to say that it collectively feels as though there will always be a "BEFORE" COVID and an "AFTER" COVID.

One notable legacy of the SARS and MERS epidemics was the push to understand and to design efficient vaccines and antiviral treatments for the viruses that cause them. Building on that wealth of knowledge and research has allowed significant breakthroughs, with highly effective vaccines for COVID-19 brought to market within a year of the start of the current pandemic. This would have been impossible had these advancements—such as the development of mRNA technology for vaccine delivery—not already been in progress, in some cases for nearly a decade (NIAID 2022).

3.5.1 Symptoms

As the names suggest, SARS (severe acute respiratory syndrome), MERS (Middle East respiratory syndrome), and COVID-19 are primarily respiratory diseases. They often begin with fever and cough, can lead to difficulty in breathing, pneumonia, and low blood oxygen levels. The effects of the diseases vary greatly in severity, depending on both the virus and the overall health of the patient. MERS differs from the other coronaviruses in its tendency to cause acute renal (kidney) failure, and has the highest CFR, at up to 30% (Hu et al. 2020; WHO 2019).

The rest of this subsection will focus exclusively on the symptoms, diagnosis, and prevention of COVID-19. Further details on MERS and SARS are available through the WHO and the CDC, with significant research available (WHO 2019; CDC 2019).

Although the medical community still has much to uncover regarding COVID-19, the virus has shown that it differs from other coronaviruses in some important ways. While COVID-19 is primarily a respiratory disease and can cause an atypical pneumonia, it can also cause an array of other symptoms that can affect almost every organ system and that can range from mild to fatal. As many as 40% of patients may have no symptoms at all but can still be highly contagious – one of the key reasons SARS-CoV-2 has managed to spread so quickly. Onset of symptoms usually occurs two days to two weeks after exposure and may include any or all of the following listed in subsection 3.5.1.1.: (NIAID 2022; Piret and Boivin 2021)

3.5.1.1 Typical Acute Symptoms

- Fever and/or chills
- Loss of taste or smell
- Cough
- Shortness of breath/difficulty in breathing
- Fatigue
- Muscle or body aches
- Headache
- Sore throat
- Congestion or runny nose
- Nausea or vomiting
- Diarrhea

The above list of acute symptoms describes a typical progression of the illness. Those patients requiring hospitalization and *intubation* – insertion of a breathing tube for mechanical ventilation – will face prolonged challenges and rehabilitation.

As mentioned above, pneumonia is common in patients with COVID-19. It is atypical from other causes of pneumonia. Many patients with COVID pneumonia may seem to be mildly short of breath and yet are actually critically ill with extremely low oxygen levels in their blood. At the beginning of the pandemic, this aspect of COVID pneumonia was a new experience for the healthcare workers caring for them, and there were many reports worldwide of the shockingly rapid decline and death of patients who only moments ago were speaking normally.

In addition, some patients who have recovered from COVID-19, even those with mild or asymptomatic cases, develop what are now known as post-COVID conditions. These symptoms can last months to at least a year (and possibly much longer). Research is ongoing, but these cases can include the following symptoms listed in subsection 3.5.1.2. (CDC 2021f)

3.5.1.2 Post-COVID Conditions

- Shortness of breath
- increased respiratory effort
- Loss of taste or smell
- Post-exertional exhaustion
- Heart palpitations/rapid heart rate
- Brain fog or cognitive impairment
- Insomnia and other sleep difficulties
- Dizziness or lightheadedness
- Impaired daily function and mobility
- Rash (e.g. urticaria)
- Menstrual cycle irregularities
- New autoimmune diagnoses

- Fatigue
- Cough
- Chest pain
- Headache
- Joint pain
- Muscle pain
- Tingling or burning pain
- Abdominal pain
- Diarrhea
- Fever
- Mood changes
- Pain

3.5.1.3 COVID-19 Multiorgan System Effects (MIS)

Further confounding recovery from COVID-19 illness in some patients is the development of multisystem inflammatory syndrome (MIS), leading to additional complications after the acute illness has resolved. MIS refers to the unexplained inflammation of any organ system, including heart, lungs, brain, kidneys, eyes, skin, or gastrointestinal tract. The hallmark of MIS is a high, persistent fever, with any of the following (CDC 2021f):

- Stomach pain
- Bloodshot eyes
- Diarrhea

- Dizziness or lightheadedness (signs of low blood pressure)
- Skin rash
- Vomiting

Effects have been documented in almost all body systems, including cardiovascular, pulmonary, renal, dermatologic, neurologic, and psychiatric. The extent and long-term impact of MIS is not fully understood and is currently being investigated. While this is a rare complication, it is a severe condition and has been reported in both children (MIS-C) and adults (MIS-A) (CDC 2021f).

3.5.2 COVID-19 DIAGNOSIS

There are two types of COVID-19 diagnostic tests used to detect a current viral infection and can be used for screening and diagnosis (CDC 2021d):

- Nucleic acid amplification tests (NAATs)
- Antigen tests

According to the CDC, anyone with signs or symptoms of COVID-19 should have diagnostic testing as soon as possible. In addition, regular screening is critical to identifying asymptomatic cases and can prevent further transmission when identified patients isolate immediately. Tests for these purposes that have been designed by the CDC include (CDC 2021d):

- *2019-Novel Coronavirus (2019-nCoV) Real-Time RT-PCR Diagnostic Panel*, which can accurately detect SARS-CoV-2 in respiratory specimens.
- *Influenza SARS-CoV-2 (Flu SC2) Multiplex Assay*, which can simultaneously identify both influenza (types A and B) and SARS-CoV-2.

In addition to antigen testing to identify current cases, antibody testing (also called serology testing) can confirm that a patient had a prior exposure to SARS-CoV-2. Antibody tests should not be used to detect current infection (CDC 2021d).

Figure 3.1 depicts the CDC recommendations for managing quarantine and isolation decisions based on antigen results.

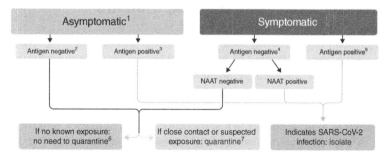

Figure 3.1 CDC SARS-CoV-2 antigen test algorithm for community settings. *Source*: CDC 2021/U.S. Department of Health and Human Services/Public Domain.

3.5.3 COVID-19 PREVENTION AND TREATMENT

WHO global COVID-19 prevention guidelines are listed below (WHO 2022c):

- All adults and children over 12 (where approved) get vaccinated as soon as possible.
- Keep physical distance of at least 1 meter.
- Avoid crowds and close contact.
- Wear a properly fitted mask when physical distancing is not possible and in poorly ventilated settings.
- Clean hands frequently with alcohol-based hand rub or soap and water.
- Cover mouth and nose with a bent elbow or tissue when coughing or sneezing.
- If symptoms develop, get tested and self-isolate.
- If one receives a positive COVID-19 test, self-isolate until recovery.

The WHO has authorized multiple vaccines for emergency use, several of which are listed below (WHO 2022a):

- AstraZeneca/Oxford vaccine (viral vector)
- Bharat Biotech (inactivated virus)
- Janssen/Johnson & Johnson (viral vector)
- Moderna (mRNA)
- Novavax (protein subunit)
- Pfizer/BioNTech (mRNA)
- Sinopharm (inactivated virus)
- Sinovac (inactivated virus)

Vaccine technologies and current and future therapies, including those currently available to protect against COVID-19, will be discussed in Chapter 4: Virus Prevention, Diagnosis, and Treatment.

3.6 CURRENT AND EMERGING VIRAL THREATS

The National Institute of Allergy and Infectious Diseases (NIAID) is charged with the investigation of all emerging infectious diseases that have the potential to threaten the health and security of the United States. The Biodefense and Emerging Infectious Diseases Surveillance program began in 2001 when, in the aftermath of the September 11th terrorist attacks, the nation was again threatened by several *anthrax* mailing attacks. While the scope of those attacks was limited, the danger that terrorists could possibly develop and attack a population with an infectious agent had become clear.

In the subsequent 20 years, NIAID has built a far-reaching infectious disease research infrastructure and has been on the front line of exploration into vaccines

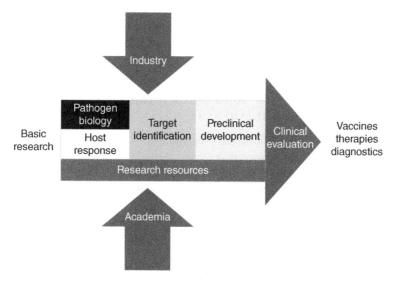

Figure 3.2 The National Institute of Allergy and Infectious Diseases Strategic Research Plan for Biodefense and Emerging Infectious Diseases. *Source*: NIAID 2018/U.S. Department of Health and Human Services/Public Domain.

and therapies against the newest and most dangerous pathogens around the globe. NIAID runs national and regional biocontainment laboratories and partners with academic and clinical centers on key research. An illustration of the comprehensive plan for supporting research in biodefense and emerging infectious diseases is shown in Figure 3.2. (NIAID 2018)

According to NIAID, emerging infectious diseases are defined as those newly appeared in a population or have existed but are rapidly increasing in incidence or geographic range, or that are caused by one of the NIAID Category A, B, or C priority pathogens. NIAID categorizes priority pathogens that have the potential to cause the most widespread damage or have the highest potential to be used as bioweapons (NIAID 2018). The following indexes of priority pathogens are cataloged in coordination with the Department of Homeland Security (DHS), which determines threat assessments, and the CDC, which is responsible for responding to emerging pathogen threats in the United States.

Category A pathogens pose the highest risk to national security and public health due to several factors, which include (NIAID 2018):

- Can be easily disseminated or transmitted from person to person
- Result in high mortality rates
- Have the potential for major public health impact
- Might cause public panic and social disruption
- Require special action for public health preparedness

Viruses included in Category A:

- Variola major (smallpox) and other related pox viruses
- Viral hemorrhagic fevers (including Ebola, Marburg, Hantavirus, Dengue, and Lassa)

Category B pathogens are the second highest priority organisms/biological agents for the following reasons (NIAID 2018):

- Require special action for public health preparedness
- They are moderately easy to disseminate
- Result in moderate morbidity rates and low mortality rates
- Require specific enhancements for diagnostic capacity and enhanced disease surveillance

Viruses included in Category B:

- Caliciviruses
- Hepatitis A
- Mosquito-borne viruses (including West Nile, equine encephalitis, yellow fever, zika, and chikungunya viruses)

Category C pathogens are the third highest priority and include emerging pathogens that could be engineered for mass dissemination in the future due to the following (NIAID 2018):

- Availability
- Ease of production and dissemination
- Potential for high morbidity and mortality rates and major health impact

Viruses included in Category C:

- Nipah and Hendra viruses
- Additional hantaviruses
- Tickborne hemorrhagic fever viruses (Bunyaviruses, Flaviviruses)
- Tickborne encephalitis complex flaviviruses/Tickborne encephalitis viruses
- Influenza virus
- Rabies virus
- Severe acute respiratory syndrome associated coronavirus (SARS-CoV), MERS-CoV, and other highly pathogenic human coronaviruses

Additional Emerging Viral Diseases:

- Australian bat lyssavirus
- BK virus
- Enterovirus 68 and 71
- Hepatitis C and E
- Human herpesvirus 6 and 8

- JC virus
- Poliovirus
- Rubeola (measles)

The research and clinical testing sites of the NIAID allow for extensive clinical trial capacity and expertise. This played a key role in testing the vaccine for the 2009 H1N1 influenza pandemic, as well as medical research on therapies against SARS and MERS. The years-long search for a vaccine is, in part, what enabled the speed with which scientists were able to develop a highly effective vaccine against COVID-19 in record time (NIAID 2022d).

While NIAID is responsible for all potential biological risks, for the purposes of this book, only viruses have been listed here. The reader is referred to the reference section for further information regarding the NIAID website.

3.7 APPLICATIONS

The following four illustrative examples are intended to complement the above material and to provide a better understanding of aspects discussed within the chapter.

Illustrative Example 3.1 Ebola
Examine the significance of biohazard level in handling Ebola virus.

Solution
Due to its virulence, Ebola virus handling requires adherence to biosafety level 4 requirements (BSL-4). BSL-4 is reserved for the most toxic and dangerous biohazards. It is the highest biohazard level, with a few specialized microbiology laboratories able to achieve such standards. Biosafety safeguards, along with proper training, are essential to protect healthcare workers, laboratory personnel, and the surrounding community.

Illustrative Example 3.2 HIV
Describe in detail HIV transmission and prevention methods.

Solution
According to the WHO, HIV main modes of transmission are through bodily fluids of infected people, including blood, semen, vaginal secretions, and breast milk. There is also risk of vertical transmission from mother to fetus.

HIV prevention includes frequent testing for both HIV and other sexually transmitted diseases, regular condom use, safe IV drug use, medical male circumcision, antiretroviral use for prevention, and elimination of mother-to-child transmission of HIV through testing and treatment (WHO 2022d).

Illustrative Example 3.3 INFLUENZA
Briefly describe both avian and swine flu subtypes (influenza A) listed below, including transmission and symptoms in humans.

- *Avian influenza virus subtypes include A(H5N1), A(H7N9), and A(H9N2).*
- *Swine influenza virus subtypes include A(H1N1), A(H1N2), and A(H3N2).*

Solution
Most zoonotic cases are a result of direct contact with infected animals and rarely involve human-to-human transmission. Infections may be mild, usually including fever and cough, but more serious cases can have varying symptoms. Severe pneumonia, septic shock, acute respiratory distress syndrome (ARDS), and death have all been reported. Other possible symptoms include encephalitis, conjunctivitis, and gastrointestinal distress.

Illustrative Example 3.4 SARS-CoV-2
Explain the significance of the basic reproduction number (R_0) and how it applies to the COVID-19 pandemic and to the other corona pandemics.

Solution
The basic reproduction number (R_0) reflects the rate of disease transmission. For a virus to cause a sustained outbreak in a population, it must be contagious enough for at least one person to spread it to one other susceptible person. Otherwise, the outbreak and the virus will die out. Therefore, if $R_0>1$, it will have the transmissibility necessary to cause a sustained epidemic among humans. The R_0 of a virus is mainly based on viral characteristics, but it is not constant and involves many variables, including the density and vulnerability of hosts in one area. SARS-CoV finally disappeared in 2004 after causing a high rate of disease and death. Its R_0 ultimately dropped below one. MERS-CoV continues to reappear because it is endemic to camels and can make the occasional species jump to humans, but its R_0 is very low. SARS-CoV-2 is less virulent than its predecessors, but it has a much higher infection rate. Its R_0 fluctuates depending on location, social distancing, and vaccine uptake. It has been estimated to be within a range of 1.4–2.6 (Achaiah et al. 2020; Hu et al. 2020).

3.8 CHAPTER SUMMARY

- For humans, there are hundreds of viral species known to cause disease.
- Viral infections are among the most challenging illnesses to treat.
- Pandemics and outbreaks have changed the course of history.
- Smallpox was the first virus to be eradicated by vaccine through an immense, well-coordinated worldwide effort.

- Several novel viruses have emerged within the last century that continue to threaten humanity.
- Viruses that cause hemorrhagic fever, such as Ebola and Marburg, from the Filoviridae family, are among the deadliest pathogens in existence.
- Global HIV cases rates have exploded over the last 30 years with more than 37 million people currently living with the virus.
- Seasonal influenza refers to the annual epidemics of influenza that emerge around the globe during the coldest months.
- On March 11, 2020, the WHO declared a global pandemic of Coronavirus Disease 2019.
- While COVID-19 is primarily a respiratory disease and can cause an atypical pneumonia, it can also cause an array of other symptoms.

3.9 PROBLEMS

1 Which type of immune cell does human immunodeficiency virus target?

2 What are four diseases that patients with late-stage AIDS are likely to acquire due to their greatly damaged immune system?

3 What are the natural reservoirs for MERS-CoV and SARS-CoV?

4 Explain the different modes of transmission for ebola and influenza.

REFERENCES

Achaiah, N.C., Subbarajasetty, S.B., and Shetty, R.M. (2020). R0 and Re of COVID-19: can we predict when the pandemic outbreak will be contained? *Indian Journal of Critical Care Medicine: Peer-Reviewed, Official Publication of Indian Society of Critical Care Medicine* 24 (11): 1125–1127. https://doi.org/10.5005/jp-journals-10071-23649.

Battistini Garcia, S.A. and Guzman, N. (2021). *Acquired Immune Deficiency Syndrome CD4+ Count*. In: *StatPearls*. Treasure Island (FL): StatPearls Publishing https://www.ncbi.nlm.nih.gov/books/NBK513289/.

Bayot, M.L. and King, K.C. (2021). *Biohazard Levels*. StatPearls Treasure Island, FL: StatPearls Publishing https://www.ncbi.nlm.nih.gov/books/NBK535351/.

Becker, T., Husni, E., Reperant, L. et al. (2021). Influenza vaccines: successes and continuing challenges. *The Journal of Infectious Diseases* 224 (Supplement_4): S405–S419. https://doi.org/10.1093/infdis/jiab269.

Bouvier, N.M. and Palese, P. (2008). The biology of influenza viruses. *Vaccine* 26 (Suppl 4): D49–D53. https://doi.org/10.1016/j.vaccine.2008.07.039.

Centers for Disease Control and Prevention (CDC) (2018). History of 1918 flu pandemic. Centers for Disease Control and Prevention. https://www.cdc.gov/flu/pandemic-resources/1918-commemoration/1918-pandemic-history.htm (accessed 19 October 2021).

Centers for Disease Control and Prevention (CDC) (2019). *About Middle East respiratory syndrome (MERS)* Centers for Disease Control and Prevention. https://www.cdc.gov/coronavirus/mers/about/index.html (accessed 8 July 2022).

Centers for Disease Control and Prevention (CDC) (2020). Influenza virus testing methods. Centers for Disease Control and Prevention. https://www.cdc.gov/flu/professionals/diagnosis/table-testing-methods.htm (accessed 19 October 2021).

Centers for Disease Control and Prevention (CDC) (2021a). History of smallpox. Centers for Disease Control and Prevention. https://www.cdc.gov/smallpox/history/history.html (accessed 20 October 2021).

Centers for Disease Control and Prevention (CDC) (2021b). Signs and symptoms. CDC. https://www.cdc.gov/vhf/ebola/symptoms/index.html (accessed 17 October 2021).

Centers for Disease Control and Prevention (CDC) (2021c). About HIV/AIDS. Centers for Disease Control and Prevention. https://www.cdc.gov/hiv/basics/whatishiv.html (accessed 17 October 2021).

Centers for Disease Control and Prevention (CDC) (2021d). CDC diagnostic tests for covid-19. Centers for Disease Control and Prevention. https://www.cdc.gov/coronavirus/2019-ncov/lab/testing.html (accessed 4 January 2022).

Centers for Disease Control and Prevention (CDC). (2021e). About flu. Centers for Disease Control and Prevention. https://www.cdc.gov/flu/about/index.html (accessed 8 July 2022).

Centers for Disease Control and Prevention (CDC). (2021f). Post-covid conditions: Information for healthcare providers. Centers for Disease Control and Prevention. https://www.cdc.gov/coronavirus/2019-ncov/hcp/clinical-care/post-covid-conditions.html (accessed 21 October 2021).

Centers for Disease Control and Prevention (CDC) (2022a). *Guidance for antigen testing for SARS-COV-2 for healthcare providers testing individuals in the community*. Centers for Disease Control and Prevention. https://www.cdc.gov/coronavirus/2019-ncov/lab/resources/antigen-tests-guidelines.html (accessed 8 July 2022).

Centers for Disease Control and Prevention (CDC) (2022b). HIV/AIDS timeline. *Centers for Disease Control and Prevention*. https://npin.cdc.gov/pages/hiv-and-aids-timeline#1980 (accessed 20 September 2022).

Hu, T., Liu, Y., Zhao, M., Zhuang, Q., Xu, L., & He, Q. (2020). A comparison of COVID-19, SARS and MERS. *PeerJ*, 8, e9725. https://doi.org/10.7717/peerj.9725.

Kramer, L. D. (2020). Overview of viruses. *Merck Manuals Professional Edition.* https://www.merckmanuals.com/professional/infectious-diseases/viruses/overview-of-viruses (accessed 23 June 2021).

Mahmoudabadi, G. and Phillips, R. (2018). A comprehensive and quantitative exploration of thousands of viral genomes. *eLife* 7: e31955. https://doi.org/10.7554/eLife.31955.

National Institute of Allergy and Infectious Diseases (NIAID) (2018). *Emerging Infectious Diseases/Pathogens.* U.S. Department of Health and Human Services (HHS). https://www.niaid.nih.gov/research/emerging-infectious-diseases-pathogens (accessed 30 October 2021).

National Institute of Allergy and Infectious Diseases (NIAID). (2020). *HIV/AIDS Treatment.* U.S. Department of Health and Human Services (HHS). https://www.niaid.nih.gov/diseases-conditions/hiv-treatment (accessed 18 October 2021).

National Institute of Allergy and Infectious Diseases (NIAID). (2022). *Coronaviruses.* U.S. Department of Health and Human Services (HHS). https://www.niaid.nih.gov/diseases-conditions/coronaviruses (accessed 20 September 2022).

Piret, J. and Boivin, G. (2021). Pandemics throughout history. *Frontiers in Microbiology* 11: 631736. https://doi.org/10.3389/fmicb.2020.631736.

Roizman, B. (1996). Multiplication. In: *Medical Microbiology*, 4e (ed. S. Baron). Galveston, TX: University of Texas Medical Branch at Galveston.

Tetteh, R.A. et al. (2017). Pre-exposure prophylaxis for HIV prevention: Safety concerns. Drug safety. https://www.ncbi.nlm.nih.gov/pmc/articles/PMC5362649/ (accessed 18 October 2021).

Thorson, A.E. et al. (2021). Persistence of Ebola virus in semen among Ebola virus disease survivors in Sierra Leone: a cohort study of frequency, duration, and risk factors. *PLoS Medicine* 18 (2): e1003273. https://doi.org/10.1371/journal.pmed.1003273.

World Health Organization (WHO) (2016). *Smallpox vaccines.* World Health Organization. https://www.who.int/news-room/feature-stories/detail/smallpox-vaccines (accessed 13 July 2022).

World Health Organization (WHO) (2018a). Influenza (seasonal). World Health Organization. https://www.who.int/news-room/fact-sheets/detail/influenza-(seasonal) (accessed 19 October 2021).

World Health Organization (WHO) (2018b). *Influenza (avian and other zoonotic).* World Health Organization. https://www.who.int/news-room/fact-sheets/detail/influenza-(avian-and-other-zoonotic) (accessed 8 July 2022).

World Health Organization (WHO) (2019). *Middle East respiratory syndrome coronavirus (MERS-COV).* World Health Organization. https://www.who.int/news-room/fact-sheets/detail/middle-east-respiratory-syndrome-coronavirus-(mers-cov) (accessed 8 July 2022).

World Health Organization (WHO) (2020). *Who director-general's opening remarks at the media briefing on COVID-19 - 29 June 2020*. World Health Organization. https://www.who.int/director-general/speeches/detail/who-director-general-s-opening-remarks-at-the-media-briefing-on-covid-19---29-june-2020 (accessed 8 July 2022).

World Health Organization (WHO) (2021a). Ebola virus disease. World Health Organization. https://www.who.int/news-room/fact-sheets/detail/ebola-virus-disease (accessed 16 October 2021).

World Health Organization (WHO) (2021b). Technical update on prevention of HIV transmission in healthcare settings. (n.d.). https://www.who.int/hiv/pub/toolkits/HIV%20transmission%20in%20health%20care%20setttings.pdf (accessed 18 October 2021).

World Health Organization (WHO) (2021c). *Immunization coverage*. World Health Organization. https://www.who.int/news-room/fact-sheets/detail/immunization-coverage (accessed 9 July 2022).

World Health Organization (WHO) (2021d). Marburg virus disease. World Health Organization. https://www.who.int/news-room/fact-sheets/detail/marburg-virus-disease (accessed 16 October 2021).

World Health Organization (WHO) (2021e). *Weekly epidemiological update on COVID-19 - 28 December 2021*. World Health Organization. https://www.who.int/publications/m/item/weekly-epidemiological-update-on-covid-19---28-december-2021 (accessed 8 July 2022).

World Health Organization (WHO) (2022a). *Coronavirus disease (covid-19): Vaccines*. World Health Organization. https://www.who.int/news-room/questions-and-answers/item/coronavirus-disease-(covid-19)-vaccines (accessed 11 August 2022).

World Health Organization (WHO) (2022b). *Weekly epidemiological update on COVID-19 - 6 July 2022*. World Health Organization https://www.who.int/publications/m/item/weekly-epidemiological-update-on-covid-19---6-july-2022 (accessed 8 July 2022).

World Health Organization (WHO) (2022c). *Advice for the public on covid-19*. World Health Organization. https://www.who.int/emergencies/diseases/novel-coronavirus-2019/advice-for-public (accessed 8 July 2022).

World Health Organization (WHO) (2022d). Severe acute respiratory syndrome (SARS). World Health Organization. https://www.who.int/health-topics/severe-acute-respiratory-syndrome#tab=tab_1 (accessed 20 September 2022).

World Health Organization (WHO). (2022d). HIV/AIDS. World Health Organization. https://www.who.int/news-room/fact-sheets/detail/hiv-aids (accessed 20 September 2022).

4

Virus Prevention, Diagnosis, and Treatment

Viral diseases are some of the most difficult diseases to treat. The most effective approach is preventing a virus from being able to take hold and cause infection in the first place. To that end, vaccines are the most important tool modern medicine has in controlling and preventing viral disease, and, therefore, averting outbreaks and epidemics. Since the arrival of the earliest vaccines 200 years ago, countless lives have been saved and countless more patients have avoided life-long illness and disability that certain viruses can cause.

Emerging and *zoonotic* diseases are an ever-present threat. This threat may increase as climate change alters ecosystems and geographical patterns of animal migration (i.e. zoonotic sources) of many viruses, such as mosquitos and other carriers. The COVID-19 pandemic has demonstrated how urgent the need is for new technologies in both prevention and treatment of emerging diseases. As the COVID-19 outbreak spreads, and the epidemic became a global pandemic, the world looked for immediate solutions for this fast-spreading and deadly pathogen. Necessity brought worldwide attention, cooperation, and resources to this life-and-death race, and research on existing and nascent technologies that could be used to fight this new menace ramped up rapidly. In December 2020, just

A Guide to Virology for Engineers and Applied Scientists: Epidemiology, Emergency Management, and Optimization, First Edition. Megan M. Reynolds and Louis Theodore.
© 2023 John Wiley & Sons, Inc. Published 2023 by John Wiley & Sons, Inc.

one year after the discovery of SARS-CoV-2 (i.e. the virus that causes COVID-19), the United Kingdom became the first country to begin vaccinating its population. The vaccine was just one of several innovative technologies developed in recent years that have been used to prevent the COVID-19. Different vaccine types will be further discussed in Section 4.2. Furthermore, developments in antivirals and therapies such as monoclonal antibodies have advanced the treatment of acute viral disease and can have a significant impact on survival.

This chapter will examine vaccine types, availability, and recommendations. In addition, successes and challenges—both technological and social—will be discussed.

4.1 VACCINATION SUCCESSES AND CHALLENGES

Thanks in part to a widespread global commitment to vaccination over the last several decades, many diseases that once caused inordinate suffering and death no longer pose the threat to public health they once did. One of the most significant successes in improving global public health to date has been the elimination of smallpox from the planet. Thanks to an unprecedented international campaign, the smallpox virus became the first to be eradicated through vaccination in 1980. Successes in national and community vaccination programs have also curbed or greatly reduced infections and deaths from other several highly infectious diseases, such as polio and yellow fever (WHO 2021c).

Poliomyelitis (polio), which is caused by the poliovirus, is a life-threatening childhood disease that typically strikes children under the age of five. The virus infects the central nervous system – specifically, the spinal cord – and can cause lifelong paralysis (CDC 2019). The polio vaccine was first introduced in the 1950s, and by 2020, the endemic, wild-type version of the disease had been eliminated in all but two countries: Pakistan and Afghanistan. This drop represents a greater than 99% reduction in overall infections. Without close surveillance and continued vaccination efforts, however, the World Health Organization has long warned that polio could return as a global health threat within a decade (CDC 2018b; WHO 2022). In fact, in September 2022, the CDC released a statement alerting healthcare professionals and the media that polio was again circulating in the United States (CDC 2022a).

Measles is an example of both achievement and shortfalls of global vaccination campaigns. The success was the development of a safe and widely available vaccine. From 2002 to 2018, extensive vaccine use reduced mortality from measles by 73% worldwide, preventing an estimated 23.2 million deaths from 2000 to 2018. Given such progress, measles had been added to the World Health Organization's (WHO) list for potential eradication. This goal, however, turned out to be unattainable, due in part to misinformation and vaccine hesitancy. In 2019, measles had once again become a global threat, killing over 200,000 people worldwide. "After a decade-long failure to reach optimal vaccination coverage," the world has seen the highest case rate in more than 20 years, according to a joint statement by the WHO and the CDC (WHO 2019; United Nations 2020).

In recent years, new vaccines have been developed to protect against one of the deadliest diseases, Ebola. The first, Ervebo, was first used to protect lives prior to its approval under Emergency Compassionate Use Authorization during the Ebola outbreaks in Guinea and the Democratic Republic of the Congo (DRC) (WHO 2021a). Ervebo is approved for the Zaire strain of Ebola. It is not effective against other strains or against Marburg viral disease. According to the FDA, Ervebo was determined to be 100% effective in preventing Ebola cases with symptom onset more than 10 days after vaccination (FDA 2019).

Although millions of lives have been saved thanks to international vaccine campaigns, many challenges remain. For public health officials, one of the most pernicious and frustrating obstacles to saving lives through vaccine programs is the growing problem of vaccine hesitancy and the threat that unvaccinated people pose to everyone.

The COVID-19 global pandemic has underscored the difficulties in public health messaging and fighting disinformation campaigns. Vaccine misinformation and distrust among the public have led to significant imbalances in regional vaccine coverage. Vaccine mandates by federal and local governments and private employers are not always popular but have shown to be effective in greatly increasing vaccine compliance.

When COVID-19 vaccines became available in 2021, inequitable distribution of resources and a lack of global collaboration resulted in many low- and middle-income countries unable to access any vaccine supply. The result was a global public health failure that increased the threat of more dangerous variants developing, potentially placing all nations at higher risk. The issue highlighted some of the failures of WHO and the need for better future cooperation, on a global scale, to protect public health in all countries.

4.2 CURRENT VACCINE TECHNOLOGY

Vaccines are designed to trigger a response from the body's immune system when a foreign substance, or *antigen,* is detected. The vaccine mimics the antigen of a specific *pathogen*—a disease-causing organism, such as a virus—so that the body will recognize and be able to mount a defense against any subsequent exposure to that same antigen. Priming the immune system in this way allows it to immediately respond and prevent an infection. There are various types of vaccines. Their formulations affect how they are used, stored, and transported, and how they are administered.

In addition, there are many vaccines in research and development. One of the most promising areas has been in nucleic acid vaccines, such as mRNA. These vaccines have recently been shown to be highly effective against COVID-19. Future mRNA vaccine technology may allow for one vaccine to provide protection against multiple diseases, thus decreasing the number of shots needed for protection against common vaccine-preventable diseases (CDC 2022b).

Similar in design, DNA vaccines have the potential for application to a wide range of infectious diseases. Although success in DNA vaccines has so far been elusive, once developed, these vaccines would, in theory, be straightforward and inexpensive to manufacture. DNA vaccines would also produce strong, durable immunity.

As of 2021, commercially available vaccines, which will be discussed in detail in the next several subsections, include: (OIDP 2021)

- Live-attenuated vaccines
- Inactivated vaccines
- Recombinant subunit vaccines
- Viral vector vaccines
- Messenger RNA (mRNA) vaccines

4.2.1 LIVE-ATTENUATED VACCINES

First developed in the 1950s, *live-attenuated* vaccines are made up of a weakened form of the disease-causing organism that is targeted by the vaccine. An advantage of live-attenuated vaccines is that, for viruses with simple structures, they can be relatively straightforward to manufacture. In addition, because the attenuated virus acts as the *antigen, i.e., as foreign to the body*, the immune response is similar to its response to a natural infection, and therefore highly robust. For some viruses, childhood immunization is enough for life-long protection from disease. The drawback is that these vaccines still deliver a *live*—though significantly weakened—version of the virus, which means that there remains a risk of illness for patients with compromised immune systems. Live-attenuated vaccines are also more complicated to transport and store, since they need to be kept refrigerated. This can limit their usefulness in rural, hard-to-access locations (NIAID 2021b; OIPD 2021).

Examples of live-attenuated include:

- Measles, mumps, and rubella (MMR combination vaccine)
- Chickenpox
- Rotavirus
- Influenza, nasal spray
- Yellow fever

- Polio (oral)
- Smallpox

4.2.2 Inactivated Vaccines

Inactivated vaccines utilize a killed version of an organism as the antigen to stimulate the immune response. Variations on this type of vaccine have been in use since the 19th century. The method used to kill the virus differs, but can be achieved through heat, radiation, or chemical inactivation. The resulting immune response is generally less durable than from the aforementioned live-attenuated vaccines, indicating that both the initial response is less effective, and that this protection will wane over time. Booster shots are usually recommended for inactivated vaccines (NIAID 2021b).

The following viruses are prevented using inactivated vaccines:

- Hepatitis A
- Influenza (injection, not nasal spray)
- Polio (injection, not oral)
- Rabies

4.2.3 Recombinant Subunit Vaccines

Subunit vaccines are comprised of only those viral components necessary to arouse an appropriate immune response from the body, as opposed to containing the entire virus. These highly specific vaccines are safe and effective, particularly because they do not deliver the organism itself in any form, unlike either the aforementioned live-attenuated or inactivated vaccines. The term *recombinant* refers to modifications made by taking genetic material from one organism and placing it into another organism (NIAID 2021b).

The antigen, as explained earlier, is the part of the pathogen that stimulates the immune system when a virus naturally attacks the body, and it is the main element of subunit vaccines. However, in certain cases, the antigen alone may not produce the optimal level of response in the body. Therefore, subunit vaccines often require additional ingredients called *adjuvants* to elicit a stronger protective reaction. Booster shots may also be necessary, as the immune protection may not be lasting. The benefit of subunit vaccines, however, is that they can limit any potential side effects (OIDP 2021).

Similar in approach to subunit vaccines, recombinant, polysaccharide, and conjugate vaccines also aim to stimulate the immune system while limiting side effects. These approaches have been highly successful for bacterial vaccines, such as (and in particular) pertussis or meningococcal meningitis (NIAID 2021b).

The following viruses are prevented using subunit vaccines:

- Hepatitis B
- HPV (human papillomavirus)
- Shingles

Toxoid vaccines, which target a toxin made by the pathogen rather than the pathogen itself, are highly useful against bacteria that produce toxins. These vaccines do not apply to viruses. However, they are mentioned here for the sake of completeness (NIAID 2021b).

4.2.4 VIRAL VECTOR VACCINES

A *viral vector* is a modified, innocuous version of a virus that has been adapted in order to produce the desired immune response. Various viruses can be used as vectors, including influenza, vesicular stomatitis virus (VSV), measles virus, and adenovirus which causes the common cold. Viral vectors have been studied since the 1970s and vaccines have been used in recent outbreaks of Ebola. The first wide-spread use of viral vector vaccines, however, began with the COVID-19 vaccination campaign. There is also significant on-going research involving these vaccines against a number of other infectious diseases such as Zika, influenza, and HIV (OIDP 2021; NIAID 2021b).

Viral vector vaccines are used to protect against:

- COVID-19
- Ebola

4.2.5 MESSENGER RNA (mRNA) VACCINES

Scientists have been investigating mRNA vaccines for decades. In 2020, the first two mRNA vaccines were introduced to protect against COVID-19. Both are highly effective against SARS-CoV-2, and demonstrated a greater than 90% reduction against the original version of the virus in serious disease, hospitalization, and death. In addition, the vaccines have remained effective to differing degrees to successive variants. These vaccines represent a particularly important breakthrough because the technology has the potential to apply to many other pathogens that have so far eluded researchers' attempts to find a suitable vaccine (OIDP 2021). Other mRNA vaccines have been studied or are in development in several disease areas, including flu, Zika, rabies, and cytomegalovirus (CMV), but it was the COVID-19 pandemic that provided a critical opportunity to overcome several obstacles that had been preventing their commercial viability. Beyond vaccines, cancer researchers are investigating the use of mRNA in eliciting an immune response against specific cancer cells (CDC 2022).

The mRNA vaccines have several advantages over other traditional types of vaccines in that they do not contain a live virus, so there is no risk of inadvertently causing disease through vaccination. They also have significantly shorter manufacturing times because the process can be scaled up to produce large quantities of vaccine more easily than with other methods (OIDP 2021).

4.3 U.S.-APPROVED VACCINES AND REQUIREMENTS

Two subsections complement the presentation for this section concerned with approved vaccines and requirements.

4.3.1 COMMERCIALLY AVAILABLE VIRAL VACCINES

The following list includes all viral illnesses that have FDA-approved vaccines available in the United States as of March 2022:

- Adenovirus
- COVID-19
- Ebola
- Hepatitis A
- Hepatitis B
- Human Papillomavirus (HPV)
- Seasonal Influenza (Flu)
- Japanese Encephalitis
- Measles, Mumps, and Rubella
- Polio
- Rabies
- Rotavirus
- Shingles
- Smallpox
- Varicella
- Yellow Fever

4.3.2 VACCINATION REQUIREMENTS

Thanks in part to public health requirements, most vaccine-preventable diseases remain at historically low levels. In the United States, all states require children to be vaccinated for communicable disease in order to attend daycare and school, whether public or private. State laws also have vaccine mandates for employees who work in those settings and for other groups, such as healthcare workers (CDC 2018a).

Table 4.1 lists the viruses and the vaccines that are currently recommended by the Centers for Disease Control (CDC) for children from birth to 18 years. A schedule detailing when each vaccine should be administered, and if it is a requirement, is available on the CDC website at *https://www.cdc.gov/vaccines/schedules/index.html*.

Table 4.1 Centers for disease control recommended vaccines for children.

Disease/vaccine	Disease symptoms	Disease complications
Chickenpox (Varicella) vaccine	Rash, tiredness, headache, fever	Infected blisters, bleeding disorders, encephalitis, pneumonia
Hepatitis A vaccine	Fever, stomach pain, loss of appetite, fatigue, vomiting, jaundice (yellowing of skin and eyes), dark urine	Liver failure, joint pain, kidney, pancreatic, and blood disorders
Hepatitis B vaccine	Asymptomatic, fever, headache, weakness, vomiting, jaundice, joint pain	Chronic liver infection, liver failure, liver cancer
Human Papillomavirus (HPV) vaccine	May be no symptoms, genital warts	Cervical, vaginal, vulvar, penile, anal, oropharyngeal cancers
Influenza (Flu) vaccine	Fever, muscle pain, sore throat, cough, extreme fatigue	Pneumonia
Measles (MMR[a] vaccine)	Rash, fever, cough, runny nose, pink eye	Encephalitis, pneumonia, death
Mumps (MMR[a] vaccine)	Swollen salivary glands (under the jaw), fever, headache, tiredness, muscle pain	Meningitis, encephalitis, inflammation of testicles or ovaries, deafness
Polio (Inactivated Poliovirus vaccine)	May be no symptoms, sore throat, fever, nausea, headache	Paralysis, death
Rotavirus (RV vaccine)	Diarrhea, fever, vomiting	Severe diarrhea, dehydration
Rubella (MMR[a] vaccine)	Sometimes rash, fever, swollen lymph nodes	Can lead to miscarriage, stillbirth, premature delivery, birth defects

a)   MMR combines protection against measles, mumps, and rubella.

4.4   VIRAL TESTING AND DIAGNOSIS

Due to the highly specific nature of antiviral treatment, proper identification of viral disease is crucial to determining the appropriate treatment. As mentioned earlier, the most effective approach to overcoming viral disease is preventing the virus from being able to take hold and cause infection in the first place. Aside from vaccination, the best way to prevent viruses from spreading is to test potentially infected individuals, and to quarantine or isolate those who test positive. This approach works both in acute situations during an outbreak or pandemic and in longer-term infections, such as HIV or HPV. The sooner the virus is identified, the sooner patients can take actions to prevent close contacts and others from harm.

The COVID-19 pandemic has proven that proper preparation is required in preventing and treating viral threats. As the outbreak grew into an epidemic, and then exploded into a global pandemic, the diagnosis and proper isolation and quarantine proved to be among the most successful approaches in controlling further infections.

Diagnostic methods for specific viruses, including Ebola, HIV, Coronaviruses, and Influenza, are reviewed in detail in Chapter 3: Pandemics, Epidemics, and Outbreaks. This section offers an overview of different diagnostic strategies and techniques.

4.4.1 Viral Testing

Viral testing involves the detection of the viral antigen or the viral nucleic acids. These can be rapid tests or laboratory tests. Rapid tests are especially useful for screening the large portions of the population, such as essential workers, healthcare workers, and travelers. Rapid tests can be performed in minutes and can include antigen and some NAATs (Nucleic Acid Amplification Testing). Rapid testing can be used effectively for serial testing (i.e., daily or weekly for in-office employees). It can also be used as an epidemiologic tool for determining the frequency of the virus in the population (CDC 2020a).

Self-tests are a subset of rapid tests which can be utilized at home, workplaces, or anywhere since they are easy to use and can determine results in minutes. Self-tests have been used in the COVID-19 pandemic as a risk-reduction strategy, along with vaccination, masking, and social distancing (CDC 2021a).

In contrast, laboratory tests can take hours or days to complete and include RT-PCR (reverse transcriptase-polymerase chain reaction) and other types of the aforementioned NAATs, such as isothermal DNA amplification. Samples sent to a laboratory can also test for mutations or variants. Due to the expense and time involved, samples are often not tested to determine specific variants, but are useful when determining the dominant variant in specific populations. During the COVID-19 pandemic, variant-specific testing has allowed changes in the viral sequence of SARS-CoV-2 to be identified earlier so that governments and the technical and medical communities can more quickly adjust to changes in virus virulence and transmissibility (CDC 2021a).

4.4.2 ANTIBODY TESTING

An antibody test (also known as a serology test) can detect viral antibodies. Serology tests look for antibodies in blood. Antibodies are proteins produced by the immune system when a person is exposed to a specific pathogen and are one of the body's main approaches for fighting off infections. For a more detailed review of what antibodies are and how they work can be found in Chapter 1: Overview of Molecular Biology.

If antibodies are found, it means there has been a previous exposure or infection. Antibody tests are not useful in diagnosing current infections but can indicate *past* exposure. There are different types of antibodies produced by the immune system. The two types that are useful in regard to infections are immunoglobulin M (IgM) and immunoglobulin G (IgG). Levels of IgM, which the body produces shortly after exposure, indicate a recent infection and only last for a certain time. IgG levels, on the other hand, can indicate an older infection, and IgG levels tend to last much longer, sometimes permanently. Antibody testing helps engineers and applied scientists learn how human immune systems defend against specific viruses, and provide insight into population-level protection. Several serologic techniques are available to determine if antibodies are present in a blood sample, and details of these are readily available in the literature (CDC 2021a).

4.5 ANTIVIRAL TREATMENT OPTIONS

As discussed in previous chapters, it is particularly challenging to design medications that are specifically targeted to stop viral infections. This difficulty stems from the fact that viral replication involves entering the host cell in order to commandeer its reproduction machinery. Advancements in the area of virology and the discovery of new viral entry pathways has opened the door to therapies for several viral diseases.

Antiviral drugs are approved to treat diseases such as herpes, influenza, hepatitis, varicella zoster (chicken pox), and COVID-19, among others. Antiretroviral therapy (ART), specific to retroviral viruses such as HIV, has also been highly effective in transforming HIV from a fatal disease to a manageable life-long illness. This section will focus on ART for the management of HIV, and antiviral options for influenza and hepatitis C.

Remdesivir, originally approved for the treatment of Hepatitis C, was the first antiviral to become FDA approved for COVID-19. Others approved (at the time of the submission of this manuscript) for emergency use include Merck's molnupiravir, and Pfizer's PAXLOVID™ (PF-07321332, ritonavir).

4.5.1 HIV

ART can control the levels of HIV virus in the body of an infected patient and prevent destruction of the immune system. ARTs are typically used in combination with drugs of different classes in order to inhibit drug resistance that the virus can develop when only one medication is taken. Viral load and CD4+ T-cell count are used to monitor the continued efficacy of drug combinations. As of 2021, more than thirty HIV medications were approved for use in the United States. (HIV.gov 2019) It should be noted that there is no cure for HIV. With lifelong antiretrovirals taken exactly as prescribed, however, patients with HIV can maintain a high quality of life. The following is a list of various classes of antiretrovirals that are approved for the treatment of HIV:

4.5.1.1 Nucleoside Reverse Transcriptase Inhibitors (NRTIs)

HIV uses reverse transcriptase to convert its RNA into DNA (reverse transcription). NRTIs bind to and inhibit this enzyme. By blocking reverse transcriptase, NRTIs prevent HIV from replicating.

4.5.1.2 Non-nucleoside Reverse Transcriptase Inhibitors (NNRTIs)

Similar to NRTIs, NNRTIs bind to and inhibit the reverse transcriptase enzyme. By blocking reverse transcriptase, NNRTIs prevent HIV from replicating.

4.5.1.3 Protease Inhibitors (PIs)

By blocking protease, an HIV enzyme, PIs prevent new HIV viruses from maturing so they cannot infect other CD4+ T cells.

4.5.1.4 Fusion Inhibitors (FIs)

Fusion inhibitors prevent HIV from entering the CD4+ T cells by thwarting fusion. Fusion normally occurs when the HIV envelope merges with the host CD4+ T-cell membrane.

4.5.1.5 Integrase Strand Transfer Inhibitors (INSTIs)

Integrase strand transfer inhibitors (INSTIs) block the HIV enzyme integrase. HIV uses integrase to integrate its viral DNA into the host CD4+ T-cell's DNA. Blocking integrase prevents HIV from replicating.

4.5.1.6 CCR5 Antagonists

CCR5 antagonists block CCR5 coreceptors on the surface of certain immune cells that HIV needs in order to enter, and infect, those cells.

4.5.1.7 Attachment Inhibitors

Attachment inhibitors bind to the gp120 protein on the outer surface of HIV, preventing HIV from entering CD4 cells.

4.5.1.8 Post-Attachment Inhibitors

Post-attachment inhibitors block CD4 receptors on the surface of certain immune cells that HIV needs to enter the cells.

4.5.1.9 Pharmacokinetic Enhancers

Pharmacokinetic enhancers are not antiretrovirals but are used in conjunction with HIV treatment to increase the effectiveness of medical therapy included in an HIV treatment regimen.

4.5.2 INFLUENZA

When started within 48 hours of symptom onset, antiviral medications for influenza can shorten the illness and make the course of the illness milder, as well as prevent potential complications, such as pneumonia.

There are currently four FDA-approved influenza antiviral drugs recommended by the CDC for use against recently circulating influenza viruses. Older drugs are no longer recommended because of widespread resistance to them. The following antiviral medications are available to treat influenza: (FDA 2020; HIV.gov 2019)

- Rapivab (peramivir)
- Relenza (zanamivir)
- Tamiflu (oseltamivir phosphate, also available as generic)
- Xofluza (baloxavir marboxil)

4.5.3 HEPATITIS C VIRUS (HCV)

Hepatitis C virus (HCV) is the virus that causes hepatitis C, a disease that causes inflammation of the liver, and can be contracted through contact with blood from an infected person. Hepatitis C can range in severity from a mild, acute illness to a serious, chronic illness that can ultimately lead to liver cirrhosis and cancer. Most patients become infected through exposure to blood or blood products. Most infections occur through unsafe injection practices, lack of proper healthcare precautions, unscreened blood transfusions or organ transplants, injection drug use, and sexual practices that lead to exposure to blood (WHO 2021b).

Globally, an estimated 58 million people have chronic HCV infection, with about 1.5 million new infections occurring per year. According to the WHO, in 2019, approximately 290,000 people died from hepatitis C, in most cases as a result of cirrhosis and liver carcinoma. There is currently no effective vaccine against hepatitis C. Antiviral medicines are highly effective and more than 95% of individuals with HCV could clear the infection within eight to 12 weeks with appropriate treatment. However, access to both diagnosis and treatment remains low (WHO 2021b; CDC 2020b).

According to the CDC, the following people are at increased risk for hepatitis C:

- People with HIV infection
- Current or former people who use injection drugs (PWID)
- People with selected medical conditions
- Patients treated for hemodialysis
- Prior recipients of transfusions or organ transplants (prior to July 1992)
- Blood donor recipients (prior to July 1992)
- Health care personnel exposed to HCV-positive blood or other fluids
- Children born to mothers with HCV infection

Currently, there are three classes of direct-acting antiviral (DAA) medications approved for use against hepatitis C. They are classified based on which mechanism they use against HCV. In addition, they can be used in combination. The following DAA classes recommended by the CDC are as follows (CDC 2020b):

- NS3/4A protease inhibitors
- NS5A polymerase inhibitors
- NS5B polymerase inhibitors

4.5.4 Other Treatment Options

Monoclonal antibodies (mAbs) are a type of therapeutic agent using pathogen-specific antibodies developed from patients who have recovered from the disease. They are genetically modified immune cells and have been used to help patients recover from infectious disease. They have also been used in autoimmune diseases and as cancer treatments (IDSA 2021).

Two monoclonal antibodies, Inmazeb and Ebanga, were approved for the treatment of Zaire ebolavirus infection in 2020. Along with early supportive care with rehydration, these therapies can significantly improve chances of survival in patients with Ebola.

Monoclonal antibodies have been approved for the treatment of SARS-CoV-2 for patients who test positive and are at a high risk for serious disease progression. Treatment, if started within 10 days of initial symptoms, can reduce the amount of viral load. A lower viral load leads to milder symptoms, thereby decreasing the likelihood of hospital stay or death (ASPA 2021).

4.6 APPLICATIONS

The following four illustrative examples are intended to compliment the above material and to provide a better understanding of the aspects discussed within the chapter.

Illustrative Example 4.1 Ebola Vaccines
Explore available vaccine options for the prevention of Ebolavirus.

Solution
The first FDA-approved vaccine for Ebola, rVSV-ZEBOV (Ervebo®) was approved as a single-dose vaccine against Zaire ebolavirus in 2019. Ervebo is a vector vaccine approved for the Zaire strain of Ebola. It is not effective against other Ebola strains or against Marburg viral disease. According to the FDA, Ervebo was determined to be 100% effective in preventing Ebola cases with symptom onset greater than 10 days after vaccination (FDA 2019).

The Advisory Committee on Immunization Practices (ACIP) recommends pre-exposure prophylaxis vaccination with rVSV-ZEBOV for adults who are at potential occupational risk of exposure to Zaire ebolavirus. This recommendation includes adults who are:

- Responding or planning to respond to an outbreak of EVD
- Laboratory staff at biosafety-level 4 facilities dealing with live Ebola virus
- Healthcare personnel working at federally designated Ebola Treatment Centers

A two-dose vaccine is also available. The vaccine consists of two separate components, Ad26.ZEBOV and MVA-BN-Filo, with the second dose required 56 days after the initial dose. It is designed to protect against the Zaire ebolavirus and was used under a research protocol in 2019 during an Ebola outbreak in the Democratic Republic of the Congo (DRC).

Illustrative Example 4.2 Adjuvants
Explain the use of adjuvants in vaccine manufacturing.

Solution
Vaccine adjuvants are components used in vaccines which accelerate, augment, and extend the immune responses triggered by antigens. Adjuvants can reduce the need for booster shots or decrease the amount of antigen required to achieve an appropriate immune reaction. In addition, elderly patients with weakened immune systems will have a larger, more protective response. Aluminum-containing adjuvants, referred to as alum, have been safely used since the 1930s (Iwasaki and Omer 2020; NIAID 2021a).

Illustrative Example 4.3   MONOCLONAL ANTIBODIES
Define monoclonal antibodies and discuss the use of monoclonal antibodies.

Solution
Monoclonal antibodies are often created by identifying patients who have recently recovered from an infection and harvesting effective antibodies from them. Monoclonal antibodies are pathogen-specific B cells designed to act against a predetermined target which can augment the patient's own immune system response to the pathogen. While monoclonal antibodies are used in several other therapeutic areas, their success in antiviral therapy has been limited. To date, monoclonal antibodies have been effective in the treatment of respiratory syncytial virus (RSV), Ebola, and COVID-19. Others have been developed but proved unsuccessful in clinical trials for viruses such as HIV, influenza, MERS-CoV, and Zika virus. Research is ongoing, with many promising candidates.

Illustrative Example 4.4   HEPATITIS VACCINES
Describe available hepatitis vaccines.

Solution
Hepatitis A (HAV) and B (HBV) are the only hepatitis viruses with available vaccines. The vaccine for hepatitis A, sold under various trade names, is an inactivated vaccine, given as two shots six months apart, with both doses necessary for long-term protection. Hepatitis B vaccine is a subunit vaccine that can be given in two, three, or four separate doses depending on the manufacturer. There is also a combination vaccine that protects against both hepatitis A and hepatitis B.

4.7 CHAPTER SUMMARY

- Vaccines are the most effective solution in preventing and controlling epidemics.
- Emerging and zoonotic diseases are an ever-present threat.
- Developments in antivirals and other therapies have advanced the treatment of acute viral disease and can have a significant impact on survival.
- Thanks to a widespread global commitment to vaccination over the last several decades, many diseases that once caused inordinate suffering and death are no longer the threat they once were.
- Successes in national and community vaccination programs have eliminated or greatly reduced infections and deaths from highly infectious diseases such as smallpox, polio, yellow fever, and measles.
- Although millions of lives have been saved thanks to international vaccine campaigns, many challenges remain.

- Vaccines mimic the antigens of the pathogens they are intended to prevent by priming the immune system.
- Advancements in virology and the discovery of new viral pathways have opened the door to therapies for several viral diseases.
- Antiviral drugs are approved for diseases such as herpes, influenza, hepatitis, varicella zoster (chicken pox), and COVID-19, among others.
- ART, specific to retroviral viruses such as HIV, has been highly effective in transforming HIV from a fatal disease to a manageable life-long illness.

4.8 PROBLEMS

1   List the advantages and drawbacks of live-attenuated vaccines.

2   What viruses can be prevented with mRNA vaccines? Name some other viral treatments that are in development.

3   Choose three classes of antiretroviral therapy (ART) and describe their mechanism for inhibiting HIV infection.

4   List four viruses that can be prevented by inactivated vaccines.

REFERENCES

Assistant Secretary for Public Affairs (ASPA) (2021). Monoclonal antibodies for high-risk COVID-19 positive patients. combatCOVID.hhs.gov. https://combatcovid. hhs.gov/i-have-covid-19-now/monoclonal-antibodies-high-risk-covid-19-positive-patients (accessed 3 November 2021).

Centers for Disease Control and Prevention (CDC) (2018a). CDC - vaccination laws - publications by topic - public health law. Centers for Disease Control and Prevention. https://www.cdc.gov/phlp/publications/topic/vaccinationlaws.html (accessed 29 October 2021).

Centers for Disease Control and Prevention (CDC) (2018b). Polio vaccination. Centers for Disease Control and Prevention. https://www.cdc.gov/vaccines/vpd/polio/index.html (accessed 29 October 2021).

Centers for Disease Control and Prevention (CDC) (2019). Vaccine (shot) for polio. Centers for Disease Control and Prevention. https://www.cdc.gov/vaccines/parents/diseases/polio.html (accessed 29 October 2021).

Centers for Disease Control and Prevention (CDC) (2020a). Influenza virus testing methods. Centers for Disease Control and Prevention. https://www.cdc.gov/flu/professionals/diagnosis/table-testing-methods.htm (accessed 19 October 2021).

Centers for Disease Control and Prevention (CDC) (2020b). Hepatitis C questions and answers for Health Professionals. Centers for Disease Control and Prevention. https://www.cdc.gov/hepatitis/hcv/hcvfaq.htm#section1 (accessed 3 November 2021).

Centers for Disease Control and Prevention (CDC) (2021a). Overview of testing for SARS-COV-2 (COVID-19). Centers for Disease Control and Prevention. https://www.cdc.gov/coronavirus/2019-ncov/hcp/testing-overview.html (accessed 4 November 2021).

Centers for Disease Control and Prevention (CDC) (2021b). Easy-to-read immunization schedule by vaccine for ages birth-6 years. Centers for Disease Control and Prevention. https://www.cdc.gov/vaccines/schedules/easy-to-read/child-easyread.html#vpd (accessed 26 October 2021).

Centers for Disease Control and Prevention. (2022a). *United States confirmed as country with circulating vaccine-derived poliovirus.* Centers for Disease Control and Prevention. https://www.cdc.gov/media/releases/2022/s0913-polio.html (accessed September 30, 2022).

Centers for Disease Control and Prevention (CDC) (2022b). Understanding mrna COVID-19 vaccines. Centers for Disease Control and Prevention. https://www.cdc.gov/coronavirus/2019-ncov/vaccines/different-vaccines/mrna.html (accessed 20 September 2022).

HIV.gov (2019). HIV treatment overview. HIV.gov. https://www.hiv.gov/hiv-basics/staying-in-hiv-care/hiv-treatment/hiv-treatment-overview. (accessed 3 November 2021).

Infectious Disease Society of America (IDSA) (2021). Immunomodulators. https://www.idsociety.org/covid-19-real-time-learning-network/therapeutics-and-interventions/immunomodulators/ (accessed 3 November 2021).

Iwasaki, A. and Omer, S.B. (2020). Why and how vaccines work. *Cell* 183 (2): 290–295. https://doi.org/10.1016/j.cell.2020.09.040.

National Institute of Allergy and Infectious Diseases (NIAID) (2021a) Vaccine adjuvants. U.S. Department of Health and Human Services (HHS). https://www.niaid.nih.gov/research/vaccine-adjuvants (accessed 3 November 2021).

National Institute of Allergy and Infectious Diseases (NIAID) (2021b) Vaccine types. U.S. Department of Health and Human Services. https://www.niaid.nih.gov/research/vaccine-types (accessed 28 October 2021).

Office of Infectious Disease and HIV/AIDS Policy (OIDP) (2021) Vaccine types. HHS.gov. https://www.hhs.gov/immunization/basics/types/index.html (accessed 26 October 2021).

U.S. Food and Drug Administration (FDA) (2019). First FDA-approved vaccine for the prevention of ebola virus disease, marking a critical milestone in public health preparedness and response. https://www.fda.gov/news-events/press-announcements/first-fda-approved-vaccine-prevention-ebola-virus-disease-marking-critical-milestone-public-health (accessed 1 November 2021).

U.S. Food and Drug Administration (FDA) (2020). Influenza (flu) antiviral drugs and related information. Center for Drug Evaluation and Research. https://www.fda.gov/drugs/information-drug-class/influenza-flu-antiviral-drugs-and-related-information (accessed 2 November 2021).

United Nations (UN) (2020). Measles cases hit 23-year high last year, killing 200,000 as vaccination stalls, WHO says | | UN News. United Nations. https://news.un.org/en/story/2020/11/1077482 (accessed 1 November 2021).

World Health Organization (WHO) (2019). Measles. World Health Organization. https://www.who.int/news-room/fact-sheets/detail/measles (accessed 1 November 2021).

World Health Organization (WHO) (2021a). Ebola virus disease. World Health Organization. https://www.who.int/news-room/fact-sheets/detail/ebola-virus-disease (accessed 16 October 2021).

World Health Organization (WHO) (2021b). Hepatitis C. World Health Organization. https://www.who.int/news-room/fact-sheets/detail/hepatitis-c (accessed 3 November 2021).

World Health Organization (WHO) (2021c). *Immunization coverage*. World Health Organization. https://www.who.int/news-room/fact-sheets/detail/immunization-coverage (accessed 9 July 2022).

World Health Organization (WHO) (2022). *Poliomyelitis*. World Health Organization https://www.who.int/news-room/fact-sheets/detail/poliomyelitis (accessed 8 July 2022).

5

Safety Protocols and Personal Protection Equipment

Contributing Author: Emma Parente

CHAPTER MENU

This chapter addresses numerous topics related to various safety protocols and personal protective equipment (PPE). PPE is defined as equipment worn to minimize exposure to hazards that cause serious workplace injuries and illnesses. Response personnel and frontline medical workers must wear protective equipment when there is a probability of contact that could affect their health or the health of others.

There are many different types of protective and safety systems put into place to protect against viruses. Antimicrobial disinfectants and other personal protective devices such as sanitizer and respirators are used and also discussed throughout the chapter.

HAZMAT protection is designed to protect all employees in the United States from injury or illness due to exposure to hazardous materials by the Occupational Health and Safety Administration (OSHA). *Biohazard levels*, on the other hand, also support the principle of biosecurity, which aims at preventing the use of microorganisms as harmful biological agents. Biosafety levels, ranging from

A Guide to Virology for Engineers and Applied Scientists: Epidemiology, Emergency Management, and Optimization, First Edition. Megan M. Reynolds and Louis Theodore.
© 2023 John Wiley & Sons, Inc. Published 2023 by John Wiley & Sons, Inc.

minimal to most complex hazards, cover four different levels. However, it is important to note that all clinical specimens may contain potentially infectious materials.

Throughout the chapter, PPE and other safety protocols are discussed, and many have become common practices throughout the COVID-19 pandemic. The Center for Disease Control (CDC) has set particular standards for practices such as mask wearing, social distancing, and transmission control.

5.1 REGULATIONS AND OVERSIGHT OF SAFETY PROTOCOLS

As mentioned earlier, the US Department of Labor's Occupational Safety and Health Administration (OSHA) assures the safe and healthful working conditions for working men and women by setting and enforcing standards and providing training, outreach, education, and assistance. OSHA requires employers to provide appropriate PPE for workers who could be exposed to blood or other infectious materials (such as bloodborne pathogens). OSHA may also require employers to provide PPE to protect against other hazards at work. Although OSHA requires the use of specific equipment, it does not regulate the marketing of these devices nor grant claims of disease prevention. These issues are overseen by the Food and Drug Administration (FDA). All PPE that is intended for use as a medical device must follow the FDA's regulations and meet specific performance standards for protection (FDA 2018).

5.2 PROTECTIVE AND SAFETY SYSTEMS

The are many different types of protective and safety systems put into place to protect against viruses. Examples of PPE requirements include mask mandates in the workplace during the COVID-19 pandemic. Different requirements regarding cleaning and disinfection are another broad category of safety procedure and are discussed below.

5.2.1 PERSONAL PROTECTIVE DEVICES AND PRACTICES

Personal protective devices are used in many different public health situations. They can be used in hospitals when treating patients, or in public to protect oneself and others from contracting and spreading contagious diseases. Three such approaches are described below (CDC 2016a):

Respirators

- There are a variety of different respirators. Respirators that cover parts of the face can remove particles, gas, and vapor contaminants from inhaled air. They can come in either half or full mask.
- Masks must be fitted for a tight seal and fit properly to have the most effectiveness.

Face Masks

- Covering the mouth and nose with a cloth face mask when going out in public to contain or protect against various infectious diseases when one is indoors.

Social Distancing

- Staying at least 6 feet away from others will protect an individual and those nearby from contagions.
- Avoid large gatherings and crowded places if a close contact has been exposed to various infectious diseases.

5.2.2 ANTIMICROBIAL SUPPRESSION AND ERADICATION

Different combinations and mixtures of substances lead to different antimicrobial suppression and eradications substances on surfaces and on skin. These are briefly defined below (CDC 2016a).

Sanitizers

- Mixture of substances that reduces the bacterial population on inanimate surfaces by about 99.9%.
- Sanitizers neither eliminate, nor destroy, all bacteria.

Disinfectants

- Substances or mixtures that destroy or inactivate bacteria, fungi, and various viruses.
- Disinfectants do not necessarily reach bacterial spores in the inanimate environment.

Sterilant

- Mixtures that destroy or eliminate all forms of microbial life in the inanimate environment. This includes all forms of vegetative bacteria, bacterial spores, fungi, fungal spores, and viruses.

Fungicide

- A substance that destroys various fungi (including yeasts) and fungal spores that are pathogenic to humans and other animals in the inanimate environment.

Microbial Water Purifier

- Any unit, water treatment product, or system that removes, kills, or inactivates all types of disease-causing microorganisms from the water, including but not limited to, bacteria, viruses, and protozoan cysts, to render the treated water safe for drinking.

Tuberculocide

- A mixture of substances that either destroys or inactivates tubercule bacilli in the inanimate environment.

Virucide

- A substance that can destroy or inactivate viruses in the inanimate environment.

5.3 DISINFECTION CATEGORIES AND PROCEDURES

Disinfection describes a process that eliminates many or all pathogenic microorganisms, except *bacterial spores*, on inanimate objects. In health-care settings, equipment is usually disinfected by liquid chemicals or through *wet pasteurization*, with which equipment is submerged in water for 30 minutes at temperatures above 70°C. Each of the various factors that affect the efficacy of disinfection can nullify or limit the efficacy of the process. These factors affecting disinfection include (CDC 2016b; EPA 2022)

Number and Location of Microorganisms

- If all other conditions remain constant, the larger number of microbes will mean more time a *germicide* needs to destroy all of them.
- Only locations and surfaces that directly contact the germicide will be disinfected.

Innate Resistance of Microorganisms

- To destroy more resistant types of microorganisms (i.e., bacterial spores), users need to employ exposure times and a concentration of germicide is needed to achieve complete destruction.

Concentration and Potency of Disinfectants

- With other variables kept constant, and with one exception – iodophors – solutions containing iodine, the more concentrated the disinfectant, the greater its efficacy and the shorter the time necessary to achieve microbial kill.

Physical and Chemical Factors

- Factors that also influence disinfectant procedures are temperature, pH, relative humidity, and water hardness.

Organic and Inorganic Matter

- Organic matter in the form of serum, blood, pus, or fecal or lubricant material can interfere with the antimicrobial activity of disinfectants.
- Protection by inorganic contaminants of microorganisms to all sterilization processes results from occlusion in salt crystals.

Duration of Exposure

- Items must be exposed to the germicide for the appropriate minimum contact time.

Biofilms

- Some bacterial species produce *biofilm,* which is a thin slimy film of bacterial cells that can reduce the efficacy of some disinfectants.

5.4 OCCUPATIONAL HEALTH AND SAFETY ADMINISTRATION HAZMAT REGULATIONS

Regardless of the type of work, all employees in the United States are protected at their place of employment by the Occupational Health and Safety Administration (OSHA), which falls under the US Department of Labor. Employees must be properly trained and fully informed of any risks to their health when there is any possibility of an exposure of hazardous material of any kind at the workplace. This is including any potential exposure to infectious diseases.

There are different levels of *HAZMAT* protection, from Levels A through D. The equipment to protect the body against contact with known or anticipated chemicals has been divided into four categories according to the degree of hazardous materials, or HAZMAT, protection afforded. Occupational Safety and Health Administration (OSHA) defines HAZMAT as any substance or chemical which is a "health" or "physical" hazard. Regardless of the type of work, all employees in the United States are protected at their place of employment by OSHA, which (as noted above) falls under the US Department of Labor.

Response personnel and frontline medical workers must wear protective equipment when there is a possibility of contact that could affect their health. This includes contact with vapors, gases, or particulates that may be generated by site activities, and direct contact with skin-affecting substances. Full-facepiece respirators protect the lungs, gastrointestinal tract, and eyes against airborne toxicants. Chemical-resistant clothing protects the skin from contact with skin destructive and absorbable chemicals. Good personal hygiene also limits or helps prevent ingestion of material (OSHA 1994b).

As noted above, equipment to protect the body against contact with known or anticipated chemicals has been divided into four categories according to the degree of HAZMAT protection afforded (OSHA 1994b):

• *Level A*

Should be worn when the highest level of respiratory, skin, and eye protection is needed.

• *Level B*

Should be worn when the highest level of respiratory protection is needed, but a lesser degree of skin protection is needed.

• *Level C*

Should be worn when a lesser level of respiratory protection is needed than Level B. Skin protection criteria are similar to Level B.

• *Level D*

Should be worn only as a work uniform and not on any site with respiratory or skin hazards. It provides no protection against infectious hazards.

The level of protection selected should be based on the risk of exposure and the type of possible exposure, as well as the type of contaminant (OSHA 1994a). Additional details on each of those four levels are provided in the next four subsections.

In situations where the type of chemical, concentration, and possibilities of contact are not known, the appropriate level of protection must be selected based on professional experience and judgment until the hazard can be better characterized. Further details on risk management are available in Part II, Chapter 11: Health and Hazard Risk Assessment (Theodore and Dupont 2012; EPA 2021).

More details on the following explanations of HAZMAT level requirements are available on OSHA's website under *Standard Number 1910.120 App B – General description and discussion of the levels of protection and protective gear* at https://www.osha.gov/laws-regs/regulations/standardnumber/1910/1910 .120AppB (OSHA 1994b).

5.4.1 HAZMAT Level A Protection

Level A offers the highest degree of protection and is to be selected when the greatest level of skin, respiratory, and eye protection is required. Level A protection should be used when:

- The hazardous substance has been identified and requires the highest level of protection for skin, eyes, and the respiratory system based on either the measured (or potential for) high concentration of atmospheric vapors, gases, or particulates.
- Site operations and work functions involving a high potential for splash, immersion, or exposure to unexpected vapors, gases, or particulates of materials that are harmful to skin or capable of being absorbed through the skin.
- Substances with a high degree of hazard to the skin are known or suspected to be present, and skin contact is possible.
- Operations that must be conducted in confined, poorly ventilated areas, and where the absence of conditions requiring Level A have not yet been determined.

The following constitute Level A equipment which may be used as appropriate.

- Positive pressure, full face-piece self-contained breathing apparatus (SCBA), or positive pressure supplied air respirator with escape SCBA, approved by the National Institute for Occupational Safety and Health (NIOSH).
- Totally encapsulating chemical-protective suit.
- Coveralls
- Long underwear
- Gloves, outer, chemical-resistant
- Gloves, inner, chemical-resistant
- Boots, chemical-resistant, steel toe, and shank
- Hard hat (under suit)
- Disposable protective suit, gloves, and boots (depending on suit construction, may be worn over totally encapsulating suit).

5.4.2 HAZMAT Level B Protection

Level B is appropriate when the highest level of respiratory protection is necessary, but a lesser level of skin protection is needed. Level B protection should be used when:

- The type and atmospheric concentration of substances have been identified and require a high level of respiratory protection, but less skin protection.
- The atmosphere contains less than 19.5% oxygen.
- The presence of incompletely identified vapors or gases is indicated by a direct-reading organic vapor detection instrument, but vapors and gases are not suspected of containing high levels of chemicals harmful to skin or capable of being absorbed through the skin.

Note: This involves atmospheres with IDLH (Immediately Dangerous to Life and Health) concentrations of specific substances that present severe inhalation

hazards and that do not represent a severe skin hazard; or, that do not meet the criteria for use of air-purifying respirators.
The following constitute Level B equipment and should be used as appropriate.

- Positive pressure, full-facepiece self-contained breathing apparatus (SCBA), or positive pressure supplied air respirator with escape SCBA (NIOSH approved)
- Hooded chemical-resistant clothing (overalls and long-sleeved jacket; coveralls; one or two-piece chemical-splash suit; and, disposable chemical-resistant overalls)
- Coveralls
- Gloves, outer, and chemical-resistant
- Gloves, inner, and chemical-resistant
- Boots, outer, chemical-resistant steel toe, and shank
- Boot-covers, outer, and chemical-resistant (disposable)
- Hard hat
- Face shield

5.4.3 LEVEL C PROTECTION

Level C is to be used when concentration(s) and type(s) of airborne substance(s) is known and the criteria for using air purifying respirators are met. Level C protection should be used when:

- The atmospheric contaminants, liquid splashes, or other direct contact will not adversely affect or be absorbed through any exposed skin.
- The types of air contaminants have been identified, concentrations measured, and an air-purifying respirator is available that can remove the contaminants.
- All criteria for the use of air-purifying respirators are met.

The following constitute Level C equipment; it may be used as appropriate.

- Full-face or half-mask, air purifying respirators (NIOSH approved)
- Hooded chemical-resistant clothing (overalls; two-piece chemical-splash suit; and disposable chemical-resistant overalls).
- Coveralls
- Gloves, outer, and chemical-resistant
- Gloves, inner, and chemical-resistant
- Boots (outer), chemical-resistant steel toe, and shank
- Boot-covers, outer, and chemical-resistant (disposable)
- Hard hat
- Escape mask
- Face shield

5.4.4 LEVEL D PROTECTION

Level D should employ affording minimal protection: used for nuisance contamination only. Level D protection should be used when:

- The atmosphere contains no known hazard.
- Work functions preclude splashes, immersion, or the potential for unexpected inhalation of or contact with hazardous levels of any chemicals.

Note: As stated before, combinations of PPE other than those described for Levels A, B, C, and D protection may be more appropriate and may be used to provide the proper level of protection.
The following constitute Level D equipment and may be used as appropriate:

- Coveralls
- Gloves
- Boots/shoes, chemical-resistant steel toe, and shank
- Boots, outer, and chemical-resistant (disposable)
- Safety glasses or chemical splash goggles
- Hard hat
- Escape mask
- Face shield

5.5 BIO LEVEL SAFETY AND SECURITY

Biohazard levels, more commonly referred to as "biological safety levels" or "biosafety levels," are classifications of safety precautions necessary to be applied in the clinical microbiology laboratory depending on specific pathogens handled when performing laboratory procedures. Developed by the Centers for Disease Control and Prevention (CDC), this principle provides a way for medical laboratory scientists and other lab personnel to identify and limit any biological hazards and further reduce the risk in the laboratory. Biohazard levels, on the other hand, also support the principle of biosecurity, which aims at preventing the use of microorganisms as harmful biological agents (Bayot 2021; CDC 2021). These levels are discussed below.

- **Biosafety level 1 (BSL-1)**
  Controls microorganisms unusually known to cause disease with "minimal hazards" to the laboratory and the community.
- **Biosafety level 2 (BSL-2)**
  Controls microorganisms generating "moderate hazards" to the laboratory and the community.

- **Biosafety level 3 (BSL-3)**
  Includes the control of infectious agents, which can cause both "serious hazards" and can cause a potentially lethal condition to the laboratory and community via the respiratory transmission of the organism.
- **Biosafety level 4 (BSL-4)**
  Is the highest and "most complex" biohazard level, involving a relatively few clinical microbiology laboratories.

It is important to note all clinical specimens may contain potentially infectious materials. Precautions should be taken when handling specimens that are suspected or confirmed to be positive for specific viruses (i.e., SARS-CoV-2). Timely communication between clinical and laboratory staff is essential to minimize the potential risk of handling specimens from patients with possible infections (Bayot 2021).

Laboratories should perform site- and activity-specific risk assessments to identify and mitigate risks. Risk assessments and mitigation measures are dependent on (CDC 2021):

- The procedures performed
- Identification of the hazards involved in the process and procedures
- The competency level of the personnel who perform the procedures
- The laboratory equipment and facility
- The resources available

5.6 COVID-RELATED SAFETY PRECAUTIONS

PPE and safety protocols have been emphasized greatly throughout the ongoing COVID-19 pandemic. Below are just a few that have been in action throughout the course of the pandemic.

5.6.1 PERSONAL PROTECTIVE EQUIPMENT

Throughout the COVID-19 pandemic, PPE and Safety Protocols have been put into place all over the world. Respirators, face masks, and social distancing are some common practices and equipment people have been accustomed to.

OSHA recommends PPE standards which require that a PPE hazard assessment be conducted to assess workplace hazards, and that PPE, such as respiratory protection, be used when necessary. When respirators are necessary to protect workers, employers must implement a comprehensive respiratory protection program in accordance with the Respiratory Protection Standard (OSHA 1998).

5.6.2 TRANSMISSION CONTROL

The CDC also encourages quarantine and isolation protocols. One may quarantine if they have been in close contact (within 6 feet of someone for a cumulative total of 15 minutes or more over a 24-hour period) with someone who has COVID-19, unless they have been fully vaccinated. During quarantine, one must stay home for 14 days after their last contact with a person who was positive with COVID-19. Monitor symptoms and if possible, limit contact with others in the household. People may be isolated when they have been infected with the COVID-19, even if they are not showing symptoms. During isolation, it is important to monitor symptoms and avoid contact with others (CDC 2020).

Transmission-based precautions can be accomplished by focusing on contact precautions, droplet precautions, and airborne precautions and are detailed below (CDC 2016b):

Contact Precautions

- Used for patients with known or suspected infections
- Ensure appropriate patient placement such as a single patient space or if available, acute care in hospitals
- Using PPE appropriately and correctly
- Limit transportation and movement of patients to decrease likelihood of spread

Droplet Precautions

- Wear a mask if infected or exposed
- Ensure patients follow respiratory hygiene/cough etiquette recommendations

Airborne Precautions

- Ensure appropriate patient placement in an airborne infection isolation room (AIIR) constructed according to the Guideline for Isolation Precautions.
- Restrict susceptible healthcare personnel from entering an infected patient's room if immune healthcare personnel are available.
- Immunize susceptible persons as soon as possible

5.7 APPLICATIONS

The following four illustrative examples are intended to compliment the above material and to provide a better understanding of aspects discussed within the chapter.

Illustrative Example 5.1   EFFECTIVE SANITIZERS FOR BACTERIA
Explore which type of sanitizer is more effective for the elimination of bacteria.

Solution

Sanitizers can be categorized into two groups: Alcohol-Based and Alcohol-Free. Alcohol-based sanitizer (HBHS) can effectively and quickly reduce microbes covering a broad germicidal spectrum without the need for water or drying with towels. Nevertheless, there are a few shortcomings with the effectiveness of ABHS, such as its short-lived antimicrobial effect and weak activity against protozoa, some non-enveloped (non-lipophilic) viruses and bacterial spores. In regard to alcohol-free sanitizer, they make use of chemicals with antiseptic properties to exert the antimicrobial effects. As they are nonflammable and often used at low concentrations, they are relatively safer to use among children as compared to alcohol-based. These different categories are available in many options including gel, foam, liquid, spray, creams, and wipes. Overall, gels and foams are more widely accepted compared to other forms of sanitizer (EPA 2022).

Illustrative Example 5.2   Covid-19

Examine what disinfectants kill the COVID-19 virus on surfaces.

Solution

Normal routine cleaning with soap and water lowers the risk of spreading the COVID-19 by removing germs and dirt from surfaces. In most situations, cleaning is enough to reduce risk. However, using EPA-Registered Disinfectants according to label directions will further lower the risk of COVID-19 on surfaces. Some examples of these disinfectants would be Hydrogen Peroxide, Quaternary Ammonium, and other disinfectants that can be found on the EPA's *List N*. More recently, ozone has also been employed as another option (EPA 2022).

Illustrative Example 5.3   Biohazard Levels

Name some examples of specific viruses that are categorized as Bio Level 4.

Solution

Ebola, Marburg Virus, Lassa Fever, Bolivian Hemorrhagic Fever, and other Hemorrhagic viruses found in the tropics are all Bio Level 4 diseases (NCBI).

Illustrative Example 5.4   HIV Safety Protocols

Describe the safety protocols used to protect against HIV in hospitals.

Solution

Standard precautions should be followed at all times. Assume that blood and other body fluids are potentially infectious. Use gloves, goggles, and other barriers when anticipating contact with blood or body fluids. Wash hands and other skin surfaces immediately after contact with blood or body fluids with soap and hot water. Be careful when handling and disposing of sharp instruments during and after use. Use safety devices to prevent needle-stick injuries. Dispose off used syringes or other sharp instruments in a sharp's container.

5.8 SUMMARY

- HAZMAT protection is designed to protect all employees in the United States from injury or illness due to exposure to hazardous materials.
- Biohazard levels are implemented to oversee the handling and use of potentially hazardous biological agents, such as viruses.
- Response personnel and frontline medical workers must wear protective equipment when there is a probability of contact that could affect their health.
- Different combinations and mixtures of substances lead to different antimicrobial suppression and eradications substances on surfaces and on skin.
- Disinfection is a process that eliminates many or all pathogenic microorganisms, except bacterial spores, on inanimate objects.
- Employees must be properly trained and fully informed of any risks to their health when there is any possibility to an exposure of hazardous material of any kind at the workplace.
- Chemical-resistant clothing protects the skin from contact with skin destructive and absorbable chemicals.
- PPE and safety precautions such as respirators, face masks, and social distancing are all safety practices used during the COVID-19 pandemic.
- The CDC encourages quarantine and isolation protocols during outbreaks and pandemics.

5.9 PROBLEMS

1 What is personal protective equipment?

2 Where/when is it appropriate to wear Level C PPE?

3 Give an example of a protective and safety system and what it does.

4 What are common safety practices used when protecting against COVID-19?

REFERENCES

Bayot, Marlon L. "Biohazard Levels." *National Center for Biotechnology Information*, U.S. National Library of Medicine, Jan. 2021, Retrieved November 30, 2021, from https://www.ncbi.nlm.nih.gov/books/NBK535351/?report=printable.

Centers for Disease Control and Prevention (CDC). (2016a, September 18). Efficacy. (CDC). Retrieved December 2, 2021, from https://www.cdc.gov/infectioncontrol/guidelines/disinfection/efficacy.html#anchor 1554391079.

Centers for Disease Control and Prevention (CDC). (7 January 2016b). "Transmission-Based Precautions." Centers for Disease Control and Prevention. Retrieved November 30, 2021, from https://www.cdc.gov/infectioncontrol/basics/transmission-based-precautions.html.

Centers for Disease Control and Prevention (CDC). (2020) "Covid-19 Quarantine and Isolation." Centers for Disease Control and Prevention. Retrieved November 30, 2021, from https://www.cdc.gov/coronavirus/2019-ncov/your-health/quarantine-isolation.html.

Centers for Disease Control and Prevention (CDC). (2021, October 28). *Interim guidelines for biosafety and covid-19*. Centers for Disease Control and Prevention. Retrieved November 30, 2021, from https://www.cdc.gov/coronavirus/2019-nCoV/lab/lab-biosafety-guidelines.html.

Environmental Protection Agency (EPA). (2021). *Personal Protective Equipment*. EPA. Retrieved December 31, 2021, from https://www.epa.gov/emergency-response/personal-protective-equipment.

Environmental Protection Agency (EPA). (2022). *About List N: Disinfectants for Coronavirus (COVID-19)*. EPA. Retrieved July 29, 2022, from https://www.epa.gov/coronavirus/about-list-n-disinfectants-coronavirus-covid-19-0.

Food and Drug Administration (FDA). (2018). *Personal Protective Equipment (PPE) and Other Government Agencies*. Center for Devices and Radiological Health. Retrieved November 30, 2021, from https://www.fda.gov/medical-devices/personal-protective-equipment-infection-control/personal-protective-equipment-ppe-and-other-government-agencies.

Occupational Safety and Health Administration (OSHA). (1994a) *1910.120 - Hazardous waste operations and emergency response*. US Department of Labor. *Retrieved December 31, 2021, from* https://www.osha.gov/laws-regs/regulations/standardnumber/1910/1910.120.

Occupational Safety and Health Administration (OSHA). (1994b). *1910.120 App B - General description and discussion of the levels of protection and protective gear*. US Department of Labor. Retrieved October 30, 2021, from https://www.osha.gov/laws-regs/regulations/standardnumber/1910/1910.120AppB.

Occupational Safety and Health Administration (OSHA) (1998). *Respiratory Protection Standard (29 CFR 1910.134 and 29 CFR 1926.103)*. US Department of Labor. https://www.osha.gov/laws-regs/regulations/standardnumber/1910/1910 .134 (accessed 31 December 2021).

Theodore, L. and Dupont, R. (2012). *Environmental Health and Hazard Risk Assessment: Principles and Calculations*. Boca Raton, FL: CRC Press/Taylor & Francis Group.

6

Epidemiology and Virus Transmission

CHAPTER MENU

The field of public health is concerned with the health and well-being of communities of people. It is a vast discipline that covers a wide array of healthcare areas, with a focus on the prevention of injuries and diseases. Epidemiology is an important part of public health. Epidemiologists examine patterns of illness or injury in order to better understand the causes or sources. Merriam-Webster succinctly defines epidemiology as "a branch of medical science that deals with the incidence, distribution, and control of disease in a population." (Merriam-Webster 2022) While epidemiology began as the study of infectious disease prevention, the science has evolved significantly in the last century, and is now much broader in scope (Frérot et al. 2018).

As the nation's principle public health institution, the Centers for Disease Control and Prevention (CDC) offers several areas in which it plays a major role: (CDC 2016)

- **_Environmental Exposures_**
  Lead and heavy metals
  Air pollutants and other asthma triggers
- **_Infectious diseases_**
  Foodborne illnesses

A Guide to Virology for Engineers and Applied Scientists: Epidemiology, Emergency Management, and Optimization, First Edition. Megan M. Reynolds and Louis Theodore.

Influenza

Pandemics, such as COVID-19

- *Injuries*

  Increased homicides in a community

  National surge in domestic violence

  Workplace hazards

- *Noninfectious diseases*

  Localized or widespread rise in a particular type of cancer

  Increase in a major birth defect

  Heart disease

- *Natural disasters*

  Hurricanes

  Tornadoes

  Forest fires

  Earthquakes

  Tsunamis

- *Terrorism*

  World Trade Center attacks (2001)

  Anthrax release (2001)

This chapter will focus solely on epidemiology in the context of outbreaks and epidemics caused by viruses. There are five key objectives of epidemiology in public health practice (CDC 2016; Brachman and Baron 1996):

- Public health surveillance
- Field investigation
- Analytical studies
- Cause evaluation
- Disease–state associations

As has been discussed in earlier chapters, viral diseases are some of the most difficult illnesses to treat. Emerging and zoonotic diseases continue to be an ever-present threat, and it is crucial for countries to be able and willing to work together when new pathogens arrive. To that end, epidemiology plays a crucial role in ensuring preparedness and planning to prevent viral disease, and, therefore, to avert outbreaks and epidemics. Countless infections could be prevented, and many lives could potentially be saved with proper global coordination.

6.1 OVERVIEW OF EPIDEMIOLOGY

Several terms used within the context of epidemiology are central to understanding its critical role in healthcare and public health. This section attempts to introduce and clarify the meaning of some of this essential vocabulary.

A more comprehensive definition of epidemiology than was presented in the introduction is offered by the CDC:

"Epidemiology is the study (scientific, systematic, and data-driven) of the *distribution* (frequency, pattern) and *determinants* (causes, risk factors) of health-related states" … [i.e., conditions, situations, or] … "events (not just diseases) in specified populations (neighborhood, school, city, state, country, global). It is also the application of this study to the control of health problems."

(CDC 2016)

Epidemiological studies seek to observe, document, describe, and finally to analyze and present collected data in order to better clarify a problem. Descriptive epidemiology specifically seeks to define individual cases of illness or conditions by person, place, and time in order to detect *patterns* in resulting data. Study designs may also be analytical or experimental in nature, including *comparative* studies which are discussed later in the chapter. The term *population* refers to the group that is being studied and can be any group of people that have something in common, whether it is geographic location, gender, or people who all have a certain disease or exposure. *Distribution* refers to the incidence rate, or pattern, that can be discerned through investigation, while *determinants,* as the CDC describes, are any cause or risk factor, the knowledge of which is usually the result of an epidemiological study (CDC 2013; Frérot et al. 2018).

With infectious disease, epidemiologists utilize an organized methodology for evaluating epidemics. An outbreak investigation includes the following major steps (CDC 2016):

- Confirm the existence of an outbreak or epidemic verifying the diagnosis and ruling out other possibilities.
- Develop an initial case definition particularly.
- Collect all pertinent data on cases.
- Analyse data by time, place, and person.
- Develop a hypothesis based on initial data, and investigate further as required.
- Implement measures to control the outbreak and prevent further cases.
- Clearly communicate any findings to all stakeholders, including the public.
- Assess effectiveness of control and preventive measures.
- Monitor any effects, intended or otherwise, due to implemented measures.
- Continually update measures and provide further direction to stakeholders based on new findings.

The magnitude of the COVID-19 global pandemic, unfortunately, underscores the difficulty of recognizing an outbreak of a *novel* virus in time to prevent it from spreading to other communities. Defining and identifying a new virus, especially a fast-moving one such as SARS-CoV-2, requires open communication and vigilance in the medical community. It is critical for public officials to balance a desire to efficiently contain an outbreak with fears of causing public panic. The fact that the present pandemic has circled the globe in multiple waves, and rages on even after two years (at the time of the submission of this manuscript), demonstrates the dangers of not moving fast enough when identifying and implementing measures to control an outbreak or epidemic before it becomes a global pandemic.

6.2 GOVERNMENT AGENCIES' CONTRIBUTIONS TO PUBLIC HEALTH

Thanks to a widespread global commitment to vaccination over the last several decades, many diseases that once caused inordinate suffering and death no longer pose the threat to public health they once did. These include examples such as smallpox, polio, measles, yellow fever, hepatitis A and B, and influenza. Measles case rates illustrate the enormous impact public health efforts can have on societies, especially when goals are aligned with healthcare innovations, such as vaccines. In the case of measles, global vaccination and public health campaigns reduced mortality by 73% worldwide and prevented an estimated 23.2 million deaths over the first two decades of the twenty-first century (WHO 2019; UN 2020).

6.2.1 THE ROLE OF THE CENTERS FOR DISEASE CONTROL AND PREVENTION (CDC)

In the United States, the CDC is responsible for maintaining and improving the nation's public health. The organization's key duty is to protect the public from infectious disease outbreaks. The CDC works together with international and local governments and other government agencies and healthcare personnel wherever in the world outbreaks occur. After initial detection of a new outbreak, the CDC sends an emergency response team comprised of epidemiologists, doctors, and supplies to respond to outbreaks at their source. This is especially critical because transmissible infections spread quickly from country to country, especially with ever-increasing international travel. Therefore, the CDC's mandate is not limited to protecting the health and lives of those in the United States, but globally. According to the CDC, "the world counts on [them] to implement appropriate and immediate early interventions, which could prevent an aggressive outbreak from becoming an epidemic and prevent an epidemic in a country or region from developing into a worldwide pandemic." (CDC 2021).

There has been significant scrutiny on the performance of the CDC during the COVID-19 pandemic. Critics point to the lack of responsiveness and the comparatively high mortality rate in the United States compared to similar, high-income countries. In its overall pandemic response, the CDC fell far short of its stated goals of protecting the country from disease-related threats. In addition, on-going deficits in and the fragmentation of the US healthcare system, along with the refusal of the previous administration to coordinate a federal response in the early stages of the virus, all helped lead to failures on multiple fronts. The development of the COVID-19 vaccines was the exception, and this stands out as a major achievement of the United States, with Operation Warp Speed contributing to such scientific innovations in record time. The director of the CDC, Dr. Rochelle P. Walensky ordered an external review of the agency's failures in April of 2022 in response to on-going criticism. In August, the agency admitted its responsibility in the many breakdowns in communication with healthcare professionals and especially with the confusing and delayed advice to the public throughout the pandemic. In addition, data was often too delayed to be of significant value for informed decision-making. According to the New York Times, Walensky stated that her goal, "is a new, public health, action-oriented culture at CDC that emphasizes accountability, collaboration, communication and timeliness." (Rochelle Walensky as quoted in The New York Times 2022).

6.2.2 The World Health Organization (WHO): Successes and Challenges

The World Health Organization (WHO) was founded by a fledgling United Nations in Geneva, Switzerland in 1948 in the wake of World War II. The WHO quickly grew to become the leading agency contributing to global public health. With 194 member countries, the "...WHO works worldwide to promote health, keep the world safe, and serve the vulnerable." (WHO n.d.)

Major global issues actively addressed by WHO include, "universal health coverage, disease prevention, health impacts of climate change, antimicrobial resistance, and the "...elimination and eradication of high-impact communicable diseases." (WHO n.d.)

The WHO prepares for public health crises, whether on a local or global scale by:

- Identifying, mitigating and managing risks
- Prevent emergencies and support development of tools necessary during outbreaks
- Detect and respond to acute health emergencies
- Support delivery of essential health services in fragile settings

As reviewed in Chapter 4: Virus Prevention, Diagnosis, and Treatment, one of the most significant successes of the WHO in global public health to date has been the eradication of smallpox in 1980 after an extensive, international campaign to vaccinate people in every affected country. Successes in national and community vaccination programs have also curbed or greatly reduced infections and deaths from several other highly infectious diseases, such as polio and yellow fever (WHO 2021a).

Millions of lives have been saved thanks to international campaigns to increase public awareness and vaccine use. However, there are still many obstacles to successfully preventing future epidemics. Lack of public awareness and misinformation regarding the safety of vaccines has led to vaccine hesitancy and thus reduced community protection against several diseases, including measles and the COVID-19. The principle behind the success of vaccination initiatives involves the idea of herd immunity threshold, which represents the percentage of the population that must be immune to the disease in order to protect vulnerable individuals. The herd immunity threshold for measles is 95%. Several communities over the last decade have fallen below that level, primarily due to vaccine hesitancy, and have led to localized outbreaks nationally and globally. Polio has a herd immunity threshold of about 80% (WHO 2020).

The most effective approach to overcoming viral disease is preventing the virus from being able to take hold and cause infection in the first place. Aside from vaccination, the best way to prevent viruses from spreading is to:

(1) test potentially infected individuals
(2) quarantine or isolate those who test positive

The sooner the virus is identified, the sooner patients can take actions to prevent close contacts and others from harm. This approach works both in acute situations during outbreaks or pandemics and in longer-term infections, such as HIV or HPV.

The COVID-19 pandemic has proven that proper preparation is required in preventing and treating viral threats. Subsequent waves of the virus, driven by new variants, have demonstrated the resilience and adaptability of the SARS-CoV-2 virus. As the outbreak grew from an isolated outbreak in Wuhan, China, into a global pandemic, early diagnosis and proper isolation and quarantine proved to be among the most successful approaches. As of in 2022, the herd immunity threshold has yet to be determined. No country has managed to reach any immunity level to SARS-CoV-2 high enough to offer the kind of protection that herd immunity provides (CDC n.d.).

6.3 EPIDEMIOLOGIC STUDY DESIGN

Scientific studies are central to the understanding of public health issues. There are two categories of epidemiologic studies: descriptive and analytical. *Descriptive* studies are conducted in order to not alter the circumstance or outcome, and can address the questions of who, what, where, and when. They are, therefore, inherently observational and can be used to describe the burden of disease or

Figure 6.1 Epidemiologic study types. Adapted from *Descriptive and analytic studies*. Source: CDC (n.d.).

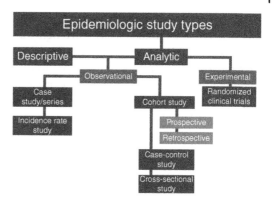

prevalence of illnesses. Types of descriptive studies include qualitative, such as case reports or cross-sectional surveys. *Analytic* studies may be either observational or experimental. *Observational* studies identify patterns in healthcare problems in order to develop a hypothesis regarding what factors led up to the issue studied and to how effective are the interventions. Analytical studies can then be used to evaluate the effectiveness of any intervention. Descriptive and analytic observational methods allow scientists to identify shared characteristics of a population in relation to certain variables in order to determine a hypothesis that explains the cause or result. *Experimental* study methods are then designed to test that hypothesis. Figure 6.1 summarizes various study design options (CDC 2012).

During infectious disease outbreaks, epidemic curves (*epi curves*) are used to interpret data on an ongoing basis. An epi curve illustrates illness progression in a population over (as a function of) time, with the x-axis depicting illness onset by days or weeks, and the y-axis representing the number of persons becoming

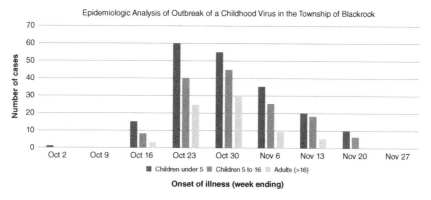

Figure 6.2 Case data stratified by age for a fictional childhood illness during a viral outbreak.

ill. Figure 6.2 represents a typical epi curve, with the onset of illness by weeks on the x-axis and number of cases in a fictional township on the y-axis. This data is also stratified by age, with three different age groups presented. The example case report in the following subsection utilizes the data presented in Figure 6.2 to portray how these epi curves can be used in real-world scenarios. The reader should note that this topic will be revisited in Part III, Chapter 16: Pandemic Health Data Models.

6.3.1 OUTBREAK CASE EXAMPLE

The following example demonstrates how epidemiologists analyze data to generate useful information and to better understand the situation during viral diseases outbreaks, as discussed in this chapter.

Case Summary

Figure 6.2 shows the epi (epidemiologic) curve of a fictional outbreak of a viral illness mainly affecting children under 5, with an *incubation period, i.e.,* time from initial exposure, until symptoms appear within 7–10 days. Patients are not contagious until symptoms appear, and these symptoms include fever, headache, and vomiting. The first case was identified on October 1. With proper *contact tracing,* this was determined to be the *index case* (first case) for this outbreak. It is unknown how the child was initially exposed. However, the family had recently returned from a trip to an island resort known for family-friendly activities. Examination of further cases revealed several immediate contacts with the index case through the child's daycare center. Siblings and parents of these cases became ill shortly afterward, along with a few staff members of the daycare center.

What conclusions can be drawn based on the case summary and the resulting epi curve chart in Figure in 6.2?

Several details can be determined from this chart. The y-axis of the graph shows incidence of new cases, with the x-axis covering the week ending October 2 through the week ending November 27. The data is stratified by age, grouping children under 5, children 5–16, and adults (>16 years). The total number of cases was 410. First, the index case becomes clear when data is depicted by week. Children under five had the highest number of cases, 196, which aligns with both the finding that this virus primarily affects young children, and that the epicenter of the outbreak was a nursery school. Older children, most likely siblings of the original cases and their immediate contacts, represented 142 cases. Fewer adults were affected (72), although an equally high number were likely exposed. Again, as mentioned in the case description, the virus mainly affects children.

In addition, it is evident from the chart that the cases in children under 5 peaked on October 23, while the cases in older children and adults peaked a week later, on October 30. One explanation for this pattern could be related to the incubation period of the virus, which is 7–10 days. Thus, the initial wave of the outbreak affected those children enrolled in the daycare center, while the secondary infections followed a week later.

It is important to note that the full picture of an outbreak is unfortunately often not evident until it is over due to reporting delays and to the often-imperfect collection of data. Viral illnesses such as COVID-19 can also be spread by asymptomatic or pre-symptomatic individuals, complicating the tracking of cases and requiring extensive diagnostic tools to gain a true understanding of infection rates in a population.

6.3.2 Clinical Trials

Clinical trials are experimental studies used for investigating new (or existing) medications or treatments. These studies involve the comparison of medical interventions to a placebo (or non-intervention) group in order to determine the efficacy of that intervention. The gold standard in clinical studies is the *randomized, double-blind, placebo-controlled trial*, which signifies:

- **Randomized**
  Patients fitting a predetermined set of parameters, are randomly assigned to one of two or more groups in the study.
- **Double-blind**
  Neither the patients nor the investigators (study coordinators, and other healthcare personnel) know which group the patient is in until the study is complete.
- **Placebo-Controlled**
  The study compares an active medication or intervention to a placebo or non-intervention group.

This study design is the most effective and accurate type of study for pharmaceutical drug trials and reduces any inherent bias that could occur if patients and investigators knew to what group the subject was assigned. Randomized, double-blind, placebo-controlled trials are required for approval by government agencies such as the Food and Drug Administration (FDA).

Any data collected from these studies require careful evaluation in order to be of any use. Epidemiology relies on statistical measures for the analysis of results. There are many different measures, depending on the type of data available. These include measures of central tendency when discussing normal distributions, which come up frequently in real-life scenarios and can also be used to simplify data when the curves are approximate. Measures of central tendency include the mean, median, and mode, as well as measures of dispersion or scatter such as the standard deviation, the range, and the standard error of the mean (Shaefer and

Theodore 2007). Statistical methods and illustrative examples are discussed in detail in Part III, Chapter 13: Probability and Statistical Principles.

6.4 VIRUS TRANSMISSION

Three conditions are required for a virus to successfully cause an active infection. The first is *transmission*, which indicates how a virus enters the body. The method of transmission is usually pathogen-specific, and one pathogen may have more than one possible route of entry. Ebola, for example, is spread through direct contact to various body fluids, such as blood or saliva. SARS-CoV-2, on the other hand, spreads through the respiratory tract and could potentially be contracted through either airborne or droplet transmission, which will be discussed later in this section (Louten 2016). The second requirement for infection to develop involves the presence of a *source*, which is any location where viruses can live outside of the body. This could include water droplets, various surfaces, or even human skin. Finally, the last requirement is *susceptible individual*, i.e., a person vulnerable to the specific virus. A person may be susceptible to a specific virus because they had never been exposed before, so that they have no acquired immunity to it. Other viruses are only dangerous to patients that have an already compromised immune system (CDC 2016).

Prior to any discussion of transmission, some terminology relating to viruses and infections should be defined. The terms *infectivity, pathogenicity,* and *virulence* are all used to indicate the severity of a particular virus, but they are each defined differently. *Infectivity* is defined as the ratio of individuals who become infected to those who were exposed, while *pathogenicity* is the ratio of those who develop clinical disease to those who become infected. The term *virulence* is defined as the ratio of clinical cases to those who become severely ill or die.

Transmission rate is dependent on several factors, some of which include the mode of transmission and the susceptibility of the population. Depending on the transmission rate, some viruses are more likely to cause an outbreak or to lead an outbreak to progress to a full-blown pandemic. According to the Principles of Epidemiology in Public Health Practice published by the CDC:

"Occasionally, the amount of disease in a community rises above the expected level. *Epidemic* refers to an increase, often sudden, in the number of cases of a disease above what is normally expected in that population in that area. *Outbreak* carries the same definition of epidemic, but is often used for a more limited geographic area. *Cluster* refers to an aggregation of cases grouped in place and time that are suspected to be greater than the number expected, even though the expected number may not be known.

Pandemic refers to an epidemic that has spread over several countries or continents, usually affecting a large number of people…" (CDC 2012)

Following an epidemic or pandemic, infectious diseases can sometimes temporarily subside, only to reemerge later, such as MERS-CoV and SARS-CoV have done since the first outbreaks. In rare situations, diseases can also be completely eradicated. However, it is more likely that they become *endemic*, which refers to a virus circulating at a constant level over time (Louten 2016; NIAID 2021).

6.4.1 MODES OF TRANSMISSION

As mentioned earlier, different viruses use different methods for entering the body. Detailed explanations for how to prevent transmission of various viruses depending on the mode of transmission is available in Chapter 5: Safety Protocols and Personal Protective Equipment. The following explains the two main types of transmission, *direct* and *indirect*.

- **Direct Transmission**
  Direct Contact: A pathogen can be transmitted from animal reservoir, contaminated surfaces, or from person to person through skin contact or sexual intercourse.
  Droplet Spread: Direct transmission can also occur via *droplet spread* through aerosols produced by an infectious person through talking, sneezing, or coughing when a susceptible person is within a few feet (CDC 2016).
- **Indirect Transmission**
  Airborne: Indirect transmission occurs by airborne particles when infectious agents are carried by dust or droplets small enough to remain suspended in air. In these cases, pathogens may remain suspended in the air long after an infected person has left the area. They can also travel longer distances through wind or air circulation. Typically, infectious airborne particles are less than 5 microns in diameter, and these can also settle on surfaces.
  Vehicles: Vehicles that may indirectly transmit an infectious agent include food, water, and biologic agents such as blood and other bodily fluids, and *inanimate objects*. In other words, a person can be infected by merely touching *fomites, i.e., inorganic surfaces* such as doorknobs or countertops where viruses and other pathogens can survive.
  Vectors: Vectors refer to animal intermediaries that can carry and transmit disease such as mosquitos, ticks, bats, or birds. Some viruses can affect multiple species, such as bird flu or coronaviruses, or vectors may be an intermediary, but the virus does not make them ill.

Examples of indirect transmission include hepatitis A, where infection usually arises from drinking contaminated water (which is a vehicle), as described earlier (Olsen 2000). Measles transmission is also usually indirect, primarily airborne, with one person spreading it to another without the two even having to meet. Just being in the same room at different times is enough to spread the disease from an infected person, as the virus can remain suspended in the air for hours (WHO 2019).

Similarly, SARS-CoV-2 can be transferred directly through airborne spread in areas with inadequate ventilation. More commonly, however, the transmission is indirect, as patients with COVID-19 pass on the SARS-CoV-2 virus through droplets in the air, which theoretically can travel up to 6 feet by an infected person talking, coughing, or singing. Initially, the virus was also thought to spread by surface contact, based on studies showing that SARS-CoV-2 can live for days on certain surfaces, but this did not result being a major mode of transmission. Further information on proper methods of ventilation can be found in Chapter 5: Safety Protocols and Personal Protective Equipment and in Part III, Chapter 15: Ventilation Calculations.

6.5 APPLICATIONS

The following four illustrative examples demonstrate the principles of epidemiology as it pertains to viral diseases and outbreaks discussed in this chapter.

Illustrative Example 6.1 EBOLA
Describe the steps of an outbreak investigation within the context of the Ebola epidemic of 2013–2016 in Western Africa.

Solution
As presented in this chapter, the major steps in investigating an outbreak are:

- Confirm the existence of an outbreak or epidemic.
- Verify the diagnosis.
- Rule out other possibilities
- Develop an initial case definition.
- Collect all pertinent data on cases.
- Analyze data by time, place, and person
- Develop a hypothesis based on initial data, and investigate further as required
- Implement measures to control the outbreak and prevent further cases
- Clearly communicate any findings to all stakeholders, including the public
- Assess effectiveness of control and preventive measures
- Monitor any effects, intended or otherwise, due to implemented measures
- Continually update measures and provide further direction to stakeholders based on new findings

Illustrative Example 6.2   HIV

Describe the evolution of the HIV pandemic from its discovery in the early 1980s until today.

Solution

As reviewed in Chapter 3: Pandemics, Epidemics, and Outbreaks, epidemiolocal studies have traced the origin of HIV to simian immunodeficiency virus (SIV) among chimpanzees, which may have crossed over to humans as early as the late seventeenth century in Central Africa through hunters who came into contact with the blood of the animals. AIDS, first recognized in 1981, reached global pandemic proportions throughout the early 1980s. In 1982, the year the term "AIDS" was coined and HIV was recognized as a global emergency. Blood was identified as a key method of transmission as the CDC identified the first case of HIV from a blood transfusion, and transmission through drug use and from exposure in healthcare settings were also reported.

By 1992, AIDS had become the number one killer of men aged 25–44 in the United States. While the first antiretroviral therapy was introduced in 1987, the standard of treatment referred to as HAART – highly active antiretroviral therapy, introduced in 1996 – changed the illness from a death sentence to a manageable lifetime disease.

Globally, case rates have exploded over the last 30 years, with more than 37 million people currently living with HIV. Although the survival rate for those living with HIV has greatly improved since the advent of antiretrovirals, an estimated 36 million people have died from AIDS (WHO 2021b).

Illustrative Example 6.3   PANDEMIC INFLUENZA

The CDC and other public health institutions have prepared for another deadly global influenza pandemic, such as occurred with the "Spanish Flu" in 1918. Discuss the need for preparedness planning in managing this emergent situation. (Further information on preparedness planning is presented in Part II, Chapter 9: Emergency Planning and Response.)

Solution

A proper emergency response plan in the event of an influenza outbreak or pandemic requires an understanding of all available data that would be able to provide healthcare workers, government agencies, and the public with the information needed to respond quickly and properly during a health crisis. Thus, successful pandemic planning begins with a thorough understanding of the infectious disease and/or the potential disaster being planned for – in this case, influenza. The impacts on public health and the environment must be estimated at an early stage.

It is critical to put science, medicine, and engineering into action to help prevent and minimize the effects of pandemics. One area that has received attention is the representation of infectious disease data in equation form for analytical and predictive purposes. This emergency planning and response (EP&R) activity offers the potential for protecting the public from illness and injury and better preparation for emergencies such as pandemics. *See also Part III, Chapter 16: Pandemic Health Data Models.*

Illustrative Example 6.4 CORONAVIRUS
List some achievements and some missteps made by public health authorities regarding vaccines during the COVID-19 pandemic.

Solution
One notable achievement in the fight against COVID-19 was the successful development of highly effective vaccines within a year of the identification of SARS-CoV-2. This was an example of the global cooperation necessary to save lives in the face of a deadly pandemic. The cooperative undertaking was in large part due to the legacy of confronting SARS and MERS epidemics over the last decade, which led to a global effort to understand and to design efficient vaccines and antiviral treatments (CDC 2022; WHO 2018; NIAID 2022).

Unlike the development of these vaccines, their global rollout highlights the failings of the WHO and world governments. Scientists had warned of the dangers of allowing SARS-CoV-2 to mutate among unvaccinated populations and called for equitable distribution of vaccines. Some feel that this plea was largely ignored by wealthy countries, who hoarded the limited vaccine doses for their own citizens, allowing a large vaccine availability gap to develop between themselves and developing countries. Once the vaccines became widely available in wealthier nations, poorer countries still had little to no access.

6.6 CHAPTER SUMMARY

- The field of public health is concerned with the health and well-being in communities and is a vast discipline that covers a wide array of healthcare areas, with focus on the prevention of injuries and diseases.
- Thanks to a widespread global commitment to vaccination over the last several decades, many diseases that once caused inordinate suffering and death no longer pose the threat to public health they once did.
- Epidemiological studies seek to observe, document, describe, and finally analyze and present collected data in order to better clarify a problem.
- In the United States, the CDC has the responsibility for maintaining and improving the nation's public health.
- Despite some major failures, millions of lives have been saved thanks to international campaigns to increase public awareness and vaccine use.

- Scientific studies are central to the understanding of public health issues.
- There are two categories of epidemiologic studies: *descriptive* and *analytical*.
- The gold standard in clinical studies is the *randomized, double-blind, placebo-controlled trial*.
- Transmission rates are dependent on several factors, some of which include the mode of transmission and the susceptibility of the population.

6.7 PROBLEMS

1 Describe the two main categories of epidemiologic studies.

2 Compare definitions of transmissibility, virulence, and pathogenicity.

3 Discuss some of the achievements and the shortfalls of public health agencies (i.e. the CDC and the WHO) over the past century.

4 List the various modes of transmission.

REFERENCES

Brachman, P. Chapter 9 Epidemiology. In Baron S, editor. (1996). *Medical Microbiology, 4*. Galveston (TX): University of Texas Medical Branch at Galveston.
Centers for Disease Control and Prevention (CDC). (2012). Principles of Epidemiology: Home Self-Study Course (archived). Atlanta: Centers for Disease Control and Prevention (CDC). Retrieved January 12, 2022, from https://www.cdc.gov/csels/dsepd/ss1978/
Centers for Disease Control and Prevention (CDC). (2013) Descriptive and analytic studies. Retrieved January 17, 2022, from https://www.cdc.gov/globalhealth/healthprotection/fetp/training_modules/19/desc-and-analytic-studies_ppt_final_09252013.pdf
Centers for Disease Control and Prevention (CDC). (2016, January 7). How infections spread. Centers for Disease Control and Prevention. Retrieved February 19, 2022, from https://www.cdc.gov/infectioncontrol/spread/index.html
Centers for Disease Control and Prevention (CDC). (2021). CDC Strategic Framework and priorities. Centers for Disease Control and Prevention. Retrieved January 16, 2022, from https://www.cdc.gov/about/organization/strategic-framework/index.html
Centers for Disease Control and Prevention (CDC). (2022). Understanding mRNA COVID-19 vaccines. Centers for Disease Control and Prevention. Retrieved

October 31, 2021, from https://www.cdc.gov/coronavirus/2019-ncov/vaccines/different-vaccines/mrna.html.

Centers for Disease Control and Prevention (CDC). (n.d.). CDC LC Quick Learn: Using an EPI curve to Determine Mode of Spread. Centers for Disease Control and Prevention. Retrieved January 28, 2022, from https://www.cdc.gov/training/quicklearns/epimode/

Epidemiology (2022) Merriam-Webster. https://www.merriam-webster.com/dictionary/epidemiology.

Frérot, M., et.al. (2018). *What is epidemiology? Changing definitions of epidemiology 1978-2017. PloS one*, 13(12), e0208442. https://doi.org/https://doi.org/10.1371/journal.pone.0208442

LaFraniere, S. (2022). Walensky, Citing Botched Pandemic Response, Calls for C.D.C. Reorganization. *The New York Times* (17 August 2022).

Louten J. (2016). Virus transmission and epidemiology. *Essential Human Virology*, 71–92. https://doi.org/https://doi.org/10.1016/B978-0-12-800947-5.00005-3

National Institute of Allergy and Infectious Diseases (NIAID). (2021). Covid-19, MERS, & SARS. U.S. Department of Health and Human Services (HHS). Retrieved October 21, 2021, from https://www.niaid.nih.gov/diseases-conditions/covid-19.

National Institute of Allergy and Infectious Diseases (NIAID). (2022). *Coronaviruses.* U.S. Department of Health and Human Services (HHS). Retrieved October 19, 2021, from https://www.niaid.nih.gov/diseases-conditions/coronaviruses.

Olsen, S.J. (2000). Surveillance for foodborne disease outbreaks — United States, 1993-1997. In: Surveillance Summaries, March 27, 2000. MMWR 2000; 49(No. SS-1):1–59. https://www.cdc.gov/csels/dsepd/ss1978/SS1978.pdf (accessed 21 October 2021).

Shaefer, S. and Theodore, L. (2007). *Probability and Statistics Applications for Environmental Science.* Boca Raton, FL: CRC Press/Taylor & Francis Group.

United Nations (UN). (2020). Measles cases hit 23-year high last year, killing 200,000 as vaccination stalls, WHO says | | UN News. United Nations. Retrieved November 1, 2021, from https://news.un.org/en/story/2020/11/1077482.

World Health Organization (WHO). (2018). Influenza (seasonal). World Health Organization. Retrieved October 19, 2021, from https://www.who.int/news-room/fact-sheets/detail/influenza-(seasonal).

World Health Organization (WHO). (2019). *Measles.* World Health Organization. https://www.who.int/news-room/fact-sheets/detail/measles (accessed 9 July 2022)

World Health Organization (WHO). (2020 December 3). Coronavirus disease (covid-19): Herd immunity, Lockdowns and covid-19. World Health Organization. Retrieved January 17, 2022, from https://www.who.int/news-room/questions-and-answers/item/herd-immunity-lockdowns-and-covid-19

World Health Organization (WHO). (2021a). *Immunization coverage*. World Health Organization. https://www.who.int/news-room/fact-sheets/detail/immunization-coverage (accessed 9 July 2022).

World Health Organization (WHO). (2021b). *HIV/AIDS*. World Health Organization. https://www.who.int/news-room/fact-sheets/detail/hiv-aids (accessed 9 July 2022)

World Health Organization (WHO). (n.d.). History. World Health Organization. Retrieved February 24, 2022, from https://www.who.int/about/who-we-are/history

Part II

Practical and Technical Considerations

Merriam-Webster defines *practical* in several ways, some of which include:

- "actively engaged in some course of action or occupation"
- "capable of being put to use or account"
- "concerned with voluntary action and ethical decisions." (Merriam-Webster 2022)

Likewise, *technical* can be defined as

- "having special and usually practical knowledge especially of a mechanical or scientific subject,"
- "of or relating to a practical subject organized on scientific principles"
- "based on or marked by a strict or legal interpretation." (Merriam-Webster 2022)

The chapters in Part II contain such practical and technical material that one might view as a prerequisite for understanding the many ramifications of a global pandemic, before reviewing the engineering applications and calculations that are addressed in Part III.

Recognizing the ethical, legal, and regulatory climate during a pandemic is essential to understanding the many repercussions of pandemics and other crises. In addition, successful emergency planning begins with a thorough appreciation of the potential health crisis and/or disaster being planned for.

A Guide to Virology for Engineers and Applied Scientists: Epidemiology, Emergency Management, and Optimization, First Edition. Megan M. Reynolds and Louis Theodore.
© 2023 John Wiley & Sons, Inc. Published 2023 by John Wiley & Sons, Inc.

There are five chapters in Part II. The chapter numbers and accompanying titles are listed below:

Chapter 7: Engineering Principles and Fundamentals
Chapter 8: Legal and Regulatory Considerations
Chapter 9: Emergency Planning and Response
Chapter 10: Ethical Considerations in Virology
Chapter 11: Health and Hazard Risk Assessment

7

Engineering Principles and Fundamentals

Contributing Author: Vishal Bhatty

In a very broad sense, engineering is a term applied to the profession in which knowledge of the mathematical and natural sciences, gained by study, experience, and practice, is applied to the efficient use of the materials and forces of nature. The term engineer refers to an individual who has received professional training in both fundamental and applied science. In addition to the professional engineer, are those individuals known as subprofessionals or paraprofessionals, who also apply scientific and engineering skills to technical problems; typical of these are engineering aides, technicians, inspectors, draftsmen, etc.

The five major branches of engineering are listed below:

- Chemical engineering
- Civil engineering
- Electrical engineering
- Environmental engineering
- Mechanical engineering

A Guide to Virology for Engineers and Applied Scientists: Epidemiology, Emergency Management, and Optimization, First Edition. Megan M. Reynolds and Louis Theodore.
© 2023 John Wiley & Sons, Inc. Published 2023 by John Wiley & Sons, Inc.

More specialized areas that fall within the focus of this book include biochemical, biomedical, healthcare, industrial, managerial, pharmaceutical, process and structural, as well as biotechnology and nanotechnology (Kunz and Theodore 2005; Theodore 2006).

7.1 HISTORY OF ENGINEERING

In terms of history, the engineering profession, as defined today, is usually considered to have originated shortly after 1800. However, many of the "processes" associated with this discipline were developed in antiquity. For example, filtration operations were carried out 5000 years ago by the Egyptians. Operations such as crystallization, precipitation, and distillation soon followed. Others evolved from a mixture of craft, mysticism, incorrect theories, and empirical guesses during this period.

The chemical industry dates back to prehistoric times when people first attempted to control and modify their environment, and it developed as did any other trade or craft. With little knowledge of science and no means of chemical analysis, the earliest "engineers" had to rely on previous art and superstition. As one would imagine, progress was slow. This changed with time. Industry in the world today is a sprawling complex of raw-material sources, manufacturing plants, and distribution facilities which supply society with thousands of chemical products, most of which were unknown over a century ago. In the latter half of the nineteenth century, an increased demand arose for individuals trained in the fundamentals of these processes. This demand was ultimately met by engineers.

The technical advances of the nineteenth century greatly broadened the field of engineering and introduced a large number of the aforementioned engineering specialties, and the rapidly changing demands of the socioeconomic environment in the twentieth and twenty-first centuries have widened the scope even further. One need only review the various branches of engineering listed above.

A related field of environmental engineering – accident and emergency management – received wide attention in the late 1970s and 1980s when the safety of nuclear reactors was questioned following accidents caused by operator errors, design failures, and malfunctioning equipment. "Human factors" engineering seeks to establish criterion for the efficient, human-centered design of, among other things, the large, complicated control panels that monitor and govern nuclear reactor operations.

Another important advancement in engineering during the middle of the twentieth century has been the adaption of modern statistical methods to the problem of quality control. For example, this area of study attempts to maintain high standards of accuracy in the manufacture of replaceable parts. By applying mathematical analysis, engineers have developed testing procedures that can improve quality and ensure uniformity.

Today, scientific methods of engineering are applied in several fields not connected directly to manufacturing and construction. Modern engineering is characterized by the broad application of what has come to be known as systems engineering principles. The systems approach is a methodology of decision-making in design, operation, or construction that adopts:

- the formal process, included in what is known as the scientific method,
- an interdisciplinary, or team, approach, using specialists from not only the various engineering disciplines but also from legal, social, aesthetic, and behavioral fields as well, and
- a formal sequence of procedures employing the principles of operations research.

7.2 PROBLEM SOLVING: THE ENGINEERING APPROACH

Engineers and scientists are known for their problem-solving abilities. It is probably this ability more than any other that has enabled many engineers to rise to positions of leadership and top management within their companies.

In problem solving, both in academia and industry, considerable importance is attached to a proper analysis of the problem, to a logical recording of the problem solution, and the overall professional appearance of the finished product calculations. Neatness and clarity of presentation should be the distinguishing marks of the work. Engineering and applied science students should always strive to practice professional habits of problem-solving analysis and to make a conscious effort to improve the appearance of each document whether it is submitted for grading or is included in a notebook.

The value of an engineer or scientist is determined by his/her ability to apply basic principles, facts, and methods in order to accomplish some useful purpose. In this modern age of industrial competition, the ultimate definition of a useful purpose is usually based on a tangible profit of monetary value. It is not sufficient, therefore, to have a knowledge and understanding of physics, chemistry, mathematics, mechanics, stoichiometry, thermodynamics, unit operations, chemical technology, and other related engineering and scientific subjects; he/she must also have the ability to apply this knowledge to practical situations, and, in making these applications, recognize the feasibility in financial terms.

Engineers and scientists who have mastered the method of problem-solving are considerably more successful in their work than those not trained in this technique. In the past, many problems were of such a routine nature that resorting to deductive reasoning would suffice, and premises of deduction could be taken from the handbooks. However, many of the engineering problems of today cannot be solved by mere "handbook techniques." Experimentation, research, and development have indeed become significant activities in today's world.

7.2.1 PROBLEM-SOLVING METHODOLOGY

If one can form good habits of problem-solving early in a career, the engineer will save considerable time and avoid many frustrations in all aspects of work in and out of school. In solving problems, one should:

- Read the available material thoroughly and understand what is required for an answer; sometimes, as in life, the major obstacle is finding out what the problem really is.
- Determine what additional data are needed, if any, and obtain this information.
- Draw a simplified picture of what is taking place and write down the available data; one may use boxes to indicate processes or equipment, and lines for the flow of streams.
- Pick a basis on which to start the problem.
- If a chemical equation is involved, write it down and make sure it is balanced.
- Problems that are long and complex should be divided into parts and attacked systematically piece by piece.

While the describing equations considered by both the engineer and the mathematician may be of identical type and form, the attitude and approach to a problem by the two groups may be markedly different. For example, the solution of a differential equation by a mathematician first requires an examination of the existence of a solution. Next, the properties of the solution are considered. Finally, the development of a suitable form for the resulting equation(s) is undertaken. Unfortunately, most mathematicians deal with x, y, z, t, etc., without considering the need of relating those variables to real terms such as pressure, temperature, time, etc. The engineer, on the other hand, when faced with a physical situation must derive first the governing differential equation(s). The actual existence of a solution to the physical problem often is taken for granted since the physical problem does exist. The existence of the proof of a solution to the mathematical model that has been selected to represent the physical problem is also often skipped by the engineer. Having obtained an answer, within the accuracy required, an engineer should then go back and be assured, finally, that the solution reasonably satisfies the physical problem originally considered. No particular form or method of presentation distinguishes an engineering problem from any other problem in applied science.

7.2.2 ENGINEERING AND SCIENTIFIC SOURCES

Virtually every university, business, government (national, state, and local) and organization has its own website, which can provide plentiful sources of information. For instance, federal agencies such as the U.S. Environmental Protection Agency (EPA), the National Institute of Health (NIH), the Center for Disease Control (CDC), and the Department of Energy (DOE) have comprehensive information available. Mapping details can be provided by several sources, from Google Earth to Geographic Information Systems (GIS) where available.

Note that some collaborative sites (such as Wikipedia) themselves are not generally advised to be used as scholarly citations in technical papers, due to their nature as crowdsourced information hubs subject to rapid changes or bias. However, this rarely detracts from their accuracy in engineering applications, and they are extremely helpful as encyclopedic information for personal use and review of information. In addition, the sources and citations that they list are normally of sufficient quality for reference.

7.3 UNITS AND CONVERSION CONSTANTS

This section introduces the reader to the general subject of units, presenting both the metric system and the International System of units (SI).

7.3.1 THE METRIC SYSTEM

The need for a single worldwide coordinated measurement system was recognized over 300 years ago. Gabriel Mouton, Vicar of St. Paul in Lyons, proposed a comprehensive decimal measurement system in 1670 based on the length of one minute of the arc of a great circle of the earth. In 1671, Jean Picard, a French astronomer, proposed the length of a pendulum swing in seconds as the unit of length. (Such a pendulum would have been easily reproducible, thus facilitating the widespread distribution of uniform standards.) Other proposals were made, but over a century elapsed before any action was taken.

In 1790, amid the French Revolution, the National Assembly of France requested the French Academy of Sciences to "deduce an invariable standard for all the measures and weights." The Commission appointed by the Academy created a system that was, at once, simple and scientific. The unit of length was to be a portion of the earth's circumference. Measures for capacity (volume) and mass (weight) were to be derived from the unit of length, thus relating the basic units of the system to each other and to nature. Furthermore, the larger and smaller versions of each unit were to be created by multiplying or dividing the basic units by 10 and its multiples. This feature provided great convenience to users of the system by eliminating the need for such calculations and divisions by 16 (to convert ounces to pounds) or by 12 (to convert inches to feet). Similar calculations in the metric system could be performed simply by shifting the decimal point. Thus, the metric system is a base-10 or decimal system.

The Commission assigned the name metre (which is now spelled meter) to the unit of length. This name was derived from the Greek word "metron" meaning "a measure." The physical standard representing the meter was to be constructed so that it would equal one 10-millionth of the distance from the north pole to the Equator along the meridian of the earth running near Dunkirk in France and Barcelona in Spain. The metric unit of mass, called the gram, was defined as the mass of one cubic centimeter (a cube that is 1/100 of a meter on each side) of water at its temperature of maximum density. The cubic decimeter (a cube 1/10 of a meter on each side) was chosen as the unit of fluid capacity. This measure was given the name liter.

Although the metric system was not accepted with enthusiasm at first, adoption by other nations occurred steadily after France made its use compulsory is 1840. The standardized character and decimal features of the metric system made it well-suited to scientific and engineering work. Consequently, it is not surprising that the rapid spread of the system coincided with an age of rapid technological development. In the United States, by an Act of Congress in 1866, it was made "lawful throughout the United States of America to employ the weights and measures of the metric system in all contracts, dealings, or court proceedings."

By the late 1860s, even better metric standards were needed to keep pace with scientific advances. In 1875, the international "Treaty of the Meter," set up well-defined metric standards for length and mass, and established permanent machinery to recommend and adopt further refinements in the metric system. This treaty, known as the Metric Convention, was signed by 17 countries, including the United States. Because of the Treaty, metric standards were constructed and distributed to each nation that ratified the Convention. Since 1893, the internationally agreed metric standards have served as the fundamental weights and measures standards of the United States.

A total of 35 nations – including the major nations of continental Europe and most of South America – had officially accepted the metric system by 1900. Today, except for the United States and a few small countries, the entire world is predominantly using the metric system or is committed to its use. In 1971, the Secretary of Commerce, in transmitting to Congress the results of a three-year study authorized by the Metric Study Act of 1968, recommended a program that the U.S. change to the predominant use of the metric system through a coordinated national program.

The International Bureau of Weights and Measures (located at Sevres, France) serves as a permanent secretariat for the Metric Convention, coordinating the exchange of information about the use and refinement of the metric system. As measurement science develops more precise and easily reproducible ways of defining the measurement units, the General Conference of Weights and Measures – the diplomatic organization made up of adherents to the Convention – meets periodically to ratify improvements in the system and the standards.

7.3.2 THE SI SYSTEM

The aforementioned General Conference adopted an extensive revision and simplification of the system in 1960. The name, Le Systeme International d'Unites (International System of Units), with the international abbreviation SI, was adopted for this modernized metric system. Further improvements in and additions to SI were later made by the General Conference in 1964, 1968, and 1971.

The basic units in the SI system are the kilogram (mass), meter (length), second (time), kelvin (temperature), ampere (electric current), candela (the unit of luminous intensity), and radian (angular measure). All are commonly used by an engineer. The celsius scale of temperature ($0\,^{\circ}C = 273.15\,K$) is commonly used with the absolute Kelvin scale. The important derived units are the newton (SI unit of force), the joule (SI unit of energy), the watt (SI unit of power), the pascal (SI unit of pressure), and the hertz (unit of frequency). There are several electrical units: coulomb (charge), farad (capacitance), henry (inductance), volt (potential), and weber (magnetic flux). As noted, one of the major advantages of the metric system is that larger and smaller units are given in powers of 10. In the SI system, a further simplification is introduced by recommending only those units with multipliers of 10^3. Thus, for lengths in engineering, the micrometer, (previously micron), millimeter, and kilometer are recommended, and the centimeter is generally avoided, A further simplification is that the decimal point may be substituted by a comma, while the other number, before and after the comma, will be separated by spaces between groups of three, e.g., one million dollars will be $100 000 000. In common practice in Europe, the comma and the decimal point are merely switched from the US standard, so that one million euros reads as €1.000.000,00. More details are provided in a later section.

7.4 DIMENSIONAL ANALYSIS

Engineers and scientists spend a surprising amount of time converting data and equations from one set of units to another. Keep in mind that a unit is defined as a measure of a physical extent while a dimension is a description of the physical extent.

Units, unlike physical laws, can be considered as either derived or basic. There is a certain latitude in choosing the basic units and, unfortunately, this free choice has resulted in the aforementioned mild form of confusion. Two systems of units have arisen: metric, the cgs, or centimeters-gram-second system and the English, the fps, or foot-pound-second system of engineering. The metric system has come to be defined as the System of International Units, or more commonly the SI system. As long as physicists and chemists make measurements in grams and centimeters while engineers employ pounds and feet, this confusion in terminology will continue to exist.

The presentation on the conversion of units is first addressed. Momentum and rate of momentum come into play in both fluid flow and the conservation law for momentum studies. The latter term is defined (in terms of units) as

$$\text{Rate of momentum} = \frac{\text{ft} \cdot \text{lb}}{\text{s}^2} \tag{7.1}$$

The above units can be converted to lb_f if multiplied by an appropriate conversion constant, a term that is used to obtain units in a more convenient form. All conversion constants have magnitude and units in the term but can also be shown to be equal to 1.0 (unity) with no units. An often-used conversion constant is

$$\frac{12 \, \text{in.}}{1 \, \text{ft}} \tag{7.2}$$

This term is obtained from the following defining equation:

$$12 \, \text{in} = 1 \, \text{ft} \tag{7.3}$$

If both sides of this equation are divided by 1 ft, one obtains:

$$12 \, \text{in/ft} = 1 \tag{7.4}$$

Note that this conversion constant, like all others, is also equal to unity without any units. Another defining equation is:

$$1 \, \text{lb}_f = 32.2 \, \text{ft} \cdot \text{lb/s}^2 \tag{7.5}$$

If this equation is divided by lb_f,

$$1 = 32.2 \ \text{ft} \cdot \text{lb/lb}_f \cdot \text{s}^2 \tag{7.6}$$

Equation (7.6) results. This serves to define the conversion constant, g_c. If the rate of momentum is divided by g_c (32.2 lb · ft/lb$_f$ · s^2), the following units result. Note that this operation is equivalent to dividing the rate of moment by 1:

$$\text{Rate of momentum} = \left(\frac{\text{ft} \cdot \text{lb}}{\text{s}^2} \right) \left(\frac{\text{lb}_f \cdot \text{s}^2}{\text{ft} \cdot \text{lb}} \right) = \text{lb}_f \tag{7.7}$$

One can conclude from the above dimensional analysis that a force is equivalent to a rate of momentum.

There are hundreds of conversion constants employed by environmental engineers and applied scientists. Some of the more common "conversion constants" were provided earlier in the chapter. Conversion of units can be accomplished by the multiplication of the quantity to be converted by appropriate unit ratios, i.e., the conversion constants. For example, suppose energy of 50 Btu must be converted to units of (ft $\cdot$ lb$_f$). One notes that to convert from Btu to (lb$_f$ $\cdot$ ft), one simply multiplies by 778. Therefore,

$$1\ \text{Btu} = 778\ \text{ft} \cdot \text{lb}_f \tag{7.8}$$

The conversion constant, or unit ratio, is

$$\frac{778\ \text{ft} \cdot \text{lb}_f}{1\ \text{Btu}} = 1 \tag{7.9}$$

The 50 Btu may be multiplied by the above conversion constant without changing its value. Therefore,

$$(50\ \text{Btu}) \left(\frac{778\ (\text{ft} \cdot \text{lb}_f)}{1\ \text{Btu}} \right) = 38\,900\ (\text{ft} \cdot \text{lb}_f) \tag{7.10}$$

with the Btu units canceling, just like numbers.

One can apply the same procedure to convert a density of 0.06 g/cm^3 to lb/ft^3. The details are provided below:

$$\left(\frac{0.06\ \text{g}}{1\ \text{cm}^3} \right) \left(\frac{\text{lb}}{454\ \text{g}} \right) \left(\frac{30.48\ \text{cm}}{1\ \text{ft}} \right)^3 = 4.0\ \text{lb/ft}^3 \tag{7.11}$$

Problems are frequently encountered in studies and other engineering work that involve several variables. Engineers and scientists are generally interested in developing functional relationships (equations) between these variables. When these variables can be grouped in such a manner that they can be used to predict the performance of similar pieces of equipment, independent of the scale or size of the operation, something very valuable has been accomplished. Details are available in the literature (Theodore 2014).

7.5 PROCESS VARIABLES

The authors originally considered the titles "State, Physical, and Chemical Properties" for this section. However, since these three properties have been used interchangeably and have come to mean different things to different people, it was decided to simply employ the title "Process Variables." The three aforementioned properties were therefore integrated into this all-purpose title and eliminated the need for differentiating between the three.

This section provides a review of some basic concepts from physics, chemistry, and engineering in preparation for material that is covered in later sections. All of these topics are vital to engineering and science applications. Because many of these topics are unrelated to each other, this section admittedly lacks the cohesiveness that one covering a single topic might have. This is usually the case when basic material from such widely differing areas of knowledge as physics, chemistry, and engineering is surveyed. Though these topics are widely divergent and covered with varying degrees of thoroughness, all of them will find later use in this book. If additional information on these review topics is needed, the reader is directed to the literature in the reference section of this chapter. A baker's dozen process variables of interest to the engineer and applied scientist are listed below:

- Temperature
- Pressure
- Moles and molecular weight
- Mass and volume
- Viscosity
- Heat capacity
- Thermal conductivity
- Reynolds number
- PH
- Vapor pressure
- Diffusion coefficient
- Surface tension
- Property estimation

There is a need to discuss the traditional difference in definition between chemical and physical properties. Every compound has a unique set of properties that allows one to recognize and distinguish it from other compounds. These properties can be grouped into two main categories: physical and chemical. Physical properties are defined as those that can be measured without changing the identity of the substance. Key properties include viscosity, density, surface tension, melting point, boiling point, etc. Chemical properties are defined as those that may be altered via reaction to form other compounds or substances. Key chemical properties include upper and lower flammability limits, enthalpy of reaction, autoignition temperature, etc. These properties may be further divided into two categories – intensive and extensive. Intensive properties are not a function of the quantity of the substance, while extensive properties do depend on the quantity of the substance.

Note that much of the material in this chapter was drawn from Kunz and Theodore in their textbook, *Nanotechnology: Environmental Implications and Solutions* (Kunz and Theodore 2005).

7.6 THE CONSERVATION LAWS

To better understand engineering, it is necessary to first understand the theory underlying it. How can one predict what viruses will be emitted from a hospital's effluent stream? At what temperature must a drug process be operated? How much energy in the form of heat is given off during a vaccine process? The answers to these questions are rooted in the various theories of thermodynamics, including thermochemistry, phase equilibrium, and chemical reaction equilibrium. However, one of the keys necessary to answer the above questions is often obtained via the application of one or more of the conservation laws, and the contents of this section deal with these laws. Topics covered include the conservation laws: the conservation law for momentum, the conservation law for mass, and the conservation law for energy. Obviously, at the heart of the section are the conservation laws for mass and energy.

Four important terms are defined below before proceeding to the conservation laws.

- A system is any portion of the universe that is set aside for study.
- Once a system has been chosen, the rest of the universe is referred to as the surroundings.
- A system is described by specifying that it is in a certain state.
- The path, or series of values certain variables assume in passing from one state to another, defines a process.

Momentum, energy, and mass are all conserved. As such, each quantity obeys the conservation law below as applied within a system.

$$
\begin{Bmatrix} \text{Quantity} \\ \text{into} \\ \text{system} \end{Bmatrix} - \begin{Bmatrix} \text{quantity} \\ \text{out of} \\ \text{system} \end{Bmatrix} + \begin{Bmatrix} \text{quantity} \\ \text{generated} \\ \text{in system} \end{Bmatrix} = \begin{Bmatrix} \text{quantity} \\ \text{accumulated} \\ \text{in system} \end{Bmatrix} \tag{7.12}
$$

The equation may also be written on a time-rate basis:

$$
\begin{Bmatrix} \text{Rate of} \\ \text{quantity} \\ \text{into} \\ \text{system} \end{Bmatrix} - \begin{Bmatrix} \text{rate of} \\ \text{quantity} \\ \text{out of} \\ \text{system} \end{Bmatrix} + \begin{Bmatrix} \text{rate of} \\ \text{quantity} \\ \text{generated} \\ \text{in system} \end{Bmatrix} = \begin{Bmatrix} \text{rate of} \\ \text{quantity} \\ \text{accumulated} \\ \text{in system} \end{Bmatrix} \tag{7.13}
$$

The conservation law may be applied at the macroscopic, microscopic, or molecular level. One can best illustrate the differences in these methods with an example. Consider a system in which a fluid is flowing through a cylindrical tube (see Figure 7.1). Define the system as the fluid contained within the tube between points 1 and 2 at any time. If one is interested in determining changes

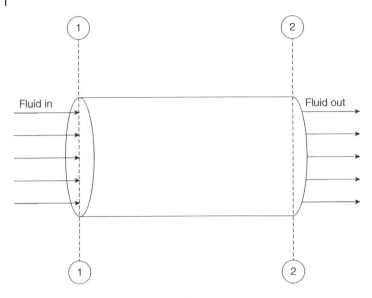

Figure 7.1 Conservation law example.

occurring at the inlet and outlet of the system, the conservation law is applied on a "macroscopic" level to the entire system. The resultant equation—usually algebraic—describes the overall changes occurring to the system without regard for internal variations within the system. This approach is usually applied by the practicing engineer.

The microscopic approach is employed when detailed information concerning the behavior within the system is required, and this is often requested of and by the engineer or applied scientist. The conservation law is then applied to a differential element within the system, which is large compared to an individual molecule, but small compared to the entire system. The resultant equation is then expanded, via an integration, to describe the behavior of the entire system. This is defined by some as the transport phenomena approach.

The molecular approach involves the application of the conservation law to individual molecules. This leads to a study of statistical and quantum mechanics – both of which are beyond the scope of this text. In any case, the description of individual molecules at the molecular level is of little value to the practicing engineer. However, the statistical averaging of molecular quantities in either a differential or finite element within a system leads to a more meaningful description of the behavior of a system. The macroscopic approach is adopted and applied in this book, and no further reference to microscopic or molecular analyses will be made.

It should be noted that the applied mathematician has developed differential equations describing the detailed behavior of systems by applying the appropriate conservation law to a differential element or shell within the system. Equations were derived with each new application. The engineer later removed the need for these tedious and error-prone derivations by developing a general set of equations that could be used to describe systems. These came to be defined as the aforementioned transport equations. Needless to say, these transport equations have proven to be an asset in describing the behavior of some systems, operations, and processes (Bird et al. 2002; Kunz and Theodore 2005; Theodore 2006).

The conservation law for mass can be applied to any process or system. The general form of this law is given as

$$\text{Mass in} - \text{mass out} + \text{mass generated} = \text{mass accumulated} \qquad (7.14)$$

or, on a time-rate basis, by

$$\left\{ \begin{array}{c} \text{Rate of} \\ \text{mass in} \end{array} \right\} - \left\{ \begin{array}{c} \text{rate of} \\ \text{mass out} \end{array} \right\} + \left\{ \begin{array}{c} \text{rate of mass} \\ \text{generated} \end{array} \right\} = \left\{ \begin{array}{c} \text{rate of mass} \\ \text{accumulated} \end{array} \right\} \qquad (7.15)$$

In engineering-related processes, it is often necessary to obtain quantitative relationships by writing mass balances on the various elements in the system. This equation may be applied either to the total mass involved or to a particular species on either a mole or mass basis. This law can be applied to steady-state or unsteady-state (transient) processes and to batch or continuous systems. As noted earlier, in order to isolate a system for study, it is separated from the surroundings by a boundary or envelope. This boundary may be real (e.g. the walls of a thermal device) or imaginary. Mass crossing the boundary and entering the system is part of the *mass in* term in Eq. (7.15), while that crossing the boundary and leaving the system is part of the *mass out* term. Equation (7.15) may be written for any compound whose quantity is not changed by chemical reaction and for any chemical element whether or not it has participated in a chemical reaction. (This is treated in even more detail in the next section.) It may be written for one piece of equipment, around several pieces of equipment, or around an entire process. It may be used to calculate an unknown quantity directly, to check the validity of experimental data, or to express one or more of the independent relationships among the unknown quantities in a particular problem situation (Fogler 2006).

A steady-state process is one in which there is no change in conditions (i.e. pressure, temperature, and composition) or rates of flow with time at any given point in the system. The accumulation term in Eq. (7.15) is then zero. (If there is no chemical or nuclear reaction, the generation term is also zero.) All other processes are unsteady-state.

In a batch process, a given quantity of reactants is placed in a container, and by chemical and/or physical means, a change is made to occur. At the end of the process, the container (or containers) to which material may have been transferred holds the product or products.

In a continuous process, reactants are continuously fed to a piece of equipment or to several pieces in series, and products are continuously removed from one or more points. A continuous process may or may not be steady-state. A coal-fired power plant, for example, operates continuously. However, because of the wide variation in power demand between peak and slack periods, there is an equally wide variation in the rate at which the coal is fired. For this reason, power plant problems may require the use of average data over long periods of time. However, most thermal operations are assumed to be steady state and continuous.

As indicated previously, Eq. (7.15) may be applied to the total mass of each stream (referred to as an overall or total material balance) or to the individual component(s) of the stream (referred to as a componential or component material balance). The primary task in preparing a material balance in engineering calculations is often to develop the quantitative relationships among the streams. The primary factors, therefore, are those that tie the streams together. An element, compound, or unreactive mass (e.g., ash) that enters or exits in a single stream or passes through a process unchanged is so convenient for this purpose that it may be considered a key to the calculations. If sufficient data are given about this component, it can be used in a component balance to determine the total masses of the entering and exiting streams. Such a component is sometimes referred to as the *key* component. Since a key component does not react in a process, it must retain its identity as it passes through the process. Obviously, except for nuclear reactions, elements may always be used as key components because they do not change identity even though they may undergo a chemical reaction (Reynolds et al. 2002).

Four important processing concepts are *bypass, recycle, purge*, and *makeup*. With bypass, part of the inlet stream is diverted around the equipment to rejoin the (main) stream after the unit (see Figure 7.2). This stream effectively moves in parallel with the stream passing through the equipment. In recycle, part of the product stream is sent back to mix with the feed. If a small quantity of nonreactive material is present in the feed to a process that includes recycle, it may be necessary to remove the nonreactive material in a purge stream to prevent it from building up above a maximum tolerable value. This can also occur in a process without

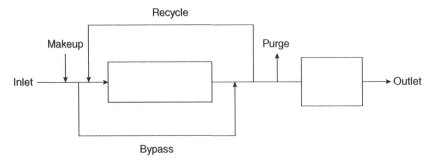

Figure 7.2 Bypass, recycle, purge, and makeup.

recycle; if a nonreactive material is added in the feed and not totally removed in the products, it will accumulate until purged. The purging process is sometimes referred to as *blowdown*. Makeup, as its name implies, involves adding or making up part of a stream that has been removed from a process. Makeup may be thought of as the opposite of purge and/or blowdown (Reynolds 1992).

7.7 THERMODYNAMICS AND KINETICS

Thermodynamics deals with the transfer of energy from one form to another. Many important practical conclusions can be derived from the two fundamental laws of thermodynamics. The energy balance, which was mentioned in the prior section, is an expression of the first law of thermodynamics. The second law of thermodynamics states that in a process involving heat transfer alone, energy may be transferred only from a higher temperature to a lower one. The thermodynamic analysis of a process leads to conclusions concerning the feasibility and efficiency of the various process steps. Thermodynamics also is useful in determining the composition of phases in equilibrium and in predicting the distribution of chemical species in reaction equilibrium (Theodore and Reynolds 1992; Theodore et al. 2010).

Kinetics considers the rate at which chemical compounds react. Data on rate of reaction are necessary in the design of the industrial chemical reactions. Chemical kinetics is the subject concerned with the study of the rates at which chemical reactions occur and the variables that affect these rates. The objective of part of this section is to develop a working understanding of this subject in order to apply it to chemical reactors. Chemical reactions and rates of reaction are treated from an engineering point of view in terms of physically measurable quantities. This information is a requisite for the study and analysis of real-world reaction systems (Theodore 2014).

All chemical reactions take place at a finite rate depending on the conditions, the most important of which are the concentration of reactants and products, temperature, and presence of a catalyst, promoter, or inhibitor. Some reactions are so rapid that they appear to be instantaneous (i.e., combustion), whereas other reactions are so slow at ordinary temperatures that no detectable change would be observed with time (i.e., the combination of two gases such as nitrogen and oxygen at ambient conditions.) Between these two extremes are many processes involving both inorganic and organic reactions (Theodore 1995).

With regard to chemical reactions, there are two important questions that are of concern to the engineer and applied scientist:

1. How far will the reaction go?
2. How fast will the reaction go?

Chemical thermodynamics provides the answer to the first question; however, it provides nothing about the second (Kunz and Theodore 2005; Theodore 2006). To illustrate the differences and importance of both of the above questions in an engineering analysis of a chemical reaction, consider the following process (Theodore 2006). Substance A, which costs 1 cent/ton, can be converted to B, which is worth $1 million/lb, by the reaction A ⇔ B. Chemical thermodynamics will provide information on the maximum amount of B that can be formed. If 99.99% of A can be converted to B, the reaction would then appear to be economically feasible, from a thermodynamic point of view. However, a kinetic analysis might indicate that the reaction is so slow that, for all practical purposes, its rate is vanishingly small. For example, it might take 10^6 years to obtain a 10^{-6} % conversion of A. The reaction is then economically unfeasible. Thus, it can be seen that both equilibrium and kinetic effects must often be considered in an overall engineering analysis of a chemical reaction.

Finally, the term *stoichiometry* has come to mean different things to different people. In a loose sense, stoichiometry involves the balancing of an equation for a chemical reaction that provides a quantitative relationship among the reactants and products. In the simplest stoichiometric situation, exact quantities of pure reactants are available, and these quantities react completely to give the desired product(s). In an industrial process, reactants usually are not pure; one reactant is usually in excess of what the reaction requires. Due to this and a host of other consideration, the desired reaction may not go to completion. In any event, the chemical equation provides a variety of qualitative and quantitative information essential for the calculation of the combining weights of materials involved in chemical changes. It deals with the combining weights of elements and compounds. The ratios obtained from the numerical coefficients in the chemical equation are the stoichiometric ratios that permit one to calculate the quantity of one substance as related to the quantity of another substance in the chemical equation in moles (Kunz and Theodore 2005; Theodore 2006; Bhatty et al. 2017).

7.8 APPLICATIONS

Four illustrative examples complement the presentation for this section concerned with principles of engineering overview.

Illustrative Example 7.1 BALANCING REACTIONS
Provide a balanced reaction equation for the overall metabolism of glucose.

Solution
The unbalanced reaction equation for the metabolism of glucose is:

$$C_6H_{12}O_6 + O_2 \rightarrow CO_2 + H_2O$$

A chemical equation provides a variety of qualitative and quantitative information essential for the calculation of the quantity of reactants reacted and products formed in a chemical process. A balanced chemical equation, as noted above, must have the same number of atoms of each type in the reactants and products. Thus, the balanced equation for glucose is:

$$C_6H_{12}O_6 + 6\,O_2 \rightarrow 6\,CO_2 + 6\,H_2O$$

Note that:

- Number of carbon atoms in reactants = number of carbon atoms in products = 6.
- Number of hydrogen atoms in reactants = number of hydrogen atoms in products = 12.
- Number of oxygen atoms in reactants = number of oxygen atoms in products = 18.
- Number of moles in reactants = 1 mol $C_6H_{12}O_6$ + 6 mol O_2 = 7 mol.
- Number of moles in products = 6 mol CO_2 + 6 mol H_2O = 12 mol.

The reader should realize that although the number of moles on both sides of the equation do *not* balance, the masses of reactants and products (in line with the conservation law for mass) *must* balance.

Illustrative Example 7.2   COLLECTION EFFICIENCY
Given the following inlet loading and outlet loading of a virus-contaminated stream of an air purification system, determine the collection efficiency of the unit.

- *Inlet loading = 0.02 ppm*
- *Outlet loading = 0.001 ppm*

Solution
Collection efficiency is a measure of the degree of performance of a control device; it specifically refers to the degree of removal of a pollutant and may be calculated through the application of the conservation law for mass. Loading refers to the concentration of pollutant usually in parts of pollutant per million (ppm) or in some form of concentration per unit volume.

The equation describing collection efficiency (fractional) E in terms of inlet and outlet loading is:

$$E = \frac{\text{Inlet loading} - \text{outlet loading}}{\text{Inlet loading}}$$

Calculating the collection efficiency of the control unit in percent for the rates provided yields:

$$E = \frac{0.02 - 0.001}{0.02} = 95\%$$

The reader should also note that the collected amount of pollutant by the control unit is the product of E and the inlet loading. The amount discharged to the atmosphere is given by the inlet loading minus the amount collected.

Illustrative Example 7.3   IDEAL GAS LAW
Given the following pressure, temperature, and molecular weight data of an ideal gas, determine its density:

Pressure = 1.0 atm
Temperature = 60 °F
Molecular weight of gas = 29 g/mol

Solution
An ideal gas is an imaginary gas that exactly obeys certain simple laws (i.e., Boyle's Law, Charles' Law, and the ideal gas law). No real gas obeys the ideal gas law exactly, although the "lighter" gases (hydrogen, oxygen, air, etc.) at ambient conditions approach ideal gas law behavior. The "heavier" gases such as sulfur dioxide and hydrocarbons, particularly at high pressures and low temperatures, can deviate considerably from the ideal gas law. Despite these deviations, the ideal gas law is routinely used in engineering calculations. The ideal gas law equation takes the form:

$$PV = nRT$$

where

P = absolute pressure
V = volume
T = absolute temperature
n = number of moles
R = ideal gas law constant

The ideal gas law in terms of density is ρ, is

$$\rho = m/V = n(\text{MW})/V$$
$$= P(\text{MW})/RT$$

where

MW = molecular weight
m = mass of gas
ρ = density of gas

Typical values of R are given below:

$$R = 10.73 \, \text{psia} \cdot \text{ft}^3/(\text{lbmol} \cdot {}^\circ\text{R})$$
$$= 1545 \, \text{psfa} \cdot \text{ft}^3/(\text{lbmol} \cdot {}^\circ\text{R})$$
$$= 0.73 \, \text{atm} \cdot \text{ft}^3/(\text{lbmol} \cdot {}^\circ\text{R})$$

$$= 555 \, \text{mmHg} \cdot \text{ft}^3/(\text{lbmol} \cdot {}^\circ\text{R})$$
$$= 82.06 \, \text{atm} \cdot \text{cm}^3/(\text{gmol} \cdot {}^\circ\text{K})$$
$$= 8.314 \, \text{kPa} \cdot \text{m}^3/(\text{kgmol} \cdot {}^\circ\text{K})$$
$$= 1.986 \, \text{cal}/(\text{gmol} \cdot {}^\circ\text{K})$$
$$= 1.986 \, \text{Btu}/(\text{lbmol} \cdot {}^\circ\text{R})$$

The choice of R is arbitrary, provided consistent units are employed. The density of the gas using the appropriate value of R may now be calculated:

$$\rho = P(\text{MW})/RT$$
$$= [(1)(29)]/[(0.73)(60 + 460)]$$
$$= 0.0764 \, \text{lb}/\text{ft}^3$$

Since the molecular weight of the given gas is 29, this calculated density may be assumed to apply to air. Also note that the effect of pressure, temperature, and molecular weight on density can be obtained directly from the ideal gas law equation. Increasing the pressure and molecular weight increases the density; increasing the temperature decreases the density.

Illustrative Example 7.4   Mass Conservation Law
An external gas stream is fed into a hospital laboratory room at a rate of 100 lb/h in the presence of 200 lb/h of air. 12.5 lb/h of cleansing agent is added to assist in treatment. Determine the rate of product gases exiting the unit in pounds per hour (lb/h). Assume steady-state conditions, implying that the process can continue indefinitely under these parameters without maintenance or removal of build-up.

Solution
As noted in the chapter, the conservations law for mass can be applied to any process or system. The general form of this law is given by:

Mass accumulated = mass in − mass out + mass generated

Apply the conservation law for mass in the room on a rate basis (effectively dividing mass by time):

Rate of mass in − rate of mass out + rate of mass generated

   = rate of mass accumulated

Rewrite this equation subject to the conditions in the problem statement:

Rate of mass in = rate of mass out

or, in engineering terms:

$$\dot{m}_{\text{in}} = \dot{m}_{\text{out}}$$

Note that mass is not generated, and steady-state conditions means no accumulation. The problem statement can provide the inlet flows:

$$\dot{m}_{in} = \dot{m}_{out} = 100\,lb/h + 200\,lb/h + 12.5\,lb/h$$
$$= 312.5\,lb/h$$

7.9 CHAPTER SUMMARY

- Engineering is a term applied to the profession in which a knowledge of the mathematical and natural sciences, gained by study, experience, and practice, is applied to the efficient use of the materials and forces of nature.
- Problems are almost always approached by checking on and reviewing all sources of information.
- In solving problems, one should:
  1. Read the available material thoroughly and understand what is required for an answer. Sometimes, as in life, the major obstacle is to find out what the problem really is.
  2. Determine what additional data are needed, if any, and obtain this information.
  3. Draw a simplified picture of what is taking place and write down the available data. One may use boxes to indicate processes or equipment, and lines for the flow of streams.
  4. Pick a basis on which to start the problem.
  5. If a chemical equation is involved, write it down and make sure it is balanced.
- The basic units in the SI system are the kilogram (mass), meter (length), second (time), kelvin (temperature), ampere (electric current), candela (the unit of luminous intensity), and radian (angular measure). All are commonly used by an engineer.
- The conservation laws may be applied at the macroscopic, microscopic, or molecular level.
- Thermodynamics deals with the transformation of energy from one form to another. Kinetics considers the rate at which chemical compounds react.

7.10 PROBLEMS

1 Convert 8.03 years to seconds.

2 Convert $0.03\,g/cm^3$ to lb/ft^3.

3 If 0.75 ft³/min of fluid with the density of water exits a system through a pipe whose cross-sectional area is 0.0025 ft², determine the mass flowrate, in lb/min and the exit velocity in ft/s.

4 A storage tank contains a gaseous mixture comprised of 15% N_2, 45% N_2O, 5% H_2O, and 35% O_2 by volume. What is the partial pressure of each component if the total pressure is 2 atm? What are their pure-component volumes if the total volume is 10 ft³?

REFERENCES

Bhatty, V., Butron, S., and Theodore, L. (2017). *Introduction to Engineering: Fundamentals, Principles, and Calculations.* New York: Theodore Tutorials and Butte, MT: Montana State University, Department of Environmental Engineering Graduate Division.

Bird, R., Stewart, W., and Lightfoot, E. (2002). *Transport Phenomena*, 2e. Hoboken, NJ: John Wiley & Sons.

Fogler, S. (2006). *Elements of Chemical Engineering*, 4e. Upper Saddle River, NJ: Prentice-Hall.

Kunz, R. and Theodore, L. (2005). *Nanotechnology: Environmental Implications and Solutions.* Hoboken, NJ: John Wiley & Sons.

Reynolds, J. (1992). *Material and Energy Balances, A Theodore Tutorial.* East Williston, NY: Theodore Tutorials (Originally published by the USEPA/APTI, RTP, NC).

Reynolds, J., Jeris, J., and Theodore, L. (2002). *Handbook of Chemical and Environmental Engineer Calculations.* Hoboken, NJ: John Wiley & Sons.

Theodore, L. (1995). *Chemical Reaction Kinetics, A Theodore Tutorial.* East Williston, NY: Theodore Tutorials (Originally published by the USEPA/APTI, RTP, NC).

Theodore, L. (2006). *Nanotechnology: Basic Calculations for Engineers and Scientists.* Hoboken, NJ: John Wiley & Sons.

Theodore, L. (2014). *Chemical Engineering: The Essential Reference.* New York, NY: McGraw-Hill.

Theodore, L. (2014). *Chemical Reactor Analysis and Applications for the Practicing Engineer.* Hoboken, NJ: John Wiley & Sons.

Theodore, L. and Reynolds, J. (1992). *Thermodynamics A Theodore Tutorial.* East Williston, NY: Theodore Tutorials (Originally published by the USEPA/APTI, RTP, NC).

Theodore, L., Ricci, F., and Van Vliet, T. (2010). *Thermodynamics for the Practicing Engineer.* Hoboken, NJ: John Wiley & Sons.

8

Legal and Regulatory Considerations

Environmental, health, and safety regulations are not simply a collection of laws. They are an organized system of statutes, regulations, and guidelines that minimize, prevent, and punish the consequences of damage to the public health and the environment. Environmental regulations deal with the problems of human activities and the effect they have over the environment, while public health regulations govern a wide-ranging collection of health issues.

There are several government agencies responsible for the health and safety of the public and the environment within the United States. These agencies implement and oversee the laws governing these matters through Federal Regulations authorized by Congress. Each agency ultimately answers to the Cabinet of the United States, of which there are several department heads appointed by the President. The three departments applicable to the health, safety, and environmental protection include the Department of Health and Human Services (HHS), Environmental Protection Agency (EPA), and the Department of Labor (DOL). These are listed below, along with the corresponding agencies they govern.

A Guide to Virology for Engineers and Applied Scientists: Epidemiology, Emergency Management, and Optimization, First Edition. Megan M. Reynolds and Louis Theodore.
© 2023 John Wiley & Sons, Inc. Published 2023 by John Wiley & Sons, Inc.

- Department of Health and Human Services (HHS)
  - Centers for Disease Control and Prevention (CDC)
    - National Institute for Occupational Safety and Health (NIOSH)
    - Coordinating Center for Infectious Diseases (CCID)
    - Healthcare Infection Control Practices Advisory Committee (HICPAC)
  - National Institutes of Health (NIH)
    - The National Institute of Allergy and Infectious Diseases (NIAID)
    - National Library of Medicine (NLM)
  - Food and Drug Administration (FDA)
  - Centers for Medicare and Medicaid Services (CMS)
  - Health Resources and Services Administration (HRSA)
  - Agency for Healthcare Research and Quality (AHRQ)
- Environmental Protection Agency (EPA)
- Department of Labor (DOL)
  - Occupational Health and Safety Administration (OSHA)

For the purposes of this chapter, CDC, FDA, EPA, and OSHA will be reviewed in detail.

8.1 THE REGULATORY SYSTEM

Over the past five decades environmental regulations have become a system in which laws, regulations, policies, and guidelines have become interrelated. The history and development of this regulatory system has led to laws that focus principally on only one environmental medium, i.e., air, water, or land. Some environmental managers feel that more needs to be done to manage all of the media simultaneously. Thankfully, the environmental regulatory system has evolved into a truly integrated, multimedia management framework.

As mentioned earlier, federal laws are the product of Congress. Regulations written to implement the law are promulgated by the executive branch of government, but until judicial decisions are made regarding the interpretations of the regulations, there may be uncertainty about what regulations mean in real situations. Until recently, environmental protection groups were most frequently the plaintiffs in cases brought to court seeking interpretation of the law. Today, industry is becoming more active in this role (W. Matistic, personal communications to L. Theodore 2021).

Enforcement approaches for environmental regulations are environmental management oriented in that they seek to remedy environmental harm, not simply a specific infraction of a given regulation. All laws in a legal system may be used in enforcement to prevent damage or threats of damage to the environment or human health and safety. Tax laws (e.g., tax incentives) and business regulatory laws (e.g., product claims and liability disclosure) are examples of laws not directly focused on environmental protection, but that may also be used to encourage compliance and discourage noncompliance with environmental regulations.

Common law also plays an important role in environmental management. It is the set of rules and principles relating to the government and the security of persons and property. Common law authority is derived from the usages and customs that are recognized and enforced by the courts. In general, no infraction of the law is necessary when establishing a common law court action. A common law "civil wrong" (e.g., environmental pollution) that is brought to court is called a tort. Environmental torts may arise because of nuisance, trespass, or negligence.

Laws tend to be general and contain uncertainties relating to the implementation of principles and concepts they contain. Regulations derived from laws may be more specific but are also frequently too broad to allow clear translation into technology practice. Permits may be used to bridge this gap and prescribe specific technical requirements concerning activities carried out by a facility.

Most major federal laws provide for citizen lawsuits. This empowers individuals to seek compliance or monetary penalties when these laws are violated, and regulatory agencies do not take enforcement action against the violator.

8.1.1 Laws, Regulations, Plans and Policy: The Differences

The following is a listing of some of the major differences between a federal *law* and a federal *regulation*, as briefly described in the previous section:

- A law (or act) is passed by both houses of Congress and signed by the President. A regulation is issued by a government agency such as the Occupational Safety and Health Administration (OSHA).
- Congress can pass a law on any subject it chooses. It is only limited by the restrictions in the Constitution. A law can be challenged in court only if it violates the Constitution. It may not be challenged if it is merely unwise, unreasonable, or even silly. If, for example, a law were passed that placed a tax on sneezing, it could not be challenged in court just because it was unenforceable. A regulation can be issued by an agency only if the agency is authorized to do so by the law passed by Congress. When Congress passes a law, it usually assigns an administrative agency to implement that law. A law regarding radio stations, for example, may be assigned to the Federal Communications Commission (FCC). Sometimes a new agency is established to implement a law. This was the case with the Consumer Product Safety Commission (CPSC). OSHA is authorized by the Occupational Safety and Health Act to issue regulations that protect workers from exposure to the hazardous chemicals they use in manufacturing processes.
- Laws can include a Congressional mandate directing government agencies to develop a comprehensive set of regulations. Regulations are issued by an agency,

such as the EPA, and translate the general mandate of a statute into a set of requirements for the agency and the regulated community.

- Regulations are developed in an open and public manner according to an established process. When a regulation is formally proposed, it is published in an official government document called the *Federal Register* to notify the public of intent to create new regulations or modify existing ones. It then provides the public, which includes the potentially regulated community, with an opportunity to submit comments. Following an established comment period, the agency may revise the proposed rule based on both an internal review process and public comments.

- The final regulation is published, or promulgated, in the Federal Register. Included with the regulation is a discussion of the agency's rationale for the regulatory approach, known as preamble language. Final regulations are compiled annually and incorporated in the Code of Federal Regulations (CFR) according to a highly structured format based on the topic of the regulation. This latter process is called codification, and each CFR title corresponds to a different regulatory authority. For example, EPA's regulations are in Title 40 of the CFR. The codified Resource Conservation and Recovery Act (RCRA) regulations can be found in Title 40 of the CFR, Parts 240-282.

- A regulation may be challenged in court on the basis that the issuing agency exceeded the mandate given to it by Congress. If the law requires the agency to consider the costs versus benefits of their regulation, the regulation could be challenged in court based on the basis argument that the cost/benefit analysis was not correctly or adequately done. If OSHA issues a regulation limiting a worker's exposure to a hazardous chemical to 1 part per million (ppm), OSHA could be called upon to prove in court that such a low limit was needed to prevent a worker from being harmed. Failure to prove this would mean that OSHA exceeded its mandate under the law, as OSHA is charged to develop standards only as stringent as those required to protect worker health and provide worker safety.

- Laws are usually brief and general. Regulations are usually lengthy and detailed. The Hazardous Materials Transportation Act, for example, is approximately 20 pages long. It speaks in general terms about the need to protect the public from the dangers associated with transporting hazardous chemicals and identifies the Department of Transportation (DOT) as the agency responsible for issuing regulations and implementing the law. The regulations issued by the DOT are several thousand pages long and are very detailed, down to the exact size, shape, design, and color of the warning placards that must be used on trucks carrying any of the thousands of regulated chemicals.

- Generally, laws are passed infrequently. Often, years pass between amendments to existing law. A completely new law on a given subject already addressed by an existing law is unusual. Laws are published as a "Public Law #_-_" and are eventually codified into the United States Code.

- Regulations are issued and amended frequently. Proposed and final new regulations and amendments to existing regulations are published daily in the Federal Register. Final regulations have the force of law when published.

8.1.2 POLICIES AND PLANS

The definitions of "policy" and "plan" can be described as:

Policy: "a high-level overall plan embracing the general goals and acceptable procedures especially of a governmental body" (Merriam-Webster 2021b).

Plan: "a method for achieving an end; an often-customary method of doing something; a detailed formulation of a program of action" (Merriam-Webster 2021a).

Policy is a general framework or plan for a course of action embodying desired principles. Policy can form the basis for law, (i.e., a nondiscrimination policy can form the basis for civil rights legislation). Laws can directly flow from policy or be the rationale for legislation. Additionally, in some cases the law is the policy, i.e., U.S. Energy Policy Act). A policy can also stand alone without a supporting law, i.e., foreign policy while setting a general direction for diplomacy (W. Matystik, personal communication to L. Theodore, Westchester, NY 2011).

Any *policy* or *plan* must first be developed and sent to Congress. Congress then passes a law that is intended to reflect the objectives of the policy. In the process, Congress must identify an agency or regulatory body to oversee and implement the law. The agency or regulatory body then may pass a regulation (or regulations) to ensure that the intent of the law is achieved. However – and here is where it gets tricky – the regulators may set a policy without going through the normal process associated with passing a regulation.

Stander and Theodore, in their 2008 book *Environmental Regulatory Calculations Handbook*, have summarized the difference between laws, regulations, and policy in the following manner:

- *Laws* provide vision, scope, and authority. They are usually enacted by a legislative body and serve as a basis for administrative implementation.
- *Regulations* establish general requirements that must be met by a regulated community. These requirements generally apply at a national, state, or local level and are usually adopted by an administrative entity that has received authorization from a law. These regulations generally interpret the enabling legislation and determine how the law is to be implemented.
- *Policy* is developed by an administrative entity to explain and further interpret how laws and regulations are to be implemented and to resolve issues and conflicts of interpretation (Stander and Theodore 2008).

8.2 THE ROLE OF INDIVIDUAL STATES

When it comes to pandemics, the role of the states is, unfortunately, confusing. The responsibilities of the states are based on the Tenth Amendment to the Constitution, which declares, "Powers not delegated to the United States by the Constitution nor prohibited by it to the States, are reserved to the States …"

This doctrine effectively grants state governments all powers not explicitly specified by the Constitution as the responsibility of the federal government. This gap can create a conflict concerning public health in part because the federal health agency, the CDC, cannot become involved in an outbreak or epidemic unless or until it crosses state lines and effects multiple states. In the case of COVID-19, however, it quickly became clear that this was not a problem restricted to certain states. COVID-19 was declared a global pandemic, with national governments around the world declaring shutdowns and restrictions. Yet, thanks in part to the patchwork of laws that make up healthcare in the United States and differences in state responses, there was a disjointed, chaotic effort to contain the disease, when a nationally coordinated effort would have been more effective. This global catastrophe has exposed inherent conflicts among groups with varied interests, socio-economic backgrounds, localities, states, and nations often forcing health authorities to wade into politics rather than focusing on what was best for public health.

In summary, the current view of state versus federal laws maintains that the states have the right to adopt and pass laws during a public health emergency such as a full-scale pandemic. However, any state laws related to the allocation of supplies must comply with federal law. In effect, limitations can be placed on individual rights and liberties during an emergency in order to benefit the entire nation.

Thus, one of the greatest challenges facing any pandemic planning is that the public health system is not a single entity but rather a loosely affiliated network of approximately 3000 federal, state, and local health agencies. Although state and local governments have responsibility toward the health of their citizens, the federal government can also play a role in protecting and improving the health of the population by setting health goals, policies, and standards, especially through its regulatory powers.

As mentioned earlier, the CDC is Health and Human Services' lead agency for ensuring that the government can not only respond to health emergencies but also oversee disease-prevention efforts. The CDC works with state health agencies, usually allowing the states to determine priorities and implement strategies through federally funded initiatives. While allocating significant financial and technical resources for these health agencies, the CDC does not have the authority to directly establish standards for a state's health-protection efforts.

Similarly, consider the Resource Conservation and Recovery Act (RCRA). Similar to most federal environmental legislation, it encourages states to develop and run their own hazardous waste programs as an alternative to EPA management. Thus, in a given state, the hazardous waste regulatory program may be run by the EPA or by a state agency. For a state to have jurisdiction over its hazardous waste program, it must receive approval from the EPA by showing that its program is at least as stringent as the EPA program. States that are authorized to operate RCRA (or other) programs oversee the hazardous waste tracking system in their state, operate the permitting system for hazardous waste facilities, and act as the enforcement arm in cases where an individual or a company practices illegal hazardous waste management. If needed, the EPA steps in to assist the states in enforcing the law. The EPA can also act directly to enforce RCRA or other laws in states that do not yet have authorized programs. The EPA and the states currently act jointly to implement and enforce the regulations.

One major problem of having fifty separate state environmental agencies is perhaps best illustrated by an initiative of the National Governor's Association (NGA) to solve it. In 1996, it recognized that "... the environmental technology industry has long been heralded as a key to enhanced environmental quality and accelerated economic development ..." and that "... [t]his diverse industry develops technologies to prevent pollution, monitor, and control pollutant emissions, detect and measure contamination, and clean up environmental pollution ..." It went on to point out, however, that "... developers and vendors of new environmental technologies are frustrated in their efforts to penetrate the enormous U.S. market ..." and that "... [m]any observers place the blame at the door of the states, where a single technology may be subject to fifty different procedures to gain a permit or approval ..." (NGA 1997). Another problem is competition for authority in particular energy situations (i.e., whether federal law or state law prevails). Solutions are governed by the often-unclear doctrine of *pre-emption* (NGA 1997).

State agencies do not operate in a vacuum. Indeed, most trace their very existence to the passage of major federal laws. While federal environmental legislation, such as the Clean Air Act (CAA), for example, established national ambient air quality standards (NAAQS), it went on to direct that these standards be met by state implementation plans (SIPs). The National Permit Discharge Elimination System (NPDES), set up by the Clean Water Act, even anticipated a gradual handing off of the control to states to administer their own permitting systems once they evidenced adequate authority. Thus, a mix of federal and state statutory laws evolved along with the appropriate bureaucracies to implement them.

One can blame this patchwork quilt of laws, regulations, and standards on the notion of federalism on which the U.S. system of governance was built. Federalism is a sharing of power between the states and the national (federal) government. In 1787, the new federal government only exercised limited or enumerated powers granted to it by the Constitution, such as making treaties and printing money. Later, the Tenth Amendment to the Constitution appears to have clarified that all other powers belong to the states, and that, "the powers not delegated to the United States by the Constitution, nor prohibited by it to the states, are reserved to the states respectively, or to the people" (W. Matystik, personal communication to L. Theodore, Westchester, NY 2011).

8.3 KEY GOVERNMENT AGENCIES

This section reviews the following four agencies:

- Environmental Protection Agency (EPA)
- Centers for Disease Control and Prevention (CDC)
- Food And Drug Administration (FDA)
- Occupational Health and Safety Administration (OSHA)

This section concludes with a discussion of legal considerations during a public health crisis.

8.3.1 Environmental Protection Agency (EPA)

The year 1970 was a cornerstone year for modern environmental policy. The Nixon Administration at that time became preoccupied with not only trying to pass more extensive environmental legislation but also implementing the laws. Nixon's White House Commission on Executive Reorganization proposed in the Reorganizational Plan No.3, of 1970 that a single, independent agency be established, separate from the Council for Environmental Quality (CEQ). The plan was sent to Congress by President Nixon on 9 July 1970, and this new U.S. Environmental Protection Agency (EPA) began operation on 2 December 1970. The EPA was officially born!

The EPA is perhaps the most far-reaching regulatory agency in the federal government because its authority is so broad. The EPA is charged with protecting the nation's land, air, and water systems. Under a mandate of national environmental laws, the EPA strives to formulate and implement actions that lead to a compatible balance between human activities and the ability of natural systems to support and nurture life (Theodore and Theodore 2021).

The EPA works with the states and local governments to develop and implement comprehensive environmental programs. Federal laws such as the Clean Air Act (CAA), the Safe Drinking Water Act (SDWA), the Resource Conservation and Recovery Act (RCRA), and the Comprehensive Environmental Response, Compensation, and Liability Act (CERCLA), etc., all mandate involvement by state and local government in the details of implementation.

8.3.2 CENTERS FOR DISEASE CONTROL AND PREVENTION (CDC)

The CDC was first established in 1946 and tasked with preventing malaria as it spread across the country. At the time, epidemiology was in its infancy. Over the decades, disease surveillance has become its fundamental mission, and the CDC has been instrumental in many developments and achievements in the area of public health. Today, according to the CDC, they are focused on health promotion, prevention, and preparedness as it plays a central role in protecting the nation from biological threats and disease, both domestic and foreign (CDC 2018). The CDC has been the focus of criticism for its handling of the COVID-19 pandemic, particularly for its lack of response time and inability to develop clear messaging to the public. The CDC's mandate over the state governments is limited, however, and the pandemic revealed the many flaws in how system works. Chapter 10 will further explore some of these limitations placed on the federal agency and of the far-reaching consequences to the nations health.

8.3.3 FOOD AND DRUG ADMINISTRATION (FDA)

As the name suggests, the FDA is responsible for oversight of the foods and drugs permitted in the United States. In its charter, the FDA is considered to be the comprehensive consumer protection agency in the country. Among other responsibilities, they are tasked with approving new drugs, vaccines, and medical devices, while monitoring all such products on the market. The predecessor to the FDA was first established in 1848 with its first mission to regulate the safety of the nation's agricultural products. In 1906, the modern FDA was founded with the passage of the Pure Food and Drugs Act. This legislation, which was finally passed after more than two decades, provided protection from counterfeit or harmful products. This offered a level of consumer protection that had, until then, been unimaginable (FDA 2021).

8.3.4 OCCUPATIONAL HEALTH AND SAFETY ADMINISTRATION (OSHA)

In addition to EPA regulations concerning human health, safety, and the environment, the OSHA regulates safety in the workplace (OSHA 2021a). It was established as part of the United States Department of Labor in 1970 by congress with the Occupational Safety and Health Act. Its mission is "to ensure safe and healthful working conditions for workers by setting and enforcing standards and by providing training, outreach, education, and assistance" (OSHA 2021b).

In November 2021, OSHA issued an Emergency Temporary Standard on COVID-19 Vaccination and Testing. According to a press release from the Department of Labor, "under this standard, covered employers must develop, implement and enforce a mandatory COVID-19 vaccination policy, unless they adopt a policy requiring employees to choose to either be vaccinated or undergo regular COVID-19 testing and wear a face covering at work" (DOL 2021). As of the time of this writing, the rule is in force until further notice, although legal challenges have put its full implementation in doubt. Other regulatory considerations regarding such issues as vaccination and quarantine which are specific to the pandemic are discussed later in the chapter.

8.3.5 Legal Considerations During a Public Health Crisis

Legal difficulties have arisen since the arrival of COVID-19. Interestingly, states have the power to adopt rules and regulations that are related to the general subject area of public health, while noting that any state initiative *must* comply with federal law(s).

One federal law – the OSHA Act – specifically states that workers/workplaces must be "free from recognized hazards that are causing or are likely to cause death or physical harm …" (OSHA 2022). How this relates to the current COVID-19 pandemic is yet to be satisfactorily determined. At this point, there is confusion regarding legal issues. Courts have historically, however, allowed states to infringe on citizens' rights if and when necessary to protect public health. Notwithstanding the above, a host of frivolous lawsuits and exaggerated claims related to COVID-19 activities have appeared during this pandemic. These reports of unprecedented litigation will ultimately create pressure on the legal system to rule in a fair manner. However, it will no doubt take many years of legal battles to resolve many of these issues.

The pandemic opened the door to litigation for individuals such as patients, doctors, employees, consumers, and citizens who claimed they were adversely affected. The future is certain to see widespread litigations at both the private and public levels.

A case in point is ventilators. It should be noted that it is usually not normal to remove a ventilator without a patient's consent. However, pandemic rules, regulations, laws and/or policies may supersede patients' rights. A host of individuals, mostly healthcare workers, has been faced with unprecedented dilemmas such as rationing care. The result is that not all patients requiring ventilators may be able to receive the care and treatment they require.

Finally, it should be noted that federal laws provide limited liability protection for health care workers and providers during a pandemic period. The Public Readiness and Emergency Preparedness (PREP) Act permits the Department of Health and Human Services (HHS) to issue a declaration that grants covered persons immunity from liability claims "resulting from the administration or use of countermeasures to prevent diseases, threats and conditions" that create "a present, or credible risk of a future public health emergency"; here, a covered person is defined as one "involved in the department, manufacture, testing, distribution, administration, and use of any countermeasures." The Volunteer Protection Act (VPA) also provided some liability relief. More specific information on the legal and regulatory implications of the pandemic is available in the next section (Theodore and Theodore 2021).

8.4   PUBLIC HEALTH EMERGENCY DECLARATIONS

Most pandemics in the past produced conflicts between states and nations as well as intergovernmental entities. Ordinarily, the responsibility for public health issues rests primarily with state and local governments, and a health crisis can often lead to inconsistent rules and confusion. At the beginning of the COVID-19 pandemic, with the federal government either unable or unwilling to lead, state governors were left with no choice but to implement their own lockdowns, and to procure their own disparately needed emergency supplies. This led to many missteps, including the states outbidding each other with suppliers for personal protective equipment (PPE) and ventilators; This action drove up prices unnecessarily.

The Public Health Service Act (PHSA) was passed in 1946. It gave the Human and Health Services (HHS) secretary responsibility for preventing the introduction, transmission, and spread of communicable diseases from foreign countries into the United States and within the United States and its territories. This authority is delegated to the Centers for Disease Control and Prevention (CDC), which is empowered to detain, medically examine, or conditionally release individuals reasonably believed to be carrying a communicable disease under the federal authorization of Title 42 – Public Health (CDC 2021). The PHSA also provides that the list of diseases for which quarantine is authorized must first be specified in an executive order of the president, on the recommendation of the HHS secretary.

The PHSA has led to legal questions regarding individual rights and restrictions, which are yet to be satisfactorily resolved. Those supporting states' rights fear that government might cut controls and limitations of personal liberties needlessly. At the time of this writing, some examples include extending the aforementioned restrictions beyond the end of the crisis or enforcing rules that do little or nothing to decrease virus transmission. These could unwittingly lead to unnecessary hardships. Negative effects on the economy associated with many restrictive policies would be borne primarily by those with the fewest financial resources, which could further increase inequality. At the same time, this population is also at the greatest risk of serious complications and death from the virus, due in part to both the lack of access to proper medical care and working in jobs in which working from home is not an option. High-quality contact tracing and free testing should also be utilized to further reduce the risk of new infections. Whether by state or federal decree, the rules and restrictions made in the interest of public health should be based on solid scientific evidence and state-to-state cooperation, rather than politics and territorial disputes.

According to the CDC, the distinction between isolation and quarantine are as follows: "*Isolation* separates sick people with a contagious disease from people who are not sick …" while, "… *quarantine* separates and restricts the movement of people who were exposed to a contagious disease to see if they become sick" (CDC 2021). Ideally, quarantine and isolation should be voluntary whenever possible and, when that is impossible, they should be enforced by the least intrusive means available which can still guarantee compliance (Theodore 2008).

Quarantine and isolation are two of the most complex and controversial public health control powers. Both involve a temporary deprivation of an individual's liberty in the name of public health and have the potential to be abused by governments. Therefore, there is a need not only to protect the health of its citizens but to also safeguard the civil liberties of individuals. It is generally accepted that large-scale public health threats can and should necessitate such extraordinary measures by the government, and quarantine and isolation may be legitimately justified if public health interests are judiciously balanced against the freedom of the individual. In this respect, the benefits to the public should outweigh any detriment that quarantine or isolation may place on individuals (L. Theodore, personal notes. East Williston, NY 2021).

The issue of civil liberties is an international concern. According to the United Nations Office of the High Commissioner on Human Rights (OHCHR), governments have the right to "adopt exceptional measures to protect public health that may restrict certain human rights …. Some rights, such as freedom of movement, freedom of expression, or freedom of peaceful assembly may be subject to restrictions for public health reasons, even in the absence of a state of emergency …. These restrictions must meet the requirements of legality, necessity, and proportionality, and be non-discriminatory" (OHCHR 2020).

No discussion on the legal role associated with pandemics would be complete without a discussion of *liability*, which is one of the legal tribulations potentially associated with the pandemic. These should not be ignored since the rationing of both healthcare equipment (e.g., ventilators) and healthcare services could easily lead to either liability claims or legal challenges. This could produce numerous frivolous lawsuits that are accompanied by inflated or exaggerated claims. However, the Volunteer Protection Act of 1997, which was amended in 2017, along with the Public Readiness and Emergency Preparedness Act, have essentially granted immunity from liability claims of any detriment caused as a result of the administration or use of countermeasures to prevent diseases, threats, and other related emergencies declarations (Chabot 2017).

The Public Health Service Act was used to declare a nationwide public health emergency (PHE) due to COVID-19 on 31 January, 2020. This allowed several government agencies more flexibility and authority beyond the scope of the legal and regulatory framework they have under normal circumstances. The PHE was subsequently renewed on 21 April, 23 July, and 2 October 2020, and 7 January, 15 April, 19 July, and 15 October 2021, with the expectation that the emergency will extend into 2022 (CMS 2021).

8.5 KEY ENVIRONMENTAL ACTS

Key environmental acts passed during the latter part of the twentieth century included the following acts:

- Clean Air Act (CAA)
- Toxic Substance Control Act (TSCA)
- Clean Water Act (CWA)
- Safe Drinking Water Act (SDWA)
- Occupational Safety and Health Act (OSH)
- Resource Conservation and Recovery Act (RCRA)
- Superfund Amendments and Reauthorization Act (SARA)
- USEPA Risk Management Program (RMP)
- Pollution Prevention Act (PPA)

For the purposes of this book, the main areas of interest are related to clean air and protection against toxic substances, including viruses. The *Clean Air Act (CAA)* will be covered in detail in Section 8.6, while regulations of dangerous elements, including the *Toxic Substance Control Act (TSCA),* will be covered in Section 8.7. More detailed information, including a discussion of the other acts mentioned, is available in the literature (Stander and Theodore 2008; Theodore and Theodore 2021).

8.6 THE CLEAN AIR ACT

The CAA defines the national policy for air pollution abatement and control in the United States. It establishes goals for protecting health and natural resources and delineates what is expected of federal, state, and local governments to achieve those goals. The CAA, which was initially enacted as the Air Pollution Control Act of 1955, has undergone several revisions over the years to meet the ever-changing needs and conditions of the nation's air quality. On 15 November 1990, the president signed the most recent amendments to the CAA, referred to as the 1990 CAA Amendments. Embodied in these amendments were several progressive and creative new themes deemed appropriate for effectively achieving the air quality goals and for reforming the air quality control regulatory process. Specifically, the amendments:

- Encouraged the use of market-based principles and other innovative approaches similar to performance-based standards and emission banking and trading.
- Promoted the use of clean low-sulfur coal and natural gas, as well as innovative technologies to clean high-sulfur coal through the acid rain program.
- Reduced energy waste and created a market for clean fuels derived from grain and natural gas to cut dependency on oil imports by one million barrels per day.
- Promoted energy conservation through an acid rain program that gave utilities the flexibility to obtain needed emission reductions through programs that encouraged customers to conserve energy.

Two of the key provisions of the Act are reviewed below (Stander and Theodore 2008; Theodore and Theodore 2021).

Although the CAA brought about significant improvements in the nation's air quality, the urban air pollution problems of ozone (smog), carbon monoxide (CO), and particulate matter (PM) persist to this day. In 1995, approximately 70 million U.S. residents were living in counties with ozone levels exceeding the EPA's current ozone standard. The CAA, as amended in 1990, established a more balanced strategy for the nation to address the problem of urban smog. Overall, the amendments revealed Congress's high expectations of the states and the federal government. While it gave states more time to meet the air quality standard (up to 20 years for ozone in Los Angeles), it also required states to make constant progress in reducing emissions. It required the federal government to reduce emissions from cars, trucks, and buses; from consumer products such as hair spray and window-washing compounds; and, from ships and barges during the loading and unloading of petroleum products. The federal government also developed the technical guidance that states need to control stationary sources. The CAA addresses the urban air pollution problems of ozone (smog), carbon monoxide, and PM. Specifically, it clarifies how areas are designated and redesignated "attainment." It also allows the EPA to define the boundaries of "nonattainment" areas, i.e., geographical areas whose air quality does not meet federal ambient air quality standards designed to protect public health. The law also establishes provisions defining when and how the federal government can impose sanctions on areas of the country that have not met certain conditions (Stander and Theodore 2008; Theodore and Theodore 2021).

Toxic air pollutants are those pollutants that are hazardous to human health or the environment. These pollutants are typically carcinogens, mutagens, and reproductive toxins. The toxic air pollution problem is widespread. The information generated in 1987 from the Superfund "Right to Know" rule (SARA Section 313) discussed earlier, indicated that more than 2.7 billion pounds of toxic air pollutants were emitted annually in the United States. The EPA studies indicated that exposure to such quantities of toxic air pollutants may result in one thousand to three thousand cancer deaths each year. Section 112 of the CAA includes a list of 189 substances that are identified as hazardous air pollutants, including a list of categories of sources that emit these pollutants. The list of source categories included major sources (those emitting 10 tons per year of any single hazardous air pollutants), and area sources, which are smaller sources, such as dry cleaners and auto body refinishing. In turn, EPA promulgated emission standards, referred to as maximum achievable control technology (MACT) standards, for each listed source category. These standards were based on the best demonstrated control technology or practices utilized by sources that make up each source category. Within eight years of promulgation of a MACT standard, EPA must evaluate the level of risk that remains (residual risk), due to exposure to emissions from a source category, and determine if the residual risk is acceptable. If the residual risks are determined to be unacceptable, additional standards are required (Stander and Theodore 2008; Theodore and Theodore 2021).

8.7 REGULATION OF TOXIC SUBSTANCES

The reader should note this chapter is meant to provide a general overview of the regulatory framework in the United States, while also attempting to identify where in this framework virus protection can be found.

People have long recognized that sulfuric acid, arsenic compounds, and other chemical substances can cause fires, explosions, or poisoning. More recently, researchers have determined that many chemical substances such as benzene and several chlorinated hydrocarbons may cause cancer, birth defects, and other long-term health effects. Today, the hazards of new kinds of substances, including genetically engineered microorganisms, including viruses, and nanoparticles are being evaluated. The EPA has several legislative tools to use in controlling the risks from toxic substances, which are listed in the table below (Table 8.1).

The Federal Insecticide, Fungicide, and Rodenticide Act of 1972 (FIFRA) encompasses all pesticides used in the United States. When first enacted in 1947, FIFRA was administered by the U.S. Department of Agriculture and was intended to protect consumers against fraudulent pesticide products. When many pesticides were registered, their potential for causing health and environmental problems was unknown. In 1970, the EPA assumed responsibility for FIFRA, which was amended in 1972 to shift emphasis to health and environmental

Table 8.1 Major toxic chemical laws administered by the EPA.

Statue	Provisions
Toxic Substances Control Act (TSCA)	Requires that the EPA be notified of any new chemical prior to its manufacture and authorizes EPA to regulate the production, use, or disposal of a chemical
Federal Insecticide, Fungicide, and Rodenticide Act (FIFRA)	Authorizes the EPA to register all pesticides and specify the terms and conditions of their use and remove unreasonably hazardous pesticides from the marketplace
Federal Food, Drug, and Cosmetic Act	Authorizes the EPA in cooperation with the FDA to establish tolerance levels for pesticide residues on food and food producers
Resource Conservation and Recovery Act (RCRA)	Authorizes the EPA to identify hazardous wastes and regulate their generation, transportation, treatment, storage, and disposal
Comprehensive Environmental Response, Compensation, and Liability Act (CERCLA)	Requires the EPA to designate hazardous substances that can present substantial danger, and authorizes the cleanup of sites contaminated with such substances
Clean Air Act (CAA)	Authorizes the EPA to set emission standards to limit the release of hazardous air pollutants
Clean Water Act (CWA)	Requires the EPA to establish a list of toxic water pollutants and set standards
Safe Drinking Water Act (SDWA)	Requires the EPA to set drinking water standards to protect public health from hazardous substances
Marine Protection, Research, and Sanctuaries Act	Regulates ocean dumping of toxic contaminants
Asbestos School Hazard Act	Authorizes the EPA to provide loans and grants to schools with financial need for abatement of severe asbestos hazards
Asbestos Hazard Emergency Response Act	Requires the EPA to establish a comprehensive regulatory framework for controlling asbestos hazards in schools
Emergency Planning and Community Right-to-Know Act	Requires states to develop programs for responding to hazardous chemical releases and requires industries to report on the presence and release of certain hazardous substances

Source: Adapted from Santoleri et al. (2000).

protection. Allowable levels of pesticides in food are specified under the authority of the Federal Food, Drug, and Cosmetic Act of 1954. Today, FIFRA contains registration and labeling requirements for pesticide products. The EPA must approve any use of a pesticide, and manufacturers must clearly state the conditions of that use on the pesticide label. Some pesticides are listed as hazardous wastes and are subject to the Resource Conservation and Recovery Act (RCRA) rules when discarded.

The Toxic Substances Control Act (TSCA) authorizes EPA to control the risks that may be posed by the thousands of commercial chemical substances and mixtures (chemicals) that are not regulated as either drugs, food additives, cosmetics, or pesticides. Under TSCA, the EPA can, among other things, regulate the manufacture and use of a chemical substance and require testing for cancer and other effects. TSCA regulates the production and distribution of new chemicals and governs the manufacture, processing, distribution, and use of existing chemicals. Among the chemicals controlled by TSCA regulations are PCBs (polychlorinated biphenyls), chlorofluorocarbons, and asbestos. In specific cases, there is an interface with RCRA regulations. For example, PCB disposal is generally regulated by TSCA. However, hazardous wastes mixed with PCBs are regulated under RCRA. Under both TSCA and FIFRA, the EPA is responsible for regulating certain biotechnology products, such as genetically engineered microorganisms designed to control pests or assist in industrial processes.

The Clean Air Act (CAA), in Section 112, listed 189 air pollutants. The CAA also requires emission standards for many types of air emission sources, including RCRA regulated incinerators and industrial boilers or furnaces.

The CWA lists substances to be regulated by effluent limitations in 21 primary industries. The CWA substances are incorporated into both RCRA and CERCLA. In addition, the CWA regulates discharges from publicly owned treatment works (POTWs) to surface waters and indirect discharges to municipal wastewater treatment systems (through the pretreatment program). Some hazardous wastewater which would generally be considered RCRA-regulated wastes are covered under the CWA because of the use of treatment tanks and a National Pollutant Discharge Elimination System (NPDES) permit to dispose of the wastewater. Sludges from these tanks, however, are subject to RCRA regulations when they are removed.

The Safe Drinking Water Act (SDWA) regulates underground injection systems, including deep-well injection systems. Prior to underground injection, a permit must be obtained which imposes conditions that must be met to prevent the endangerment of underground sources of drinking water.

It should be noted that, while there is a list of major air toxins, no equivalent list exists for water. Rather, the practitioner has several options for specifying those substances that could be classified as *water toxins*. Included in these eligible candidates are the Priority Water Pollutants (PWP), chemicals that appear under the primary drinking water standards of the SDWA, and what has come to be defined by many as *unregulated contaminants*. The latter term falls within the domain of EPA's Contaminated Candidate List (CCL), a process that is employed to determine toxic substances.

The Safe Drinking Water Act (SDWA) includes a process where new contaminants are identified that may require regulation in the future with a primary standard. The EPA is required to periodically release the aforementioned Contaminant Candidate List (CCL), which is used to prioritize research and data collection efforts to help determine whether a specific contaminant should be regulated. On 2 March 1998, the first Drinking Water CCL as part of the Unregulated Contaminant Monitoring Rule (UMCR 1) was released which contained 60 contaminants (10 microbiological contaminants and 50 chemical contaminants). The EPA has since amended the Contaminant Candidate List several times to include more toxins each time. CCL 4, which included 97 chemicals or chemical groups and 12 microbial contaminants, was finalized in 2016. The list includes, among others, chemicals used in commerce, pesticides, biological toxins, disinfection by-products, pharmaceuticals, and waterborne pathogens.

8.7.1 Toxic Water Pollutants: Control and Classification

Since the early 1980s, EPA's water quality standards guidance placed increasing importance on toxic pollutant control. The Agency urged states to adopt the criteria into their standards for priority toxic pollutants, particularly those for which EPA had published criteria guidance. EPA also provided guidance to help and support state adoption of toxic pollutant standards with the *Water Quality Standards Handbook* (1983) and the Technical Support Document for Water Quality Toxics Control (1985 and 1991).

Despite EPA's urging and guidance, the state response was disappointing. A few states adopted large numbers of numeric toxic pollutant criteria, primarily for the protection of aquatic life. Most other states adopted few or no water quality criteria for priority toxic pollutants. Some relied on "free from toxicity" criteria and the so-called action levels for toxic pollutants or occasionally calculated site-specific criteria. Few states addressed the protection of human health by adopting numeric human health criteria.

State development of case-by-case effluent limits using procedures that did not rely on the statewide adoption of numeric criteria for the priority toxic pollutants frustrated Congress. Congress perceived that states were failing to aggressively address toxics and that EPA was not using its oversight role to push the states to move more quickly and comprehensively. Many in Congress believed that these delays undermined the effectiveness of the Act's framework.

8.7.2 Drinking Water

As mentioned earlier in the section, The Safe Drinking Water Act (SDWA) is overseen by the EPA. Contaminants are characterized as any physical, chemical, biological, or radiological substance present in water. In this broad definition, contaminants can include anything other than an H_2O molecule. Drinking water, therefore, contains at least minimal amounts of contaminants. While some contaminants could potentially be harmful at high enough levels, most common contaminants do not pose a significant health risk (EPA 2021a, b). Of particular interest for the purposes of this book are the biological contaminants, which are any microbes, or microbiological contaminants, present in the water. This includes any viruses, bacteria, mold, protozoa, or parasites (EPA 2021a, b).

Contaminant Candidate List (CCL), as discussed earlier, serves as the first level of evaluation for unregulated drinking water contaminants that may need further investigation of potential health effects and the levels at which they are found in drinking water" (EPA 2021a, b). EPA recently began the development of the fifth Contaminant Candidate List (CCL 5). On 4 October 2018, EPA requested nominations of chemicals, microbes, or other materials for consideration on the CCL 5. The public was able to nominate contaminants by following the instructions contained in the Federal Register notice for CCL 5 nominations. The deadline for nominations was 4 December 2018, so is now closed. EPA is evaluating the nominations and other contaminant data and information and plans to publish a draft of the CCL 5 for public review and comment, in the near future. Table 8.2 includes the microbial contaminants on the current standards in CCL 4 (EPA 2021a, b).

For drinking water, there is no specified limit for viruses or other microbial contaminants but there are recommended Maximum Contaminant Level Goals (MCLGs) that finished water should reach. For the microbial contaminants on the National Primary Drinking Water Regulations list for Microorganisms (*Cryptosporidium, Giardia*, Heterotrophic Plate count, *Legionella*, enteric viruses) the MCLG is zero, and the way a treatment plant gets there is through minimum treatment technology (TT) standards. So, EPA has established a minimum level of treatment technology (filtration, disinfection) that is required to reach minimal levels of these microbial contaminants in finished drinking water. These TT standards are a bit different for surface water versus groundwater but are designed to provide 99.9% removal of the protozoan, 99.99% removal of viruses, etc. For coliform bacteria, no more than 5% of samples analyzed in a month can be total coliform positive (R.R. Dupont to L. Theodore, personal communication 2020).

8.7.3 Surface Water Treatment Rules (SWTR)

The principal objective of the Surface Water Treatment Rules (SWTRs) is to minimize pathogens that could transmit disease directly through drinking water. Examples of such microbes include *Legionella, Giardia lamblia, Cryptosporidium*, and enteric viruses. The SWTRs direct water treatment facilities to filter and disinfect surface water sources (EPA 2021a, b).

Table 8.2 Microbial contaminants – CCL 4.

Microbial contaminant	Contaminant type	Diseases and infections
Adenovirus	Virus	Respiratory illness and occasionally gastrointestinal illness
Caliciviruses	Virus (includes Norovirus)	Mild self-limiting gastrointestinal illness
Campylobacter jejuni	Bacteria	Mild self-limiting gastrointestinal illness
Enterovirus	Viruses including polioviruses, coxsackieviruses, and echoviruses	Mild respiratory illness
Escherichia coli (0157)	Bacteria	Gastrointestinal illness and kidney failure
Helicobacter pylori	Bacteria	Capable of colonizing human gut that can cause ulcers and cancer
Hepatitis A virus	Virus	Liver disease and jaundice
Legionella pneumophila	Bacteria	Found in the environment including hot water systems causing lung diseases when inhaled
Mycobacterium avium	Bacteria	Lung infection in those with underlying lung disease, and disseminated infection in the severely *immunocompromised*
Naegleria fowleri	Protozoan	Parasite found in shallow, warm surface, and ground water causing primary amebic meningoencephalitis
Salmonella enterica	Bacteria	Mild self-limiting gastrointestinal illness
Shigella sonnei	Bacteria	Mild self-limiting gastrointestinal illness and bloody diarrhea

Source: Adapted from EPA (2021a, b).

The reader should note that there are no specific federal effluent discharge standards for coliform bacteria from a wastewater treatment plant, but it is up to the States to determine acceptable levels based on where their water is being discharged to. For most states, the discharge standard for these coliform bacteria,

Table 8.3 Microorganism surface water treatment rules summary.

Contaminant	MCLG (mg/l)	MCL or TT (mg/l)	Potential health effects from long-term exposure above the MCL
Cryptosporidium	Zero	TT	Gastrointestinal illness (such as diarrhea, vomiting, and cramps)
Giardia lamblia	Zero	TT	Gastrointestinal illness (such as diarrhea, vomiting, and cramps)
Heterotrophic plate count (HPC)	N/a	TT	HPC has no health effects; it is an analytic method used to measure the variety of bacteria that are common in water; the lower the concentration of bacteria in drinking water, the better maintained the water system is
Legionella	Zero	TT	Legionnaire's disease, a type of pneumonia
Total Coliforms (including fecal coliform and *E. coli*)	Zero	5.0%	Not a health threat in itself; it is used to indicate whether other potentially harmful bacteria may be present
Turbidity	N/a	TT	Higher turbidity levels are often associated with higher levels of disease-causing microorganisms such as viruses, parasites, and some bacteria; these organisms can cause symptoms such as nausea, cramps, diarrhea, and associated headaches
Viruses (enteric)	Zero	TT	Gastrointestinal illness (such as diarrhea, vomiting, and cramps)

Source: Adapted from EPA (2016).

indicators of fecal contamination, is based on recreational standards, i.e., what level is acceptable for people to come in contact with that prevents them from becoming infected. Table 8.3 lists the contaminants covered under the National Primary Drinking Water Regulations (EPA 2021a, b).

8.8 REGULATIONS GOVERNING INFECTIOUS DISEASES

Rules governing the implementation of quarantines delegate the responsibility and authority to the Secretary of HHS, which, as discussed earlier in this chapter, is a cabinet-level office. HHS determines when legally mandated quarantines are necessary. According to the CDC, the "Secretary of the Department of Health and Human Services has statutory responsibility for preventing the introduction, transmission, and spread of communicable diseases in the United States" (CDC 2021). Under delegated authority from the Commerce Clause of the U.S. Constitution, and US Code Titles 8 and 42, the Division of Global Migration and Quarantine oversees several related activities as listed below:

- Installing and supervising Quarantine Stations at ports of entry.
- Governing standards for medical examination of persons prior to arrival in the United States.
- Administration of all interstate and foreign quarantine regulations, "which govern the international and interstate movement of persons, animals, and cargo" (CDC 2021).

According to the CDC, "The United States Code is a consolidation and codification by subject matter of the general and permanent laws of the United States. Sections 264-272 of the following portion of the code apply: Title 42 – The Public Health and Welfare, Chapter 6A – Public Health Service, Subchapter II – General Powers and Duties, Part G – Quarantine and Inspection" (CDC 2021). These are listed below.

- 42 USC Part G – Quarantine and Inspection
- Sec. 264. Regulations to control communicable diseases
- Sec. 265. Suspension of entries and imports from designated places to prevent the spread of communicable diseases
- Sec. 266. Special quarantine powers in time of war
- Sec. 267. Quarantine stations, grounds, and anchorages
- Sec. 268. Quarantine duties of consular and other officers
- Sec. 269. Bills of health
- Sec. 270. Quarantine regulations governing civil air navigation and civil aircraft
- Sec. 271. Penalties for violation of quarantine laws
- Sec. 272 Administration of oaths by quarantine officers

The CFR is the official and complete version of all rules within the Federal Register, as established by the executive governmental departments and agencies. As discussed earlier in the chapter, regulations related to the CDC's authority are part of Title 42: Public Health.

As described by the CDC, "Executive Orders specify the list of diseases for which federal quarantine is authorized, which is required by the Public Health Service Act. On the recommendation of the HHS Secretary, the President may amend this list whenever necessary to add new communicable diseases, including emerging diseases that are a threat to public health" (CDC 2021).

8.8.1 VACCINATION LAWS

With the exception of the COVID-19 pandemic, cases of vaccine-preventable diseases have been at all-time lows. CDC, in coordination with other public health agencies and private corporations and health institutes, has improved immunization coverage for most communicable diseases and closely monitors any outbreaks. Of the approaches used to sustain low rates of vaccine-preventable disease, the most successful is the implementation of vaccination laws. State vaccine laws include mandates for children attending school and for healthcare workers and patients in certain facilities. These laws help improve access to vaccination services by determining which vaccines are necessary for public health and should be paid for by the state or federal government (CDC 2021).

8.8.2 STATE HEALTHCARE WORKER AND PATIENT VACCINATION LAWS

In most states vaccines for healthcare workers are mandated, which is especially important, not only for public health but also to protect at-risk populations such as immunocompromised or elderly. The following viral illnesses are commonly required by state vaccine laws governing healthcare facilities:

- Hepatitis B
- Influenza
- Measles, mumps, rubella (MMR)
- Varicella

8.8.3 STATE-MANDATED CHILDHOOD VACCINATIONS

The CDC states that, "all states require children to be vaccinated against certain communicable diseases as a condition for school attendance." This includes both private and public schools and includes some religious or health-related exemptions. All states also establish vaccination requirements for children as a condition for childcare attendance. The Public Health Law Program (PHLP) offers legal resources for public health practitioners regarding state school vaccination laws. The EPA also shares some of the responsibility for implementation and provides for more training for state and federal personnel in emergency preparedness, disaster response, and hazard mitigation (CDC 2021).

8.9 APPLICATIONS

Four illustrative examples complement the presentation for this chapter on regulatory concerns.

Illustrative Example 8.1 TOXIC SUBSTANCES CONTROL ACT (TSCA) HISTORY
Provide an overview and history of TSCA.

Solution
In 1970, the President's Council on Environmental Quality developed a legislative proposal to address the increasing problems of toxic substances. After six years of public hearings and debate, Congress enacted the Toxic Substances Control Act (TSCA) in the fall of 1976. EPA/OPPT (Office of Pollution Prevention and Toxics) is charged with implementing TSCA. TSCA (Title I) does not provide opportunities for EPA to authorize state programs to operate in lieu of the federal program, although the office actively collaborates with regions, states, and tribal governments. Through the provisions of TSCA, EPA can collect or require the development of information about the toxicity of particular chemicals and the extent to which people and the environment are exposed to them. Such information allows EPA to assess whether the chemicals pose unreasonable risks to humans and the environment, and TSCA provides tools instituting appropriate control actions. TSCA provides the basis for EPA's programs on New and Existing Chemicals (Stander and Theodore 2008).

Illustrative Example 8.2 TSCA SECTION 8(D) REVISIONS
Discuss TSCA Section 8(d) Revisions concerned with unpublished health and safety studies.

Solution
Key elements of these revisions are:

- Persons who must report (unless otherwise specified) include chemical producers and importers under the NAICS Codes Subsection 325 (chemical manufacturing and allied products) and Industry Group 32411 (petroleum refiners).
- The reporting period for studies for a listed chemical substance or listed mixture will terminate 60 days after the effective date of the listing.
- Studies to be reported will be specified by EPA to include the specific type(s) of health and safety data needed; the chemical grade/purity of the test material (studies involving mixtures are not required unless otherwise specified).
- Initiated studies are reportable only for study initiation that occurs during the 60-day reporting period.
- Adequate file searches encompass reportable information data on or after 1 January 1977 (unless otherwise specified).

Under TSCA Section 8(d), EPA has the authority to promulgate rules to require producers, importers, and processors to submit lists and/or copies of ongoing and completed, unpublished health and safety studies. EPA's TSCA

Section 8(d) "Health & Safety Data Reporting Rule" was developed to gather health and safety information on chemical substances and mixtures needed by EPA to carry out its TSCA mandates (e.g. to support OPPT's Existing Chemicals Program and Chemical Testing Program and to set priorities for TSCA risk assessment/management activities). EPA has also used its TSCA Section 8(d) authority to gather information needed by other EPA Program Offices and other Federal Agencies. Chemicals that are designated or recommended for testing by the TSCA Interagency Testing Committee (ITC) may be added to the rule via immediate final rulemaking (up to 50 substances/year). Non-ITC chemicals can be added to the Section 8(d) rule via notice and comment rulemaking. Further information is available under 40 CFR Part 716.

Illustrative Example 8.3 Particle Size

Refer to Figure 8.1, and comment on the shape of the curve for particle sizes below 0.1 microns.

Solution

The particle size at 0.1 μm (microns) is 100 nanometers (nm); at 0.01 μm, it is 10 nm; and at 0.001 μm it is 1.0 nm. There are various statements in the literature regarding the size of viruses. The most widely accepted size range of human viruses is generally from 5.0 to 500 nm.

Concerning collection efficiency in Figure 8.1, the behavior of these nanoparticles in the above range appears to indicate that the curve continues to rise, approaching 100%. There is limited data available for sizes below 10 nm, but both authors believe the efficiency will continue to increase. (Theodore 2006; Theodore 2008; M. Reynolds, personal notes, Freiburg, Germany 2021).

Figure 8.1 Size-efficiency curve.

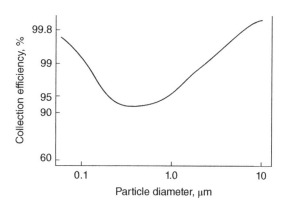

Illustrative Example 8.4 POLLUTANTS

Atmospheric concentrations of toxic and virus pollutants are usually reported using two types of units: (i) mass of pollutant per volume of air, i.e., mg/m^3, $\mu g/m^3$, ng/m^3, etc.; or (ii) parts of pollutant per parts of air (by volume), i.e., ppmv, ppbv, pptv, etc. To compare data collected at different conditions, actual concentrations are often converted to standard temperature and pressure (STP). According to EPA, STP conditions for atmospheric or ambient sampling are often either 0 or 25°C and 1 atm.

Assume the concentration of chlorine vapor is measured to be 15 mg/m^3 at a pressure of 600 mmHg and at a temperature of 10°C.

Convert the concentration units to ppmv.

Solution

To convert the concentration units to ppmv, the molecular weight of chlorine must first be obtained. Since chlorine gas is Cl_2, with 2 gmol of Cl atoms per Cl_2 molecule, it has a molecular weight of:

$$MW = 2(35.45 \text{ g/gmol}) = 70.9 \text{ g/gmol}$$

The absolute temperature of the gas is:

$$T = 10°C + 273.2°C = 283.2 \text{ K}$$

The pressure of the gas in atmospheres is:

$$P = 600 \text{ mmHg}/760 \text{ mmHg} = 0.789 \text{ atm}$$

The concentration of chlorine gas in ppmv can be determined by taking the mass per unit volume value, converting the chlorine gas mass to volume, and then expressing this concentration as volume per million volumes as follows. Take as a basis 1.0 m^3 of volume of gas and note that 15 mg = 0.015 g. Thus,

$$Cl_2(V) = \frac{(Cl_2 \text{Mass})(R)(T)}{(MW)(P)}; \quad R = \text{ideal gas law constant}$$
$$= \frac{(0.015 \text{ g})(0.082 \text{ atm} \cdot \text{l/gmol} \cdot \text{K})(283.2 \text{ K})}{(70.9 \text{ g/gmol})(600/760 \text{ atm})}$$
$$= 0.00623 \text{ l} = 6.23 \text{ ml}$$

Since there are 10^6 ml in a m^3, the concentration in ppmv is expressed as ml/m^3 or ml 10^6 ml. Thus:

$$\text{Concentration} = (6.23 \text{ ml})/1 \text{ m}^3 = 6.23 \text{ ppmv}.$$

8.10 CHAPTER SUMMARY

- Environmental regulations is an organized system of statutes, regulations, and guidelines that minimize, prevent, and punish the consequences of damage to the environment. The recent popularity of environmental issues has brought about changes in legislation and subsequent advances in technology.
- The Clean Air Act Amendments (CAA) of 1990 built upon the regulatory framework of previous CAA programs and expands their coverage to many more industrial and commercial facilities.
- Under the Clean Water Act (CWA), the permit program known as the NPDES, requires dischargers to disclose the volume and nature of their discharges as well as monitor and report the results to the authorizing agency.
- The Occupational Safety and Health Act (OSH) was established to address safety in the workplace, which is limited to conditions that exist within the workplace, where its jurisdiction covers both safety and health.
- The Toxic Substance Control Act (TSCA) of 1976 provides EPA with the authority to control the risks of thousands of chemical substances – both new and old – that are not regulated as drugs, food additives, cosmetics, or pesticides.
- Risk management has become a top priority for industry and federal officials. The CAA Amendments require industries to communicate the likelihood and degree of a chemical accident to the public.

8.11 PROBLEMS

1 Many acronyms are associated with the Clean Air Act (CAA) and the corresponding amendments. Indicate what each of the following six acronyms stands for:
- NAAQS
- NSR
- RFP
- SIP
- BACT
- VOC

2 List the viruses regarded as microbial contaminants by the CCL-4.

3 Describe the dioxin/furan family of chemicals.

4 Discuss continuous emission monitoring.

REFERENCES

Centers for Disease Control and Prevention (CDC) (2018). Our history – our story. https://www.cdc.gov/about/history/index.html (accessed 28 November 2022).

Centers for Disease Control and Prevention (CDC) (2021). Specific laws and regulations governing the control of communicable diseases. Centers for Disease Control and Prevention. https://www.cdc.gov/quarantine/specificlawsregulations .html (accessed 6 December 2021).

Centers for Medicare and Medicaid Services (CMS) (2021). Current emergencies. https://www.cms.gov/About-CMS/Agency-Information/Emergency/EPRO/ Current-Emergencies/Current-Emergencies-page (accessed 4 December 2021).

Chabot, S. (2017). H.R.2432 – Volunteer Organization Protection Act of 2017. Congress.gov. https://www.congress.gov/bill/115th-congress/house-bill/2432/text (accessed 2 December 2021).

Environmental Protection Agency (EPA) (2016). National primary drinking water regulations. https://19january2021snapshot.epa.gov/sites/static/files/2016-06/ documents/npwdr_complete_table.pdf (accessed 4 December 2021).

Environmental Protection Agency (EPA) (2021a). Drinking Water Contaminant Candidate List (CCL) and regulatory determination. EPA. https://www.epa.gov/ccl (accessed 4 December 2021).

Environmental Protection Agency (EPA) (2021b). EPA. https://www.epa.gov/ccl/ types-drinking-water-contaminants (Retrieved 4 December 2021).

Food and Drug Administration (2021) About FDA. https://www.fda.gov/about-fda (accessed 1 December 2021).

Merriam-Webster (2021a). Plan definition & meaning. Merriam-Webster. https:// www.merriam-webster.com/dictionary/plan (accessed 5 December 2021).

Merriam-Webster (2021b). Policy definition & meaning. Merriam-Webster. https:// www.merriam-webster.com/dictionary/policy (accessed 5 December 2021).

National Governor's Association (NGA), Natural Resources Policy Studies Division (1997). Interstate cooperation to speed multistate acceptance of environmental technologies. http://www.nga.org/Pubs/IssueBriefs/1997/970910Enviro Tech.asp (accessed 1 December 2021).

Occupational Safety and Health Administration (OSHA) (2021a) *US Department of Labor*. US Department of Labor issues emergency temporary standard to protect workers from coronavirus. https://www.osha.gov/news/newsreleases/national/ 11042021 (accessed 6 December 2021).

Occupational Safety and Health Administration (OSHA) (2021b). About OSHA|Occupational Safety and Health Administration US Department of Labor. https://www.osha.gov/aboutosha (accessed 6 December 2021).

Occupational Safety and Health Administration (OSHA) (2022). *OSH Act of 1970*. US Department of Labor. https://www.osha.gov/laws-regs/oshact/section5-duties (accessed 28 October 2022).

Santoleri, J.J., Theodore, L., and Reynolds, J.P. (2000). *Introduction to Hazardous Waste Incineration*, 2e. Hoboken, NJ: Wiley-Interscience.

Stander, L. and Theodore, L. (2008). *Environmental Regulatory Calculations Handbook*. Hoboken, NJ: John Wiley & Sons.

Theodore, L. (2006). *Nanotechnology: Basic Calculations for Engineers and Scientists*. Hoboken, NJ: John Wiley & Sons.

Theodore, L. (2008). *Air Pollution Control Equipment Calculations*. Hoboken, NJ: John Wiley & Sons.

Theodore, M.K. and Theodore, L. (2021). *Introduction to Environmental Management*, 2e. Boca Raton, FL: CRC Press/Taylor & Francis Group.

United Nations Human Rights, Office of the High Commissioner (2020). Emergency measures and covid-19: Guidance – OHCHR. https://www.ohchr.org/Documents/ Events/EmergencyMeasures_COVID18.pdf (accessed 29 November 2021).

9

Emergency Planning and Response

The extent of the need for emergency planning is significant and continues to expand as new laws and regulations on safety are introduced. Preparedness for any emergency is essential and is usually based on worst-case scenarios. When a crisis begins, the response time to implement any plan(s) will be, in large part, based on the level of planning, preparation, and training of emergency response teams, healthcare personnel, police, and government agencies. Effective cooperation among various responders can be one of the most important factors and can not only increase speed and efficiency but also save lives. The first line of defense against a public health emergency such as a pandemic begins at an early stage. It is much easier to prevent a public health problem rather than to try and rectify the situation once a problem has occurred.

The main objective for any plan should be to prepare a procedure to make maximum use of the combined resources of the community to accomplish the following (Theodore and Theodore 2021):

- Safeguard people during the event
- Minimize damage to property and the environment

A Guide to Virology for Engineers and Applied Scientists: Epidemiology, Emergency Management, and Optimization, First Edition. Megan M. Reynolds and Louis Theodore.
© 2023 John Wiley & Sons, Inc. Published 2023 by John Wiley & Sons, Inc.

- Initially contain and ultimately bring the situation under control
- Effect appropriate treatment and management of casualties
- Provide authoritative information to the news media which can then communicate the facts to those affected
- Secure the safe rehabilitation of the affected individuals and/or community

While the priority of this chapter is emergency planning and response as it applies to viral outbreaks and epidemics, other catastrophic events, such as industrial accidents or natural disasters are also highly relevant. These events often leave survivors and local communities without access to proper shelter, clean water, or functioning sanitation for weeks, months, or even years. Dire situations that follow such disasters can leave people vulnerable to many health issues, not the least of which are communicable diseases. Each type of crisis calls for unique approaches to lessen the harm and misery of survivors. Thus, this chapter will also review the various methodologies which can be used for events such as hazardous chemical or nuclear accidents, natural disasters, and outbreaks and pandemics.

9.1 THE IMPORTANCE OF EMERGENCY PLANNING AND RESPONSE

Regarding public health problems, exposure concerns will be briefly introduced at this time, although the topic will be discussed in detail in Chapter 11.

Health disorders are usually categorized as either acute or chronic based on the timeframe of the condition or exposure. These are classified as (Theodore and Theodore 2021):

- *Chronic:* Continuous exposure occurs over long timeframes, generally several months to years. Concentrations of inhaled viruses and contaminants are usually relatively low. Direct skin contact by immersion, splash, or by contaminated air involves contact with substances exhibiting low dermal activity.
- *Acute:* Exposures occur for relatively short time periods, generally minutes to one to two days. The concentration of air contaminants is usually high relative to their protection criteria. In addition to inhalation, airborne substances such as viruses might directly contact the skin, or liquids and sludges may be splashed on the skin or into the eyes, leading to negative health effects.

In general, acute exposures to chemicals released in the air or water are more typically the result of high contaminant concentrations which usually do not persist for long durations. (The resulting injuries, however, could be long-term or even permanent.) Exposures from splashes of liquids or from the release of vapors, gases, or particulates that might be generated can occur. Acute skin exposure may also occur. Chronic exposures, on the other hand, are usually associated with longer time periods or repeated over months or years. Unlike most acute emergencies, chronic exposures can often remain hidden or unknown for years or even decades. These types of exposures may not be included in emergency planning, and some are discovered too late for an appropriate response. One example of long-term consequences is the development of cancer years after exposure to radiation from nuclear accidents.

Emergencies such as pandemics, industrial accidents, and natural disasters have occurred in the past and will continue to occur in the future. A few of the many commonsensical reasons to plan ahead are provided in the following (Krikorian 1982):

- Emergencies will happen; it is only a question of time.
- When emergencies occur, the minimization of loss and the protection of people, property, and the environment can be achieved through the implementation of an appropriate emergency response plan.
- Minimizing the losses caused by an emergency requires planned procedures, understood responsibility, designated authority, accepted accountability, and trained and experienced people. With a fully implemented plan, these goals can almost always be achieved.
- If an emergency such as a pandemic occurs, it may be too late to plan; a lack of preplanning can turn an emergency into a disaster.

Whatever the nature of the crisis, a well-designed emergency plan that is understood by the individuals responsible for action, as well as by the public, can ease concerns over emergencies and reduce the inherent distress and fear that accompanies such situations. People will react during an emergency; how they react can be somewhat controlled through education. When confusion and ignorance are pervasive, the likely behavior during an emergency is panic (Theodore and Theodore 2021).

An emergency plan can minimize loss by helping to assure the proper responses. "Accidents, such as public health problems become crises when subsequent events, and the actions of people and organizations with a stake in the outcome, combine in unpredictable ways to threaten the social structures involved" (Shrivastava 1987). The wrong response can turn an accident into a disaster as easily as no response. For example, if a chemical fire is doused with water, which causes the emission of toxic fumes, it would have been better to let the fire burn itself out. For another example, suppose people are evacuated from a building into the path of a toxic vapor cloud; they might well have been safer staying indoors with closed windows. Still another example is offered by members of a virus rescue team who become victims because they were not wearing proper breathing protection. Thus, the proper response to an emergency requires an understanding of the various hazards. A plan can provide the right people with the information they need to respond properly during an emergency.

9.2 PLANNING FOR EMERGENCIES

In order to estimate the impact on the public or the environment, an emergency must be studied in depth. A hazardous gas leak, fire, or explosion may cause a toxic cloud to spread over a great distance, as it did in Bhopal nearly a half-century ago. An accurate estimation of the areas that could potentially be affected is critical. Depending on the type of emergency, this could include an assessment of the area and population size that would require evacuation (Krikorian 1982).

9.2.1 PREPAREDNESS TRAINING

All personnel involved in responding to environmental incidents, and those who could be exposed to hazardous substances, health hazards, or safety hazards, must receive safety training before carrying out their response functions. Health and safety training must, as a minimum, include:

- Use of personal protective equipment (i.e., respiratory protective apparatus and protective clothing)
- Safe work practices, engineering controls, and standard operating safety procedures
- Hazard recognition and evaluation
- Medical surveillance requirements, symptoms that might indicate medical problems, and first aid
- Site safety plans and plan development
- Site control and decontamination
- Use of monitoring equipment, if applicable

Training must be as practical as possible and include hands-on use of equipment and exercises designed to demonstrate and practice classroom instruction. Formal training should be followed by at least three days of on-the-job experience working under the guidance of an experienced, trained supervisor. All employers and hospital employees should, as a minimum, complete an eight-hour safety refresher course annually. Health and safety training must comply with OSHA's training requirements as defined in 29 CFR1910.20 (OSHA 2021).

The personnel at an industrial plant, particularly the operators, are trained in the operation of the plant. These people are critical to proper emergency response. They must be taught to recognize abnormalities in operations and report them immediately. Plant operators should also be taught how to respond to various types of accidents. Emergency squads at plants can also be trained to contain an emergency until outside help arrives, or, if possible, to terminate the emergency. Safety procedures are especially important when training plant personnel. In addition, similar training should be offered to hospital personnel in close proximity to the plant or manufacturing facility.

Training is essential for the emergency teams to ensure that their roles are clearly understood and that any health problem can be addressed safely and correctly without delay. The emergency teams include police, firefighters, medical people, and volunteers who will be required to take action during an emergency. These people must be knowledgeable about the potential hazards. For example, specific antidotes for different types of medical problems must be known by medical personnel. The entire emergency team must also be taught the use of personal protective equipment.

Training for emergencies should be done routinely in the following situations:

- When a new member is added to the group
- When someone is assigned a new responsibility within the community
- When new equipment or materials are acquired for use in emergency response
- When emergency procedures are revised
- When a practice drill shows inadequacies in the performance of duties
- At least once annually

Emergency procedures need to be performed as planned. As noted earlier, this requires regular training to ensure that people understand and remember how to react. The best plan on paper is likely to fail if the persons involved are reading it for the first time as an emergency is occurring. People must be trained *before* an emergency happens.

9.3 PLAN IMPLEMENTATION

Once an emergency plan has been developed, its successful implementation can be assured only through constant review and revision. Helpful ongoing procedures are:

- Routine checks of medical equipment inventory, the status of personnel, the status of hazards, and population densities
- Auditing of the emergency procedures
- Routine training exercises
- Practice drills

The emergency coordinator must assure that the emergency equipment is always in a state of readiness. Siting the control center and locating its equipment are also the coordinator's responsibility. There should be both the main control center and an alternate one in carefully chosen locations. The following items should be present at the control center:

- Copies of the current emergency plan
- Maps and diagrams of the area
- Names and addresses of key functional personnel
- The means to initiate alarm signals in the event of a power outage
- Communication equipment (e.g., phones, radio, TV, and two-way radios)
- Emergency generators and lights
- Evacuation routes detailed on the area map
- Self-contained breathing equipment for possible use by the control center crew
- Miscellaneous items, including beds/cots, ventilators, etc.

Inspection of emergency equipment such as fire trucks, police cars, medical vehicles, personal safety equipment, and alarms should be performed routinely.

The plan should be audited regularly, at least annually, to ensure that it is current. Items to be updated include the list of potential hazards and emergency procedures (adapted to any newly developed technology). A guideline of general questions for auditing the emergency response plan, adapted from literature published by the Chemical Manufacturers Association, is presented in question format as follows (Stander and Theodore 2007):

- What types of emergencies have occurred since the last audit?
- Are all potential emergency types covered by the plan?
- Who is responsible for maintaining the written plan?
- Who is authorized to activate the plan?
- When was the last revision?

9.3.1 Notification of Public and Regulatory Officials

Notifying the public of an emergency such as a virus pandemic is a task that must be accomplished with caution. People will react in different ways when receiving notification of an emergency, particularly an epidemic or pandemic. Many will simply not know what to do, some will not take the warning seriously, and others will panic. Proper training in each community, as discussed in the previous section, can help minimize panic and can condition the public to make the correct response in a time of stress.

Methods of communicating an emergency will differ from community to community, depending on its size and resources. Some techniques for notifying the public are:

- The sounding of fire department alarms in different ways to indicate emergencies of certain kinds and levels

- Chain phone calls (this method usually works well in small towns)
- Announcements made through loudspeakers from police cars or the vehicles of volunteer teams

Once the emergency has been communicated, an appropriate response by the public must be evoked. For this to occur, an accepted plan that people know and understand must be put into effect. Since an emergency can quickly become a disaster if panic ensues, the plan should include the appropriate countermeasures to bring the situation back under control.

Information reported to the emergency coordinator must be carefully screened. A suspected "crank call" should be checked out before an alarm is sounded. By taking no immediate action, however, the team runs the obvious risk that the plan will not be implemented in time. Therefore, if a call cannot be verified as bogus, a response must begin, and local police should be dispatched quickly to the scene of the reported emergency to provide firsthand information about the actual situation.

The print and broadcast media can be a major resource for communication, and one job of the emergency coordinator is to prepare information for reporters. The emergency plan should include a procedure to pass along information to the media promptly and accurately.

9.4 EP&R FOR EPIDEMICS AND PANDEMICS

Epidemics and pandemics have occurred in the past and will continue to occur in the future. Scientists have long warned of the possibility of an influenza pandemic to rival that of the Spanish Flu of 1918, which killed an estimated 50 million people worldwide (CDC 2018).

The proper response requires an understanding of all the data that is available, and a plan can provide the practitioner with the information needed to respond quickly and properly during a health crisis. The potential for pandemics, based on the history and knowledge of the region, should also be considered early in any response. Thus, successful pandemic planning begins with a thorough understanding of the infectious disease and/or the potential disaster being planned for. The impacts on public health and the environment must be estimated at an early stage.

As noted earlier, the wrong response can turn an outbreak into an epidemic or even a pandemic, as easily as no response. The application of the above has recently come under increased scrutiny in the public health field with the pandemic that began in the last days of 2019 with the arrival of the SARS-CoV-2 virus. It became more important than ever to put science and engineering into action to try to help prevent and minimize the effects of pandemics.

One area that has received significant attention is the representation of infectious disease data in equation form for analytical and predictive purposes. This emergency planning and response (EP&R) activity offers the potential for protecting the public from illness and injury, and better preparing for emergencies such as pandemics. Accurate data and analysis can help authorities better allocate emergency funds and other resources. The material in Part III, Chapter 16 attempts to provide the technical community with the tools to generate mathematical models to depict health and disease data.

9.4.1 FEDERAL PUBLIC HEALTH AND MEDICAL EMERGENCY PREPAREDNESS

Similar to EPA preparedness programs, which will be discussed in Section 9.5 – EP&R for Industrial Accidents--the Department of Health and Human Services (HHS) and the Centers for Disease Control and Preparedness (CDC) coordinate the public health emergency response. Falling under the purview of the HHS, the Office of the Assistant Secretary for Preparedness and Response (ASPR) oversees Public Health Emergency (PHE) responses, which includes a framework for inter-agency and private sector cooperation. According to the PHE/ASPR website:

> This framework includes the National Incident Management System (NIMS) and the National Response Framework (NRF) [which] integrate the capabilities and resources of government, emergency response disciplines, Non-Governmental Organizations (NGOs) and the private sector … to work effectively and efficiently together to prepare for, respond to, and recover from domestic incidents regardless of cause, size, or complexity.
>
> (Source: PHE 2019).

In addition to these measures, the Occupational Safety and Health Administration (OSHA) has set forth standards and guidelines governing public health issues in the workplace. In November 2021, for example, OSHA issued an Emergency Temporary Standard on COVID-19 Vaccination and Testing (OSHA 2021).

9.4.2 EMERGENCY OPERATIONS CENTER

Emergency operations centers (EOCs) are established during crises of various types (CDC 2020b). They are located as close to the site as is safely possible, or which may be virtual, and act as a headquarters during a crisis. EOCs are designed to support the emergency response to enhance communication and centralize decision-making. A primary EOC should be that it supports the following incident management functions: (Ready.gov 2021)

- Activation – bring knowledge and expertise together to deal with events that threaten the business
- Situation analysis – gather information to determine what is happening and to identify potential impacts
- Incident briefing – efficiently share information among team members
- Incident action plan – provide a single point for decision-making and decide on a course of action for the current situation
- Resource management – provide a single point of contact to identify, procure, and allocate resources
- Incident management – monitor actions, capture event data, and adjust strategies as needed

Among other agencies, the CDC activates its EOCs during crises ranging from disease outbreaks to hurricanes, as well as other crises immediately following other natural disasters, in order to support the health of victims and survivors. In the event of an epidemic for example this allows the agency to provide increased operational support on the gound for the response team to meet the outbreak's evolving challenges. Agency subject matter experts will continue to lead the CDC response with enhanced support from additional CDC and EOC staff (CDC 2020a). The following recent EOC activations are listed below:

- 2019 Novel Coronavirus (EOC Activation: January 2020–present)
  The CDC activated its Emergency Operations Center (EOC) on 20 January 2020, in response to the outbreak caused by a novel (new) coronavirus first identified in Wuhan, Hubei Province, China.
- 2018 Ebola Response (EOC Activation: June 2019–June 2020)
  The CDC activated its Emergency Operations Center on Thursday, 13 June 2019, to support the interagency response to the Ebola outbreak in eastern Democratic Republic of the Congo (DRC). The CDC assisted the DRC and Ugandan governments, and countries bordering the outbreak area.
- 2016 Zika Virus Disease Outbreak (January 2016–30 September 2017)
  Outbreaks of Zika are occurring in many countries and territories. Zika infection during pregnancy can cause microcephaly, a serious birth defect of the brain. CDC posted travel notices and worked with governments and other partners around the world to gain better insight into the illness and to control Zika.
- 2014 Ebola Outbreak in West Africa (March 2014–December 2015)
  In March 2014, CDC, along with domestic and international governments and partners, responded to stop an outbreak of Ebola virus disease (EVD) in West Africa.
- Chikungunya in the Americas (June 2014–present)
  Local transmission of chikungunya has been reported in 43 countries and territories throughout the Americas as well as in Florida, Puerto Rico, and the

US Virgin Islands, since late 2013. Chikungunya virus disease cases have been reported among US travelers returning from affected areas.

Since early 2020, the COVID-19 pandemic has dominated official discourse regarding emergency response. Likewise, the ongoing crisis response has also become a relentless topic of debate among both regulators and the public. Once limited to the realms of public health medicine, vocabulary such as "viral variants," "genetic mutations," "mRNA vaccines," and "personal protective equipment," have now been readily adopted into everyday language.

9.4.3 DISEASE CONTAINMENT

An integral part of disease contamination should include both medical intervention and public health intervention. Medical interventions employed – particularly as it applies to viral pandemics such as influenza in the past include (CDC 2018):

- Isolation
- Quarantines
- Social distancing
- Mask requirements
- Hygienic measures

As noted earlier, controlling the spread of disease should be performed in accordance with scientific data, laws, and policies, and with consideration to individual rights and liberties. Quarantine and social distancing – both individual and community measures that reduce the frequency of human contact – have proven effective in reducing the transmission of infectious diseases. To maximize the benefits of these policies, such measures must be planned, constrained, and instituted as early as possible during a pandemic. Some of these measures date as far back as the 1918 Influenza Pandemic (CDC 2018).

Workplace and school closings present particularly difficult ethical issues. Although effective when used judicially, issues focus on the subject of *distributive justice.* Workplaces are vital to the livelihoods of both employers and employees, so closing them can cause severe financial hardships. At the international level, there should also be at least a serious effort to weigh the risks to health and welfare from workplace closings and other social-distancing measures against those risks of disease transmission that the closings might mitigate. In different locations, the balance of risks may be resolved differently, depending on resources and the number of people living at or below a subsistence level. Thus, behavioral strategies for disease containment must also be compatible with modern legal thinking, with its emphasis on objective standards and fair procedures. Further details on ethical dilemmas unearthed during epidemics is available in Chapter 10.

Hygienic measures to prevent the spread of respiratory infections are broadly accepted and have been widely used in the past but with uncertain benefits. These hygienic methods include:

- Handwashing
- Disinfection
- The use of personal protective equipment (PPE) such as masks, gloves, gowns, and eye protection
- Respiratory hygiene, such as the use of proper etiquette for coughing, sneezing, and spitting.

It is also important that the public be informed of the need for hygienic measures, and that accurate information, including the uncertainty of the effectiveness of the recommended interventions, be provided.

9.4.4 PUBLIC NOTIFICATION OF PANDEMIC QUARANTINES AND LOCKDOWNS

Caution must be used when notifying the public of an emergency. Reactions change depending on the type of emergency, and public panic often leads to more problematic situations. At the start of the COVID-19 pandemic, several governments made numerous miscalculations while announcing closed borders and lockdowns. Careless reporting and inconsistent messaging led to confusion and panic in many countries. This led to a rush to leave (or enter) countries and states prior to announced deadlines, which in turn led to dangerous crowd conditions, overwhelming officials, and leading to larger outbreaks and super-spreading events at borders, airports, and train stations.

The public must receive accurate information from a reliable and trustworthy source. If incorrect or distorted information about an emergency is disseminated, misperceptions and rumors can easily result. In the United States, conflicting announcements regarding the COVID-19 pandemic from official sources, including White House officials, the CDC, WHO, and various media outlets led to even further confusion. In the absence of clear recommendations, people turned to social media, allowing speculation and misinformation to spread. This exposed a highly discordant national emergency response. Governments around the world must identify the missteps made during the COVID-19 crisis, so that the global response to the next pandemic be swift and effective, hopefully saving millions of lives (WHO 2022).

9.4.5 THE NATIONAL STRATEGY FOR PANDEMIC INFLUENZA (NSPI)

Public health officials in many countries have predicted that another influenza outbreak is impending and that it could potentially lead to severe epidemics or even a global pandemic, similar to the pandemic caused by SARS-CoV-2. The National Strategy for Pandemic Influenza (NSPI), created by the Homeland Security Council (HSC) as well as the detailed NSPI implementation plan, was released in May 2006. The NSPI provides an integrated framework for national planning efforts across all levels of government and in all sectors of society outside of government. The integrated response was based on the following five principles:

- The federal government will use all instruments of national power to address the pandemic threat.
- States and communities should have credible pandemic preparedness plans to respond to an outbreak within their jurisdictions.
- The private sector should play an integral role in preparedness before a pandemic begins and should be part of the national response.
- Individual citizens should be prepared for an influenza pandemic and be educated about individual responsibility to limit the spread of infection if they or their family members become ill.
- Global partnerships will be leveraged to address the pandemic threat.

The NSPI implementation plan was organized in stages that correspond to the pandemic phases detailed in the World Health Organization (WHO) global framework for pandemic influenza (WHO 2022).

9.5 EP&R FOR INDUSTRIAL ACCIDENTS

Much of the attention on emergency planning, along with the resulting promise of new laws, are a reaction to large-scale calamities including nuclear disasters like Chernobyl in 1986 and Fukushima, Japan in 2011, and toxic chemical releases such as occurred in Bhopal, India, nearly half a century ago. In 1986, Congress approved the Emergency Planning and Community Right-to-Know Act (EPCRA) in response to the nearly 7000 chemical accidents that occurred in the United States in the five years leading up to its authorization (Diamond 1985; Theodore and Theodore 2021).

Large industries have long had emergency plans designed for on-site personnel. The protection of people, property, and livelihoods has made emergency plans and prevention methods common in industry. On-site emergency plans are also often required by insurance companies. One way to minimize the effort required for emergency planning at the industrial level is to expand existing industry plans to include all significant hazards and all people in a given community or within a certain geographical area. Successful emergency planning for industrial accidents and other manmade disasters begins with a thorough understanding of the potential event or disaster being planned for. The impacts on public health and the environment must also be estimated. Some of the types of crises that can occur and should be included in the plan are (Michael et al. 1986):

- Explosions and fires
- Hazardous chemical leaks, gas, or liquid
- Power or utility failures
- Radiation incidents
- Transportation accidents
- Natural disasters causing other hazards, such as power failures or natural gas explosions

9.5.1 EMERGENCY PLANNING AND COMMUNITY RIGHT-TO-KNOW ACT (EPCRA)

As mentioned above, Congress, recognizing the need for better preparation to deal with chemical emergencies, approved the *Emergency Planning and Community Right-to-Know Act (EPCRA)*, officially designated as *Title III of the Superfund Amendments and Reauthorization Act (SARA)*, in 1986. This act requires federal, state, and local governments and industry to work together in developing emergency plans and community "right-to-know" reporting on hazardous chemicals. These requirements build on EPA's Chemical Emergency Preparedness Program and numerous state and local programs that are aimed at helping communities deal with potential chemical emergencies (EPA 2021b; Theodore and Theodore 2021).

The EPCRA, requires federal, state, and local governments and industry to work together in developing emergency preparedness plans. Under EPCRA, the community "right-to-know" signifies that the public must be notified of any potential dangers due to industrial accidents that may affect them. Communities must be made aware of risks such as nuclear or hazardous chemical plants in the immediate area. This would include any laboratories handling biological hazards.

Certain types of emergencies must be reported to government agencies; it is not always sufficient to notify just the response team. For example, state and federal laws require the reporting of hazardous releases and nuclear power plant problems. There are also more specific requirements under EPCRA for reporting chemical releases. Facilities that produce, store, or use a listed hazardous substance must immediately notify the *Local Emergency Planning Committee (LEPC)* and the *State Emergency Response Commission (SERC)* if there is a release of one or more substances specifically listed in the SARA. These substances include 402 extremely hazardous chemicals on the list prepared by the Chemical Emergency Preparedness Program and chemicals subject to the reportable quantity requirements of the original Superfund. The initial notification can be made by telephone, radio, or in person. Emergency notification requirements involving transportation incidents can be satisfied by dialing 911 or calling the operator. The emergency planning committee should provide a means of reporting information on transportation accidents quickly to the plan coordinator (Stander and Theodore 2007). EPCRA requires that the notification of an industrial emergency include:

- The name of the chemical released
- Whether it is known to be acutely toxic
- An estimate of the quantity of the chemical released into the environment
- The time and duration of the release
- Where the chemical was released (e.g. air, water, and land)
- Known health risks and necessary medical attention that will be required
- Proper precautions, such as evacuations.
- The name and telephone number of the contact person at the plant or facility at which the release occurred

As soon as is practical after the release, there must be a written follow-up emergency notice, updating the initial information and giving additional information on response actions already taken, known or anticipated health risks, and advice on medical attention. The law has required reporting and written notices since 1986.

The likely "emergency zone" must be studied to estimate the potential impact on the public or the environment of problems of different types. For example, a hazardous gas leak, fire, or explosion may cause a toxic cloud to spread over a great distance. The minimum affected area contaminated, and thus the area to be evacuated should be estimated based on an atmospheric dispersion model (Theodore 2008). Various models can be used; the more complex models often produce more realistic results, but the simple and faster models may provide adequate data for planning purposes. A more thorough discussion of atmospheric dispersion is available within the literature (Theodore and Theodore 2021; Theodore 2008; Theodore and Dupont 2012).

In formulating the plan, some general assumptions may be made:

- Organizations do a good job when they have specific assignments.
- The various resources will need coordination.
- Most of the necessary resources are likely to be already available in the community (in plants or city departments).
- People react more rationally when they have been apprised of a situation.
- Coordination is a social process, not a legal one.
- Disorganization and reorganization are common in a large group.
- Flexibility and adaptability are basic requirements for a coordinated team.

The objective of this plan should be a procedure that uses the combined resources of the community in a way that will (EPA 2021a):

- Safeguard people during any health or environmental crisis.
- Minimize damage to property and the environment.
- Initially contain the incident and ultimately bring it under control.
- Effect the rescue and treatment of casualties.
- Provide authoritative information to the news media (for transmission to the public).
- Secure the safe rehabilitation of the affected area.
- Preserve relevant records and equipment for subsequent inquiry into causes and circumstances.

During the development of the plan, the assumptions employed and the overall objectives should be kept in mind. Although prevention is an important goal in accident and emergency management, it is not the objective of this plan. The plan should also focus on minimizing damage when problems do occur. Key components of the emergency action plan include the following (Stander and Theodore 2007):

- Emergency actions other than evacuation
- Escape procedures when necessary
- Escape routes clearly marked on a site map, and perhaps also on the roads
- A method of accounting for people after evacuation
- Description and assignment of rescue and medical duties
- A system for reporting health data to the proper regulatory agencies
- A means of notification of the public by an alarm system
- Responsibilities of contact and coordination person

9.5.2 The Planning Committee

Emergency planning should grow out of a team process coordinated by a leader. The team may be the best vehicle for including people representing various areas of expertise in the planning process, thus producing a more meaningful and complete plan. The team approach also encourages planning that will reflect a consensus of the entire community. Some individual communities and areas that included several communities had formed advisory councils before EPA and other agencies implemented their requirements. These councils can serve as an excellent resource for the planning team in charge of emergency planning activities, including accidents, natural disasters, and pandemics (Beranek et al. 1987).

By law, through EPCRA, the planning committee should include the following (EPA 2001):

- Elected and state officials
- Civil defense personnel

- First aid personnel
- Local environmental personnel
- Transportation personnel
- Owners and operators of facilities subject to the SARA
- Law enforcement personnel
- Firefighting personnel
- Public personnel
- Hospital personnel
- Broadcast and print media
- Community groups

Other individuals who could also serve the community well and should be a part of the committee include technical professionals, city planners, educators, academic and university researchers, and local volunteer help organizations (Schulze 1987).

EPCRA originally called for community groups, as designated by state governors, to have a plan by 1988. Specific requirements included the following (Stander and Theodore 2007; Theodore and Dupont 2012):

- Identification of all facilities as well as transportation routes for extremely hazardous substances (EHSs).
- Establishment of emergency response procedures, both on and off plant sites (facility owner and operator actions, as well as the actions of local emergency and medical personnel).
- Establishment of methods of determining when releases occur and what areas and populations may be affected.
- Listing of community and industry emergency equipment and facilities, along with the names of those responsible for the equipment and its upkeep.
- Description and schedule of a training program to teach methods for responding to chemical emergencies.
- Establishment of methods and schedules for exercises or drills to test emergency response plans.
- Designation of a community coordinator and a facility coordinator to implement the plan.
- Designation of facilities that are subject to added risk, and provision for their protection.

While many individuals have an interest in reducing risks, their differing economic, political, and social perspectives may cause them to favor different means of promoting safety. For example, people who live near an industrial facility that manufactures, uses, or emits public health-related materials are likely to be greatly concerned about avoiding threats to their lives. They are less likely to be concerned about the costs of developing accident prevention and response measures than some of the other team members. Others in the community, e.g., those representing industry or the budgeting group, are likely to be more sensitive to costs. They may be more anxious to avoid expenditures for unnecessarily elaborate prevention and response measures. Also, industrial facility managers, although concerned with reducing risks, may be reluctant, for proprietary reasons, to disclose materials and process information beyond what is required by law. These differences can be balanced by a well-coordinated team that is responsive to the needs of its community. (Flynn and Theodore 2002).

Agencies and organizations bearing emergency response responsibilities may have differing views about the role they should play in case of a problem. The public health agency, local fire department, and an emergency management agency are all likely to have some responsibilities during a problem. However, each of these organizations might envision a very different set of ideas for what the emergency response will look like and even have differing views on the gravity of a situation. A comprehensive plan will serve to detail the necessary actions of each response group.

In organizing the community to address the problems associated with emergency planning, it is important to bear in mind that all affected parties have a legitimate interest in the choices among planning alternatives. Therefore, strong efforts should be made to ensure that all such groups are included in the planning process. The need for unity of the committee during both the planning and the implementation processes increases for larger numbers of different community groups. Each group has a right to participate in the planning, and a well-structured, well-organized planning committee should serve the entire community.

The local government has a great share of the responsibility for emergency response within its community. The official who has the power to order an evacuation, fund fire and emergency units, and educate the public is key to emergency planning and the resulting response effort. For example, an entire plan might fail if a necessary evacuation were not ordered on time. Although politics should be disassociated from technical decisions, such linkage is inevitable in emergency planning. Unpleasant options that require political courage are often necessary. For example, the decision to evacuate a town immediately in the event of an industrial accident may be unpopular with the public, but could avoid the worst-case scenario of mass fatalities. Public officials must be able to act with proper authority, given as much available information as possible. An effective plan can save elected officials hours of media criticism after a public health crisis because the details of a response were organized properly.

9.6 EP&R FOR NATURAL DISASTERS

Unlike certain industrial accidents, natural disasters cannot always be prevented. However, detailed EP&R can mitigate the aftereffects of these disasters and reduce the resultant suffering and death. Natural disasters such as hurricanes, forest fires, and flooding occur frequently around the world and are following a disturbing upward trend as a result of climate change.

In addition, as mentioned at the beginning of this chapter, natural disasters of many kinds can lead to further crises, such as disease outbreaks, dehydration, and starvation, leading to further misery long after the initial emergency is over. Unsanitary conditions, crowding, and/or lack of shelter are frequent results of such calamities and are one of the major drivers of infectious outbreaks. Diseases such as measles and hepatitis A are common examples, although such outbreaks are dependent on the baseline vaccination rates of the affected population. For example, after a volcanic eruption in the Philippines in 1991, a measles epidemic broke out among displaced individuals with over 18,000 cases (Watson et al. 2007).

Vector-borne illnesses, including those carried by mosquitos, such as malaria and dengue fever, are also potential issues for displaced persons in warmer climates, primarily due to the standing water that can result after hurricanes, tsunamis, or flooding. This scenario transpired after a severe 2016 earthquake in Ecuador, where there was a surge in Zika virus cases in the immediate aftermath of the event (Watson et al. 2007; Ortiz et al. 2017).

In the United States, the Federal Emergency Management Agency (FEMA) oversees the coordination of national disaster relief. FEMA, under the direction of the Department of Homeland Security, has several separate programs focusing on different aspects of EP&R. These include (DHS 2018):

- Preparedness and protection
- Mitigation
- Response and recovery
- Regional operations
- Mission support

FEMA, as it is today, was created in 1979 by an executive order from President Carter, and was originally tasked with both disaster relief and national defense. Disaster relief by order of Congress dates back to 1803 following a devastating fire in Portsmouth, New Hampshire (FEMA 2021).

The COVID-19 pandemic response by FEMA and other government agencies has been authorized under the federal disaster declaration of 5 April 2020. This allowed FEMA to provide emergency funding to states and local governments and health departments. Funds can be used in a wide variety of ways, which include supporting hospitals, the purchase of necessary healthcare equipment, setting up testing centers, and even funeral assistance (FEMA 2022).

9.7 CURRENT AND FUTURE TRENDS

As evident in the lessons from the past regarding pandemics and other large-scale crises, it is essential to abide by stringent public health safety procedures. Regulations such as the EPCRA detailed in SARA Title III, help to ensure that safety practices are up to standard. However, these regulations should only provide a minimum standard. It should be up to hospitals, other medical facilities, corporations, and industry to see that every possible measure is taken to ensure the safety and well-being of the employees, local community, and environment in that area. It is also up to the community itself to prepare for any problems that might arise. However, both authors believe that the next infectious disease pandemic almost certainly will once again overwhelm the US medical system (Theodore and Theodore 2021).

The future promises to bring more attention to the topics discussed in the above paragraph. In addition, it appears that there will be more research in the area of risk assessment, including fault-free, event-free, and cause–consequence analysis. Extensive details on these topics are beyond the scope of this chapter, but some information is available in Chapter 11, and in the literature. (Shaefer and Theodore 2007; Theodore and Dupont 2012).

In conclusion, considering the many tragic consequences that resulted from the missteps during the COVID-19 pandemic, it is now clear that there are several weaknesses inherent in both state and federal health systems. The threat of another pandemic, whether from a new type of influenza or an other viral pathogen, highlights the need for increased global coordination. It will take strong determination and willingness to collaborate among countries to work fast, and efficiently, and to overcome disparities. International scientific cooperation only serves to strengthen global public health, and represents one of the most important contributions that could, ultimately, save many lives.

9.8 APPLICATIONS

Four illustrative examples complement the presentation for this chapter on emergency planning and response.

Illustrative Example 9.1 PREPAREDNESS PLANNING
Briefly describe at least four specific features of an emergency preparedness plan that would be put in place to respond to a major accidental release of a virus.

Solution

- Evacuation plans and routes for nearby dwellings, schools, offices, and industries
- Access routes and predicted response times for emergency teams
- Ability to predict the trajectory and concentration of the releases of airborne or water-borne
- Plans for mobilization of area medical personnel to treat casualties associated with major airborne accidental chemical releases

Illustrative Example 9.2 Emergency Response Program

Describe in general terms, how the Emergency Response Program (ERP) works for CPI.

Solution

When a release or spill occurs, the company responsible for the release, its response contractors, the local fire and police departments, and the local emergency response personnel provide the first line of defense. If needed, a variety of state agencies should stand ready to support, assist, or take over response operations if an incident is beyond local capabilities. In cases where a local government or Native American tribe conducts temporary emergency measures in response to a hazardous substance release, but does not have emergency response funds budgeted, EPA operates the Local Governments Reimbursement program that will reimburse local governments or Native American tribes up to $25,000 per incident.

If the amount of a hazardous substance release or oil spill exceeds the established reporting trigger, the organization responsible for the release or spill is required by law to notify the federal government's National Response Center (NRC). Once a report is made, the NRC immediately notifies a pre-designated EPA or U.S. Coast Guard On-Scene Coordinator (OSC) based on the location of the spill. The procedure for determining the lead agency is clearly defined so there is no confusion about who is in charge during a response. The OSC determines the status of the local response and monitors the situation to determine whether, or how much, federal involvement is necessary. It is the OSC's job to ensure that the cleanup, whether accomplished by industry, local, state, or federal officials, is appropriate, timely, and minimizes human and environmental damage.

The OSC may determine that the local action is sufficient and that no additional federal action is required. If the incident is large or complex, the federal OSC may remain on the scene to monitor the response and advise on the deployment of personnel and equipment. However, the federal OSC will take command of the response in the following situations:

- If the party responsible for the chemical release or oil spill is unknown or is not cooperative
- If the OSC determines that the spill or release is beyond the capacity of the company, local, or state responders to manage
- For oil spills, if the incident is determined to present a substantial threat to public health or welfare due to the size or character of the spill

The OSC may request additional support to respond to a release or spill, such as additional contractors, technical support from EPA's Environmental Response Team, or Scientific Support Coordinators from EPA or the National Oceanic and Atmospheric Administration. The OSC also may seek support from the Regional Response Team (RRT) to access special expertise or to provide additional logistical support. In addition, the National Response Team should stand ready to provide backup policy and logistical support to the OSC and the RRT during an incident.

The federal government will remain involved at the oil spill site following response actions to undertake several activities, including assessing damages, supporting restoration efforts, recovering response costs from the parties responsible for the spill, and, if necessary, enforcing the liability and penalty provisions of the Clean Water Act, as amended by the Oil Pollution Act of 1990.

Illustrative Example 9.3   Emergency Response Plan Audit
Provide some appropriate questions that could be posed during an audit of an emergency response plan as applies.

Solution
- What types of emergencies have occurred since the last audit?
- Are all potential emergency types covered by the plan?
- Who is responsible for maintaining the written plan?
- Who is authorized to activate the plan?
- When was the last revision?

Illustrative Example 9.4   Risk Calculations
On 2 January 1988, a fuel tank at an Ashland Oil terminal in Pennsylvania ruptured, and a 35 ft high wave of 600,000 gallons of number 2 distillate fuel oil surged out over a containment dike into the Monongahela River, creating massive health problems. In this case, the containment dike was breached by the violence of the release of oil that surged over the dike. Assume a slightly different case in which a similar tank (containing 3.9 million gallons of fuel) ruptures slowly and the entire contents of the tank are to be retained by a circular 5 ft high dike. If the radius of the dike is 192 ft, how far from the top of the dike will the level of the oil be?

Solution

The depth of the fuel, h, held in the dike is calculated based on the volume of the spill divided by the cross-sectional area of the dike:

$$h = \frac{V}{A} = \pi r^2$$
$$= (3.9 \times 10^6 \text{ gal})(0.1337 \text{ ft}^3/\text{gal})/[\pi(192 \text{ ft})^2]$$
$$= 4.5 \text{ ft}$$

The oil will therefore rise to a level that is approximately 0.5 ft below the top of the dike, which is five feet tall.

9.9 CHAPTER SUMMARY

- Toxic and chemically active substances present special concern because they can be dangerous when inhaled, ingested, or absorbed through the skin.
- Although pandemics cannot be completely prevented, careful planning and stringent safety procedures can significantly lower the potential risk that a pandemic will occur.
- Emergency planning is essential in preventing a potential public health problem and in foreseeing what possible incidents might occur.
- OSHA has guidelines and regulations for the safe operation of industrial plants and the handling of emergencies.
- Safety and health training for personnel is essential in preventing accidents; workers must know what they are dealing with and understand the consequences.
- Natural disasters such as hurricanes, forest fires, and flooding occur frequently around the world and are following a disturbing upward trend as a result of climate change.
- In the future, more stringent regulations, and hopefully better safety techniques, will help minimize both the fallout from pandemics and industrial accidents.

9.10 PROBLEMS

1   Provide at least five questions that should be raised in evaluating emergency planning and response drills.

2   Discuss the role of personal health resiliency as is applied to pandemics.

3 Develop an original problem concerning emergency preparedness and response.

4 Refer to Illustrative Example 4. The problem in this example stems from the fact that rain begins to fall just after the rupture of the tank. The heaviest daily rainfall recorded in this region of Pennsylvania over a recent 30-year period is 5.68 in. The fuel is immiscible with water and has a lower density. Assume the worst foreseeable case (record-breaking rainfall for a long period in the future) and calculate how long it will take for the oil level to reach the top of the dike.

REFERENCES

Beranek, W., Mccullough, J.P., Pine, S.H., and Soulen, R.L. (1987). ACS comment: getting involved in community right-to know. *Chemical & Engineering News Archive* 65 (43): 62. https://doi.org/10.1021/cenv065n043.p062.

Centers for Disease Control and Prevention (CDC) (2018). History of 1918 flu pandemic. Centers for Disease Control and Prevention. https://www.cdc.gov/flu/pandemic-resources/1918-commemoration/1918-pandemic-history.htm (accessed 12 December 2021).

Centers for Disease Control and Prevention (CDC) (2020a). Emergency preparedness and response. Centers for Disease Control and Prevention. https://emergency.cdc.gov/ (accessed 1 March 2022).

Centers for Disease Control and Prevention (CDC) (2020b). CDC emergency operations center activations. Centers for Disease Control and Prevention. https://emergency.cdc.gov/recentincidents/ (accessed 1 March 2022)

Department of Homeland Security (DHS) (2018). Department of Homeland Security Federal Emergency Management Agency (FEMA) budget overview, congressional justification, FY 2018. https://www.dhs.gov/sites/default/files/publications/FEMA%20FY18%20Budget.pdf (28 February 2022).

Diamond, S. (1985). U.S. toxic mishaps in chemicals put at 6,298 in 5 Years. The New York Times. https://www.nytimes.com/1985/10/03/us/us-toxic-mishaps-in-chemicals-put-at-6298-in-5-years.html (accessed 11 December 2021).

Environmental Protection Agency (EPA) (2001). *Hazardous Materials Emergency Planning Guide*. Washington, DC: The National Response Team (NRT).

Environmental Protection Agency (EPA) (2021a). EPA. https://www.epa.gov/ccl/types-drinking-water-contaminants (accessed 4 December 2021).

Environmental Protection Agency (EPA) (2021b). Emergency planning and community right-to-know act (EPCRA). EPA. https://www.epa.gov/epcra (accessed 7 December 2021).

FEMA.gov (2021). History of FEMA. https://www.fema.gov/about/history (accessed 28 February 2022).

FEMA.gov (2022). FEMA covid-19 funeral assistance state-by-state breakdown as of Feb. 7, 2022. https://www.fema.gov/press-release/20220207/fema-covid-19-funeral-assistance-state-state-breakdown (accessed 1 March 2022).

Flynn, A. and Theodore, L. (2002). *Accident and Emergency Management for the Chemical Process Industries*. Boca Raton, FL: CRC Press (originally published by Marcel Dekker).

Krikorian, M. (1982). *Disaster and Emergency Planning*. Loganville, AL: Institute Press.

Michael, E., Bell, O., and Wilson, J. (1986). *Emergency Planning Considerations for Specialty Chemical Plants*. Boston, MA: Stone and Webster Engineering Corporation.

Occupational Safety and Health Administration (OSHA) (2021). US Department of Labor issues emergency temporary standard to protect workers from coronavirus. US Department of Labor. https://www.osha.gov/news/newsreleases/national/11042021 (accessed 6 December 2021).

Ortiz, R.M., Le, N.K., Sharma, V. et al. (2017). Post-earthquake Zika virus surge: disaster and public health threat amid climatic conduciveness. *Scientific Reports* 7: 15408. https://doi.org/10.1038/s41598-017-15706-w.

Public Health Emergency (PHE) (2019 Federal emergency preparedness. https://www.phe.gov/Preparedness/support/emergencypreparedness/Pages/default.aspx (accessed 7 December 2021).

Ready.gov (2021). Incident management. https://www.ready.gov/incident-management (accessed 1 March 2022).

Schulze, R. (1987). *Superfund Amendments and Reauthorization Act of 1986 (SARA Title III)*. Richardson, TX: Trinity Consultants Incorporated.

Shaefer, S. and Theodore, L. (2007). *Probability and Statistics Applications in Environmental Science*. Boca Raton, FL: CRC Press/Taylor & Francis Group.

Shrivastava, P. (1987). *Bhopal: Anatomy of A Crisis*. Cambridge, MA: Ballinger.

Stander, L. and Theodore, L. (2007). *Environmental Regulatory Calculations Handbook*. Hoboken, NJ: John Wiley & Sons.

Theodore, L. (2008). *Air Pollution Control Equipment Calculations*. Hoboken, NJ: John Wiley & Sons.

Theodore, L. and Dupont, R. (2012). *Environmental Health and Hazard Risk Assessment: Principles and Calculations*. Boca Raton, FL: CRC Press/Taylor & Francis Group.

Theodore, M.K. and Theodore, L. (2021). *Introduction to Environmental Management*. 2nd ed. Boca Raton, FL: CRC Press/Taylor & Francis Group.

Watson, J.T., Gayer, M., and Connolly, M.A. (2007). Epidemics after natural disasters. *Emerging Infectious Diseases* 13 (1): 1–5. https://doi.org/10.3201/eid1301.060779.

World Health Organization (WHO) (2022). Strategy and planning. https://www.who.int/emergencies/diseases/novel-coronavirus-2019/strategies-and-plans (accessed 1 March 2022).

10

Ethical Considerations within Virology
Contributing Author: Paul DiGaetano, Jr.

CHAPTER MENU

When it comes to ethical considerations within healthcare, and particularly, in the field of virology, many areas must be considered. Ethics generally involves considering how one ought to live, act, or think. When it comes to decision-making, there is the inherent question of what is "the right thing to do." According to the World Health Organization (WHO), in *Guidance for Managing Ethical Issues in Infectious Disease Outbreaks*, the process of ethical analysis starts with identifying relevant principles and then applying them to a particular situation. Judgments then arise when one must weigh competing principles and make decisions or compromise on which to follow, particularly in difficult circumstances when it is not possible to satisfy every side of an issue (WHO 2016).

Ethics plays a unique role in virology. This chapter focuses on ethical considerations within the fields of applied sciences and medicine, and more narrowly, ethics related to medical research concerning pandemics and other healthcare crises. For a broader overview of scientific ethical considerations, the interested reader may choose to review the classic 1998 work by Wilcox and Theodore, which contains over 100 case studies relating to ethical situations. (Wilcox and Theodore 1998).

A Guide to Virology for Engineers and Applied Scientists: Epidemiology, Emergency Management, and Optimization, First Edition. Megan M. Reynolds and Louis Theodore.
© 2023 John Wiley & Sons, Inc. Published 2023 by John Wiley & Sons, Inc.

10.1 CORE ETHICS PRINCIPLES

To begin any ethical discussion, it is important to understand the main principles or pillars of ethics, and for this chapter, how they specifically tie into aspects related to viral diseases. The first of the principles is *justice* which, in itself, deals with both equity and procedural justice. *Equity* refers to fairness in the distribution of available resources, opportunities, and outcomes, as well as the equal and appropriate treatment of all people involved (WHO 2016). It is also important under this principle to avoid discrimination and ensure that vulnerable populations and those who might not have equal access to protections or resources are not only considered, but accommodated, in decision-making. On the other hand, *procedural justice* deals with ensuring every person have equal access to both resources and information, that accurate information is shared freely and openly, and that it can be easily understood by the general public once it is disseminated.

The next principle, *beneficence,* addresses the obligation of scientific professionals – and more specifically, medical personnel – to always act in the best interests of the people they are trying to serve and help. While this principle most commonly applies in terms of medical ethics and decision-making, it should be a consideration in the public health sector to ensure that the basic needs of individuals and communities are sufficiently met and fulfilled, particularly in the face of a public health crisis. There is a duty to ensure that the public has equal and sufficient access to basic needs such as nourishment, safety, and shelter and that social support structures are in place, particularly amidst the chaos that often arises during a pandemic or an epidemic. Lastly, the principle of beneficence must always be balanced and weighed against the principle of *nonmaleficence* – the concept of doing no harm. While the debate of this principle is, once again, most often found in the medical ethics sector, it can be an important ethical consideration in the public health arena when assessing the benefits versus risks of a particular treatment or course of action, or even a vaccine as seen in the case of COVID-19. This will be discussed later in this chapter.

The last core principle of ethics that requires discussion is *autonomy,* which recognizes individuals as capable decisionmakers and respects their inherent rights regardless of the choices they make for themselves. In the public health arena, autonomy suggests that an individual can choose to accept or refuse any medical treatment or course of action, such as medications, vaccines, or even medical advice. However, it must be noted that this principle requires that the patient not only has access to the pertinent information needed to make the decision but also to have an understanding of the matters at hand and the consequences of whatever action they choose. To that end, the agencies that govern healthcare and safety guidelines during a public health crisis must effectively communicate with the public, to explain the utility of their decisions, and why a particular course of action is or is not in the best interest of individuals or the population as a whole. It is also critical that the public understands on a basic level that the decisions that are being made are based on scientific evidence or at least a preponderance of scientific knowledge and reasoning that allows decisionmakers to reach conclusions regarding any potential benefits or harms that could occur. Finally, it must be noted that while individuals have the autonomy to make independent decisions regarding their health, individuals' actions also have an impact on the community as a whole. In public health, the "common good" dictates that autonomy does not give an individual the right to cause harm to others. Thus, during a public health crisis, individual autonomy has limits.

It is paramount that the public can to place trust in both decisionmakers and the processes they rely on during an ongoing health crisis. Whether it be governmental agencies, or elected officials at the state or federal levels, or local response workers, *transparency* and the equal and swift dissemination of accurate information is key to building this trust. It is also important that there be periodic reviews once new information is learned or obtained, and an openness to making necessary adjustments, particularly for communities with different needs or outcomes. Any agency or elected official must consider all the principles discussed above in decision-making, and they must make every effort to account for a comprehensive review of all available information. It is important during a global health crisis to share such information on a large scale, whether globally or locally, to not only learn swiftly but also tailor approaches for more vulnerable or at-risk communities (WHO 2016).

10.2 IMPORTANT TENETS OF ETHICAL RESEARCH

In the early stages of a viral outbreak, usually little is known about many aspects of the virus, as seen in recent outbreaks of novel strains of Ebola and Coronavirus. For example, important factors such as virulence, transmissibility, vulnerable populations, response to potential treatments, etc. are often only discovered as the virus makes its way through a population. Further, as has been apparent in the most recent COVID-19 pandemic, viruses often mutate, rapidly changing the disease course, and therefore challenging the healthcare infrastructure and framework put in place. Thus, one can see the urgency of ongoing clinical research regarding viruses, particularly in the realm of outbreaks. However, important ethical considerations emerge during the time of an ongoing health crisis, especially given that the outbreak itself may be the only period when researchers have adequate numbers of people infected to allow for effective studies and trials of potential treatments and vaccines (Edwards and Kochhar 2020).

10.2.1 CONDUCTING RESEARCH DURING A HEALTH CRISIS

Scientists and researchers alike often struggle with the obligation to gather information without delay to begin trials for therapeutics, to inform the public, and to help mitigate the spread of the virus and its effects on the population. But there are definite benefits to moving as quickly as possible. In gleaning as much as possible about a new virus and innovative ideas for prevention and treatment, scientists can prevent (or at least prepare for) future outbreaks, as was the case with the novel creation of COVID-19 vaccines using already-established theories (WHO 2016).

Any research opportunity, whether conducted during a time of outbreak or not, must do its best to make certain that the perceived benefits always outweigh the risks, and that the risks are somewhat reasonable and will not cause any additional harm to the subjects enrolled (Theodore and Dupont 2012). It is also important that any participants are enrolled in the study of their own volition and fully understand any pertinent information regarding the study and its purpose and potential benefits. A study should always intend to add some value to society, even if it is just to increase knowledge in a given area. Local research, whenever possible, is a valuable asset that can help ensure that studies can effectively represent the needs and situations of a given population and that the studies can be implemented in a manner that does not affect or jeopardize any local emergency response (WHO 2016).

10.2.2 SCIENTIFIC COOPERATION DURING A HEALTH CRISIS

Certain areas may lack access to resources or the capacity to conduct much-needed research, particularly during an outbreak when available resources (and researchers with necessary expertise) are already stretched, when findings are time-sensitive, and when quick action is critical. It is vital to ensure cooperation and collaboration among all those conducting research to share findings and foster widespread implementation, particularly in hard-hit or vulnerable communities. Here again, local researchers are a critical link in the sharing of information, so that locales facing similar conditions or vulnerabilities can take advantage of existing knowledge and act quickly to respond.

This combined and coordinated effort can help to clarify social, economic, and health disparities within a population, while also encouraging the dissemination of fact-based information, and providing equitable access to pertinent healthcare guidelines or scientific discoveries that could impact health outcomes. Shared information adds to the establishment of a broader set of overarching priorities that are in line with disease response efforts overall and safeguards against unnecessary duplication of research efforts or wasted resources (WHO 2019).

10.2.3 Fair and Ethical Study Design and Implementation

Of course, no scientific or medical research should ever expose study participants to unnecessary or undue risk. The study design and research methodology must be carefully and rigorously developed. Furthermore, in the case of a quickly emerging epidemic, it is critically important that research does not usurp vital resources such as personnel, equipment, and healthcare facilities if they are essential to the local response to whatever health crisis the locality is experiencing (WHO 2019). Further details on study design are available in Part I, Chapter 6.

Overall, thoughtful considerations of the above factors are crucial to establishing and maintaining the public's trust in both the scientific community and the local and global communities where medical research is taking place. Distrust in a study's design or methodology could lead to inadequate recruitment or other obstacles to successful completion. Additionally, doubts about the integrity of the research during or after a study could result in an unwillingness to follow its recommendations and could even lead authorities to reject potential treatments or interventions, even if they were proven effective by the results of the study (WHO 2016). As such, researchers should conduct studies as transparently as possible to establish trust with participants and with the public and to ensure that demonstrated societal need from study design through completion of the study and recommendations or interventions that are being suggested.

10.3 ETHICAL DILEMMAS IN PUBLIC HEALTH

As mentioned, during a public health crisis, time and data are paramount in the race to protect people and save lives. *Public health surveillance* is necessary to gain as much information regarding the spread of disease, and this introduces ethical questions as will be described below. Similarly, *nonpharmaceutical interventions (NPI)* and quarantine requirements can cause ethical issues to be raised. Both dilemmas are discussed in the next two subsections.

10.3.1 Public Health Surveillance

Whether or not public health surveillance activities are characterized for research purposes, it is still important to evaluate any measures taken and to ensure ethical concerns are considered. This ethical analysis is multi-pronged, ranging from protecting and safeguarding personal information to ensuring that any resulting interventions and public policies are enacted equitably.

Related to public health research and the findings obtained as a result of it, public health surveillance also presents ethical considerations. According to the CDC, public health surveillance is defined as "the ongoing, systematic collection, analysis, and interpretation of health-related data essential to planning, implementation, and evaluation of public health practice" (CDC 2018). The information acquired by public health surveillance can be a powerful tool for evaluating current practices, and for influencing public health action, including the implementation of policies and programs in response to a health crisis. The CDC further states that the uses of public health surveillance include (CDC 2018):

- Identifying patients and their contacts for treatment and intervention
- Detecting epidemics, health problems, and changes in behavior
- Estimating the magnitude and scope of health problems
- Measuring trends and characterizing disease
- Assessing the effectiveness of programs and control measures implemented
- Aiding research by developing hypotheses that will then be tested and proven either right or wrong

Within the umbrella term of public health surveillance, the CDC enumerates multiple types: *Passive, Active, Sentinel*, and *Syndromic*, as described below (CDC 2018):

- *Passive* surveillance refers to healthcare providers reporting cases of disease to appropriate authorities.
- *Active* surveillance involves a particular health agency requesting case information from healthcare providers.
- *Sentinel* surveillance describes situations in which health professionals designated to represent and respond for a particular geographic area or a specific reporting group are reporting the health event.
- *Syndromic* surveillance focuses on the reporting of a particular symptom or symptoms instead of focusing on a particular diagnosable disease.

A potential issue with regard to public health surveillance is *jurisdiction*, and whether the populations involved are effectively represented and protected by the governing authorities. It is important to note that any laws or mandates regarding reporting of disease cases are determined at the state level with healthcare personnel with hospitals and similar institutions required to report to their local health department. It is the local health department that is then tasked with delving into each case further as necessary and then reporting it to the state health department. Within the United States, health decisions are the responsibility of the individual states. The CDC, meanwhile, only gets involved when their help is requested by a state, or if the disease or condition spreads beyond state boundaries and has larger implications. Furthermore, since public health surveillance usually will have to occur over many jurisdictions, it could also call into question the reliability and quality of the data since different areas can have varying levels of resources available (CDC 2018).

As stated above, personal and identifying information of affected individuals collected during a disease outbreak must be protected, since any dissemination of that data could pose a substantial risk to the individuals and their safety. It is important to ensure that whatever jurisdiction is charged with the public health surveillance has significant measures in place to protect and uphold patient confidentiality, and access to the information is limited only to authorized and appropriate personnel. In many instances, it is the job of research ethics committees to safeguard and oversee this process (CDC 2018).

Individual participation in public health surveillance presents another ethical consideration. In most cases, according to the WHO, public health surveillance is conducted on a mandatory basis, and an individual does not have the option to refuse public health surveillance measures. This practice may be justified with the argument that public health surveillance is in the interest of the common good by a given government authority, and that participation can be compelled to achieve the desired public health outcome. Thus, the governing agency responsible for the surveillance also has an ethical obligation to weigh public risks against potential benefits, evaluating any potential harms from not allowing individuals to "opt-out" of surveillance measures. Here again, transparency is critical, so that authorities effectively communicate what information will be gathered, how this information will be used, and with whom the information may be shared.

As stated above, it is important during any disease outbreak or epidemic for information to be shared accurately and promptly. Data gathered from public health surveillance is often key to inform mitigation efforts and to predict how a disease might spread. Observing commonalities among the collected information such as symptoms, treatments, and any responses to treatments or interventions – can be crucial in disease response and can ultimately save lives (CDC 2018).

10.3.2 ETHICAL EVALUATION OF NONPHARMACEUTICAL INTERVENTIONS

As stated above, one of the main goals of public health surveillance and research and clinical trials is to evaluate interventions to slow the spread of disease during an outbreak. These may include nonpharmaceutical interventions (NPI), measures that can mitigate the effects of a disease on individual patients and the community as a whole. These interventions often precede the availability of medical treatments, antiviral therapies, or vaccines, and may continue to be used once those treatments are accessible to the public (WHO 2019).

Examples of NPI include many that have been implemented during the COVID-19 pandemic, such as (WHO 2022):

- Quarantining infected or exposed individuals
- Social distancing measures to keep individuals physically separated from one another to slow the spread of disease

- Mask requirements to prevent the spread of the virus indoors or in close outdoor spaces
- Limiting the number of people in a given physical space at one time, such as large public events (both indoors and outdoors), workplace and school closures, and restrictions placed on movement and transportation.

The central issue with such types of intervention is how to maximize the common good while also attempting to maintain the rights and health of the individual. In essence, when managing public health, it is important to ensure that each person has the autonomy and resources to act in their best interests, while also working to protect the health and safety of everyone in the community, including its most vulnerable members (Stern and Markel 2008).

Authorities must remain transparent about why suggested mandates are necessary and how they will be implemented, while also demonstrating a thoughtful consideration of how implementation might affect individuals in various circumstances. Public trust in the governing authority can improve community acceptance and adoption of NPI, and thus strengthens efforts to mitigate the effects of the disease early (Stern and Markel 2008).

Historically, NPI have been implemented often during viral disease outbreaks, either as voluntary efforts adopted by a community of affected people, or as a set of rules mandated by a governing authority to mitigate the spread of disease. Before the COVID-19 pandemic, a notable example would be the 1918 influenza epidemic. This epidemic was one of the deadliest infectious disease outbreaks globally, with approximately 40 million deaths worldwide. Thus, it provides a useful model to assess the efficacy of mandatory and voluntary NPI. In general, it may seem logical that NPI would achieve the best results and meet the least resistance when they are voluntary (Stern and Markel 2008).

Indeed, as historical record and the Hastings Center Briefing further shows, there have been multiple instances in prior epidemics where people willingly volunteered to follow quarantining and isolation measures, citing "civic duty" as their main reason for doing so. For example, studies of the 2003 SARS epidemic showed that the majority of people voluntarily consented to home and work quarantine to protect the health of others. Further, a 2006 Harvard School of Public Health survey showed that the majority of people would implement NPI and voluntarily make major changes to their lives for up to one month following an influenza epidemic (Stern and Markel 2008). However, while public health surveillance and NPI can empower individuals to take such actions voluntarily, challenges can arise, and although some may feel that their autonomy is being disregarded, it may sometimes be necessary for authorities to mandate such policies. There is an inherent risk in authorizing mandatory restrictions over long periods, as public sentiment could turn against such mandates. Resultant political unrest not only risks public safety but also could in turn further spread the contagion (Stern and Markel 2008).

According to the 2017 guidelines on community outbreak mitigation, the CDC recommends that implementation of mandatory NPI should only apply in situations where the case fatality ratio for a particular infectious disease outbreak *exceeds 1%*. The establishment of such standards helps the public be more amenable to NPI and thus could result in overall better compliance when crises due occur, leading to better health outcomes. It is crucial that all NPI put in place allow people to act in whichever way they deem best for themselves as long as it does not cause harm to somebody else. Any mandatory NPI implemented should conform to international standards established for ethics requiring that (CDC 2020)

> … coercive public health measures be legitimate, legal, necessary, non-discriminatory and represent the least restrictive means appropriate to the reasonable achievement of public health goals.
>
> (Stern and Markel 2008)

Further, the implementation of NPI must not be in any way discriminatory. The right to due process must be protected in that NPI should be reviewed by accredited and reputable agencies, or institutions and individuals should be able to challenge mandatory NPI. It is imperative that NPI do not go beyond their means and are not found to be excessive infringements on individual rights as compared to the proposed benefit or level of response that is deemed necessary to have the desired effect which is described by the ethical principle of proportionality (Stern and Markel 2008; CDC 2020).

10.3.3 Ethical Consideration Involving Restrictions of Movement

As stated above, one example of NPI is the broad category of restrictions of movement which includes isolation and quarantine, travel advisories, travel restrictions, school or workplace closures, and limitations on gatherings. Because of their potential for significant social disruption, it is imperative that decisions restricting movement be rooted in the best scientific knowledge available and would lead to a measurable decrease in the transmission of disease. As new evidence regarding transmissibility emerges, restrictions and related guidelines should be reviewed and updated. Authorities should aim to impose the *least restrictive measures* that will achieve the desired health outcomes to minimize the negative consequences of limiting social contact and keeping people away from work, school, and regular daily life.

Any restrictions put in place should be distributed in an equitable manner such that every person, regardless of means and status, should be required to abide by the same laws and restrictions, including the officials who make these decisions. It is not unheard of during pandemics for government leaders, such as mayors, governors, or even heads of state fail to follow their own rules, as was the case with this recent pandemic. This situation erodes public trust and can make lockdowns much more difficult to enforce. Furthermore, agencies have an ethical obligation to make sure "individuals should not be subject to greater or lesser restrictions for reasons unrelated to the risks they may pose to others" and that the restrictions do not disproportionately impose any additional burden on already vulnerable populations (WHO 2019).

When requiring isolation or quarantine of infected or exposed individuals, questions arise about the availability of important resources. If patients are required to stay at a healthcare facility, the available capacity and supplies must be considered so that staffing and resources are not overwhelmed. Inadequate healthcare resources would not only be detrimental to the overall response and containment of disease, but it could also prevent people from receiving quality healthcare for other conditions unrelated to the outbreak.

Requiring people to isolate or stay at home for long periods can have other significant implications for individuals and their livelihoods. Authorities imposing the NPI should take a comprehensive approach to evaluating how restrictions of movement will affect the community, including workers and employers, schools, transit systems, and local businesses. Additional policies should be put into place to mitigate negative consequences. For instance, systems and infrastructure must be in place to ensure that individuals have adequate access to basic needs, such as food and safe water, medical care, and adequate shelter and facilities. It is also important to consider whether imposed confinement puts an undue burden on individuals in closely shared spaces or, conversely, further isolates people who are alone in confinement. Planning for these likelihoods can help safeguard impacted individuals' health and dignity, and also helps to lessen the psychological burdens of the epidemic, with mental health being an important factor (WHO 2019).

As was evident during the COVID-19 pandemic, restrictions can have adverse effects on financial situations both for the individual and the economy. If companies must shut down or people are not allowed to go to work, this can lead to both employees and their companies struggling to survive. Also, uncertain times, and the inability of companies to run efficiently, can bring on supply chain issues, further raising prices and leading to inflation; this effect then trickles down onto the individual consumer, many of whom are already adversely affected financially.

10.4 ETHICAL CONSIDERATIONS REGARDING MEDICAL INTERVENTIONS

To begin the discussion of any medical intervention it is important to note that, regardless of whether or not the purpose of the intervention is for diagnosis, treatment, or prevention, the intervention should be applied to the public following the highest of medical standards and protocols with patient well-being remaining a paramount concern. All individuals should be given the same level of care and be made aware of the risks and benefits of the medical intervention as well as alternatives that may prove effective. In addition, it is the ethical duty of the researchers or the medical professionals to ensure that individuals are able to understand those risks and benefits. If it is determined that a person does not have sufficient mental capacity, ethically and legally speaking, the decision shall fall on a designated health care proxy, who is charged with representing what the proxy believes to be the wishes of the patient.

According to the 2016 WHO guidelines for managing ethical issues during a disease outbreak, there are multiple factors to consider in a decision to override a patient's refusal. First, there must be a compelling reason to believe that a patient's refusal would amount to a greater risk to overall public health and that the intervention would substantially lessen that risk to others (WHO 2016).

The question of whether vaccine mandates are ethical can be complicated when it comes to the issue of vaccinating children. Children may not be fully capable of understanding all the necessary information about the vaccine and might not be able to fully assess benefits and risks, so it is the ethical responsibility of the healthcare professional to balance both the autonomy of the parent to decide what is best for the children and the benefits to public health (Hendrix et al. 2016). This conflict became even more evident in recent years with the reemergence of almost-eradicated, vaccine-preventable diseases and the issues regarding school vaccine mandates (CDC 2020).

Healthcare personnel are on the frontlines of treating ill and contagious patients during a pandemic including the treatment of already-vulnerable populations, so it could be argued that it is justifiable to require them to be vaccinated, overriding refusals for the benefit of public health (Gur-Arie et al. 2021). Requiring them to be vaccinated will help limit the spread of disease, especially during an outbreak when healthcare facilities are inundated with infected patients. In cases where healthcare personnel might refuse, it is the ethical obligation of the institution to ensure that there is open dialogue and that other safety precautions and protocols are in place and are above medical standards. These measures may help reduce resistance by increasing trust among healthcare workers in their particular institutions and increasing voluntary vaccination (Gur-Arie et al. 2021). As stated above, it is crucial for any discussion of ethics that any interventions or processes be reviewed and that people can challenge the fairness of any intervention suggested to or imposed on them.

10.4.1 Emergency Use Of Medical Interventions

In most cases, vaccines and treatments are developed and given FDA approval after rigorous research and clinical trials, typically with years of data available. However, during a public health emergency, such a long timeframe may not be feasible, and an Emergency Use Authorization (EUA) may be provided, as was the case for vaccines developed to prevent COVID-19. According to the FDA, an EUA allows for medical countermeasures like vaccines to become readily available for use, even if the products are not fully approved or if that particular use for a medical product has not yet been approved. During an emergent health crisis, the use of such interventions allows patients to be treated to help prevent serious disease or disease complications. The FDA will only issue an EUA for a vaccine when there is sufficient information regarding quality and consistency, and when the benefits generally are greater than the potential risks. With these protocols, a vaccine that is granted FDA approval or has been issued an EUA is generally considered extremely safe and effective and thus can be used to bolster the argument of vaccine mandates.

Some viral pathogens do not have any proven effective treatment or require interventions that have shown efficacy in a laboratory setting but have not been fully assessed for efficacy or safety in humans (WHO 2016). During an emergent health crisis, it might not be possible to properly and fully conduct these much-needed human clinical trials. Thus, the World Health Organization (WHO) 2016 guidelines on managing infectious disease during an outbreak cite certain instances when it would be ethically acceptable to give these experimental interventions on an emergency basis. For the use of these experimental treatments, guidelines specify that (WHO 2016):

- no alternative treatment exists that has proven effective
- there must not be sufficient time to begin and complete clinical trials
- some data is available that demonstrates or at least suggests that the intervention will prove safe and effective at achieving the desired health outcome
- sufficient resources are available to the patient to ensure that any and all risks are significantly minimized
- the agency administering the treatment must obtain informed consent of the patient and ensure that the patient will not only be monitored but that the data will be shared safely and quickly with the greater medical community to hopefully benefit others
- ensure that only pertinent information is shared and that it is only shared with other officials when deemed appropriate.

Overall, whether those are issues arising from research or from the implementation of interventions, there are a lot of factors to consider when trying to make ethical distinctions and decisions. These ethical queries are especially critical during an emergent health crisis when a swift coordinated response is essential.

10.5 APPLICATIONS

Four illustrative examples complement the presentation for this chapter on ethical considerations, particularly during pandemics.

Illustrative Example 10.1 CORE ETHICS PRINCIPLES
How can the core ethical principles of justice and beneficence apply to the HIV pandemic?

Solution
Justice deals with both equity and procedural justice. Throughout the HIV pandemic, equity has been an issue with the, sometimes unsuccessful, attempts to fairly distribute antiretroviral drugs nationally and globally. Wealthy countries are able to provide these lifesaving medicines to their citizens, while low- and middle- income countries struggle to acquire enough supply. This is particularly challenging in many countries in Africa with high rates of infection. Beneficence, the ethical principle of doing "good" and acting for the benefit of the patients was evident, particularly at the beginning of the epidemic when HIV and its mode of transmission was poorly understood. While HIV patients were shunned by many, dedicated healthcare workers were willing to risk their lives, *placing the importance of treating patients above their own health and safety. It was later discovered that casual contact did not put people at risk of contracting the disease. Healthcare workers, however, faced real risk of contagion due to the risk of working with contaminated needles.*

Illustrative Example 10.2 PUBLIC HEALTH SURVEILLANCE
Discuss the legal basis for the use of local health departments' public health surveillance.

Solution
Local health departments are legally authorized to act on behalf of public health within their jurisdiction. They are also responsible for health surveillance and for reporting to the individual state health department. This authority is granted by the constitution in two separate clauses, the *general welfare* clause and *the interstate commerce* clause.

Illustrative Example 10.3 Nonpharmaceutical Interventions
Discuss nonpharmaceutical interventions and how they were implemented during the 1918 influenza pandemic.

Solution
Historically, nonpharmaceutical interventions have been used to control outbreaks and pandemics. Public measures such as mask requirements, isolation and quarantine, and travel restrictions are typical of any pandemic response and were all used during the 1918 influenza response.

Illustrative Example 10.4 Emergency Use Authorization
Under what circumstances are Emergency Use Authorizations utilized?

Solution
During a public health emergency, an Emergency Use Authorization (EUA) allows for the use of medical countermeasures such as vaccines for use prior to full FDA approval. Such interventions help prevent serious disease or disease complications. Importantly, the FDA will only issue an EUA for use when there is sufficient safety and efficacy data, and when the benefits outweigh any potential risks. EUAs were used in 2020 to allow for several COVID-19 vaccines to be used as soon as possible. EUAs were also used during the 2014–2016 West African Ebola outbreak, which accelerated ongoing clinical trials for an Ebola virus vaccine – later approved as Ervebo.

10.6 CHAPTER SUMMARY

- Ethics generally involves considering how one ought to live, act, or think. When it comes to decision-making, there is the inherent question of what is "the right thing to do."
- To begin any ethical discussion, it is important to understand the main principles or pillars of ethics and how they specifically tie into aspects related to viral diseases.
- It is paramount that the public be able to place trust in both decisionmakers and the processes they rely on during an ongoing health crisis.
- Scientists and researchers alike often struggle with the obligation to gather information without delay to begin trials for therapeutics, to inform the public, and to help mitigate the spread of the virus and its effects on the population.
- One of the main goals of public health surveillance and of research and clinical trials is to evaluate interventions to slow the spread of disease, particularly during an outbreak.

- Because of their potential for significant social disruption, it is imperative that decisions restricting movement be rooted in the best scientific knowledge available and would lead to a measurable decrease in the transmission of disease.
- During a public health crisis, an Emergency Use Authorization (EUA) may be provided, as was the case for vaccines developed to prevent COVID-19.

10.7 PROBLEMS

1 What are the core principles of ethics?

2 What are some uses of public health surveillance?

3 Describe the four different types of health surveillance.

4 What are some examples of the use of nonpharmaceutical interventions during the COVID-19 pandemic?

REFERENCES

Centers for Disease Control and Prevention (CDC) (2018). Introduction to public health surveillance|public health 101 series. Centers for Disease Control and Prevention. https://www.cdc.gov/training/publichealth101/surveillance.html (accessed 8 March 2022).

Centers for Disease Control and Prevention (CDC) (2020). Community mitigation guidelines to prevent pandemic influenza – United States, 2017. Centers for Disease Control and Prevention. https://www.cdc.gov/mmwr/volumes/66/rr/rr6601a1.htm (accessed 11 March 2022).

Edwards, K.M. and Kochhar, S. (2020). Ethics of conducting clinical research in an outbreak setting. *Annual Review of Virology* 7 (1): 475–494. https://doi.org/10.1146/annurev-virology-013120-013123.

Gur-Arie, R., Jamrozik, E., and Kingori, P. (2021). No jab, No job? Ethical issues in mandatory COVID-19 vaccination of healthcare personnel. *BMJ Global Health* 6 (2): e004877. https://doi.org/10.1136/bmjgh-2020-004877.

Hendrix, K.S., Sturm, L.A., Zimet, G.D., and Meslin, E.M. (2016). Ethics and Childhood Vaccination Policy in the United States. *The American Journal of Public Health* 106 (2): 273–278. https://doi.org/10.2105/AJPH.2015.302952.

Stern, A.M. and Markel, H. (2008). Influenza pandemic. In: *From Birth to Death and Bench to Clinic: The Hastings Center Bioethics Briefing Book for Journalists, Policymakers, and Campaigns* (ed. M. Crowley), 89–92. Garrison, NY: The Hastings Center.

Theodore, L. and Dupont, R. (2012). *Environmental Health and Hazard Risk Assessment: Principles and Calculations*. Boca Raton, FL: CRC Press/Taylor & Francis Group.

Wilcox, J.R. and Theodore, L. (1998). *Engineering an Environmental Ethics: A Case Study Approach*. Hoboken, NJ: John Wiley & Sons.

World Health Organization (WHO) (2016). Guidance for managing ethical issues in infectious disease outbreaks. World Health Organization. https://www.who.int/publications/i/item/9789241549837 (accessed 1 March 2022).

World Health Organization (WHO) (2019). Non-pharmaceutical public health measures for mitigating the risk and impact of epidemic and pandemic influenza. World Health Organization. https://www.who.int/publications/i/item/non-pharmaceutical-public-health-measuresfor-mitigating-the-risk-and-impact-of-epidemic-and-pandemic-influenza (accessed 11 March 2022).

World Health Organization (WHO) (2022). Ethics and covid-19. World Health Organization. https://www.who.int/teams/health-ethics-governance/diseases/covid-19 (accessed 11 March 2022).

11

Health and Hazard Risk Assessment

People face all kinds of risks every day, some voluntarily and others not only involuntarily, but often unidentified. Therefore, risk plays a very important role in today's world. Studies determining the many causes of cancer, for example, caused a turning point in the world of risk because it opened the eyes of risk engineers, scientists, and health professionals to the world of risk assessment.

Both health risk assessment (HRA) and hazard risk assessment (HZRA) are ultimately concerned with characterizing risk. (Theodore and Dupont 2012).

Unfortunately, the word *risk* has come to mean different things to different people. Merriam-Webster defines risk as "… the possibility of loss or injury … something that creates … a hazard" (Merriam-Webster 2015). Stander and Theodore have defined it as "a combination of uncertainty and change" (Stander and Theodore 2008).

In relation to health and the environment, the U.S. Environmental Protection Agency (EPA) considers risk "to be the chance of harmful effects to human health

A Guide to Virology for Engineers and Applied Scientists: Epidemiology, Emergency Management, and Optimization, First Edition. Megan M. Reynolds and Louis Theodore.
© 2023 John Wiley & Sons, Inc. Published 2023 by John Wiley & Sons, Inc.

or to ecological systems resulting from exposure to an environmental stressor" (EPA 2021). EPA uses risk assessments to characterize the type and scale of health risks to humans and ecosystems from certain *stressors* that may be present. A stressor is defined as a physical, chemical, or biological entity that can cause harm (EPA 2021). Environmental risk assessments typically fall into one of two areas: human health and ecological. This chapter will focus mainly on risks to human health.

Risk is closely related to probability. For example, probability of unity, that is, 1.0, indicates that a health/hazard problem will occur, alternately, a probability of zero, i.e., 0.0, indicates that there is no risk and that a health/hazard problem definitely will not occur (Paustenbach 1989); Theodore and Dupont 2012.) Irrespective of the category of the applicable risks for a system, total risk (R_T) is given by the summation of the risk from all n events/scenarios, as illustrated in Eq. 11.1. In addition, the magnitude of each risk can be noted relative to the total risk.

$$R_T = \sum_{i=1}^{n} R_i \qquad (11.1)$$

To compound confusion about the precise meaning of risk, there are also two types of risk that environmental professionals are concerned with: *health risk* and *hazard risk*. However, these two classes of risk have been used interchangeably by practitioners, researchers, and regulators. Because of this confusion, one of the main objectives of this chapter is to both define and clarify the differences between these two risks (Stander and Theodore 2008). Details on these two categories of risk are provided below.

Regarding health risks, concern arises because contaminants can elude natural defense mechanisms upon entering the human body. Exposure to toxins or microorganisms such as viruses can lead to various pathways of entry into the human body. These include inhalation, skin absorption (absorption), and ingestion (digestion system). It is fair to say that the dominant route of human exposure to hazardous substances is via inhalation. This includes microbial contaminants such as viruses, bacteria, and mold. Note, also, that two types of potential exposure exist relative to the concentration and duration of exposure. These are categorized as acute or chronic, as described below:

- *Acute:* Exposures that occur for relatively short timeperiods, generally minutes to one to two days, at high exposure concentrations. In addition to inhalation, high concentrations of airborne substances might directly contact the skin, or liquids and sludges may be splashed onto the skin or into the eyes, leading to acutely toxic effects.
- *Chronic:* Continuous exposure occurring over long periods, generally months to years, at relatively low contaminant concentrations. Direct skin contact by

immersion, splash, or contaminated air at chronic exposure levels involves contact with substances exhibiting low dermal activity.

Alternatively, *hazard risk*, which is classified in the *acute* category, can be described as any of the following (Burke et al. 2000):

- A ratio of hazards (e.g., an explosion) to failures of safeguards.
- An acute exposure to a pathogen
- A triple combination of event, probability, and consequences.
- A measure of economic loss or human injury in terms of both an incident likelihood and the magnitude of the loss or injury.

Health risk and its assessments are addressed, in part, under the Clean Air Act (CAA) Section 112(d) and (f), where "EPA must promulgate (along with methods of calculating) residual risk standards for the source category as necessary to provide an ample margin of safety to protect public health." In the case of hazard risk, the CAA's Section 112(r), the "Risk Management Program" (RMP) rule (40 CFR Part 68), is designed to reduce the risk of accidental releases of acutely toxic, flammable, and explosive substances (Stander and Theodore 2008).

11.1 INTRODUCTION TO RISK ASSESSMENT

Risk-based decision-making and risk-based corrective action (RBCA) are decision-making processes for assessing and responding to a chemical release. These processes take into account the potential effects of chemical exposure on human health and the environment, which vary greatly in terms of complexity, physical and chemical characteristics, and in the level of risk that they may pose. RBCA was initially introduced by the American Society for Testing and Materials (ASTM) to assess petroleum releases, but the process may be tailored for use with any chemical release. For example, in the 1980s, to satisfy the need to start corrective action programs quickly, many regulatory agencies decided to uniformly apply, at underground storage tank (UST) cleanup sites, regulatory cleanup standards developed for other purposes. It became increasingly apparent that applying such standards without considering the extent of actual or potential human and environmental exposure was an incomplete means of providing adequate protection against the risks associated with UST releases. EPA now believes that risk-based corrective-action processes are tools that can facilitate efforts to clean up sites expeditiously, as necessary, while still ensuring the protection of human health and the environment (Stander and Theodore 2008).

EPA and several state environmental agencies have developed similar decision-making tools. EPA refers to the process as the aforementioned "risk-based decision-making." While the ASTM RBCA standard deals exclusively with human health risks, EPA advises that, in some cases, ecological goals must also be considered in establishing cleanup goals (Theodore and Theodore 2021).

Health risk is the probability that persons will suffer adverse health consequences as a result of exposure to a substance. The amount of risk is determined by a combination of the concentration the individual is exposed to, the rate of intake, or dose of the substance, and the toxicity of the substance. HRA is the procedure used to attempt to quantify or estimate this risk. Risk-based decision-making also distinguishes between the terms *point of exposure* and *point of compliance*. The point of exposure is the point at which the environment or the individual comes into contact with the problem. An individual may be exposed by methods such as inhalation of vapors, as well as physical contact with the substance. The point of compliance is a point between the point of release of the chemical or organism (i.e., the source) and the point of exposure. The point of compliance is selected to provide a safety buffer for affected individuals and/or environments (Theodore and Morris 1998).

Since 1970, the field of HRA has received widespread attention within both scientific and regulatory communities. It has also attracted the attention of the public. Properly conducted risk assessments have received fairly broad acceptance, in part, because they put into perspective the terms toxic, hazard, and risk. Toxicity is an inherent property of all substances. It states that all chemical and physical agents can produce adverse health effects at some dose or under specific exposure conditions. In contrast, exposure to a chemical that can produce a particular type of adverse effect represents a hazard. Risk, however, is the probability or likelihood that an adverse outcome will occur in a person or a group that is exposed to a particular concentration or dose of the hazardous agent. Therefore, risk is generally a function of exposure and/or dose. Consequently, HRA is defined as the process or procedure used to estimate the likelihood that humans or ecological systems will be adversely affected by a chemical or physical agent under a specific set of conditions (Theodore 2008).

A HZRA is a systematic examination to identify things, situations, processes, accidents, etc., that may cause harm, particularly to people. After identification of the risk, the next steps should be to analyze and evaluate the likeliness and severity of the risk. After that determination is made, controls and measures must be put in place to effectively eliminate or reduce the harm from happening. Risk assessments are very important as they form an integral part of an occupational health and safety management plan. Risk assessments are intended to CCOHS (2021):

- Create awareness of hazards and risk.
- Identify who may be at risk (e.g. employees, cleaners, visitors, contractors, the public, etc.).
- Determine whether a control program is required for a particular hazard.
- Determine if existing control measures are adequate or if more should be done.
- Prevent injuries or illnesses, especially when done at the design or planning stage.
- Prioritize hazards and control measures.
- Meet legal requirements where applicable.

Understanding risk communication dynamics is also essential to successful risk communication efforts. Two-way communication with stakeholders, including regulatory agencies, residents, and employees, prevents costly rework and permit delays, and provides information useful for prioritizing risk management efforts. As communities have become more attentive to, and concerned about health and environmental issues, the role of the environmental manager has expanded to include communications with key audiences. Interest and concern are certain to expand in the future. In addition to addressing the technical aspects of environmental and health risk, efforts to address process, health, and lifestyle concerns have become more critical to the success of environmental projects and risk management (Burke et al. 2000; EPA 1990).

11.2 THE HEALTH RISK ASSESSMENT PROCESS

HRA provides an orderly, explicit, and consistent way to deal with scientific issues in evaluating whether a health problem exists and what the magnitude of the problem may be. This evaluation typically involves large uncertainties because the available scientific data are limited and the mechanisms for adverse health impacts or environmental damage are only imperfectly understood.

Most human or environmental health problems can be evaluated by breaking down the analysis into four parts: *problem identification, dose–response assessment, exposure assessment,* and *risk characterization* (See Figure 11.1). This four-step framework has been widely adopted for U.S. federal and state agencies plus international organizations that assess and manage health and environmental issues.

Regarding health risk identification, a problem may be defined generally as a specific toxic or infectious agent, or a set of health conditions that have been detected at a certain level and that have the potential to cause adverse effects to human health or the environment. For some perceived problems, the risk

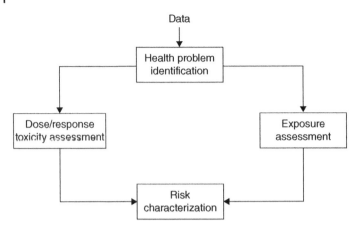

Figure 11.1 Health risk assessment flowchart. Source: Compliments of Abhishek Garg, adapted from Flynn and Theodore (2002).

assessment might stop with the first step in the process, that is, problem identification, if no adverse effect is identified or if an agency elects to take regulatory action without further analysis (Paustenbach 1989).

Problem identification involves an evaluation of various forms of information to identify different health concerns. Atmospheric air, indoor air, water, land, hazardous substances, and toxic pollutants are some of the environmental media or agents that could be identified in this step in the HRA process. *Dose–response or toxicity assessment* is the next step required in an overall HRA. Responses/effects can be different since contaminants vary in their capacity to cause adverse effects. This step frequently requires that assumptions be made to relate experimental data for animals to humans. *Exposure assessment* is the determination of the magnitude, frequency, duration, and routes of exposure of human populations and ecosystems to hazardous agents. Finally, in *risk characterization*, the aforementioned toxicology and exposure data/information are combined to provide information on effects and obtain a qualitative or quantitative expression of the health risk posed by the identified problem in response to the assumed exposure. HRA thus involves the integration of information and analyses associated with these four steps to provide a complete characterization of the nature and magnitude of risk and the degree of confidence associated with this characterization. A critical component of the assessment is the full elucidation of the uncertainties associated with each of the major steps. (See also Section 11.5 Hazard Risk versus Health Risk.) All of the essential problems of toxicology are encompassed under this broad concept of risk assessment since it takes into account available dose–response data.

It should treat uncertainty not by the application of arbitrary safety factors but by stating them in quantitatively and/or qualitatively explicit terms so that they are not hidden from decision-makers. Additional details are provided in the next section.

11.3 DOSE–RESPONSE ASSESSMENT

Dose–response assessment – perhaps the most important step in the analysis presented above in Section 11.2 – is the process of characterizing the relationship between the quantity of toxin administered or received and the incidence of an adverse health effect in exposed populations and estimating the incidence of the effect as a function of exposure to the agent. This process considers such important factors as the intensity of exposure, the age pattern of exposure, and possibly other variables that might affect response, such as sex, lifestyle, and other modifying factors. A dose–response assessment usually requires extrapolation from high to low doses and extrapolation from animals to humans, or one laboratory animal species to a wildlife species. A dose–response assessment should describe and justify the methods of extrapolation used to predict incidence, and it should characterize the statistical and biological uncertainties in these methods. When possible, the uncertainties should be described numerically rather than qualitatively.

Toxicologists, for instance, tend to focus their attention primarily on extrapolations from cancer bioassays. However, there is also a need to evaluate the risks of lower doses to see how they affect the various organs and systems in the body. Many scientific papers focused on the use of a safety factor or uncertainty factor approach since all adverse effects other than cancer and mutation-based developmental effects are believed to have a threshold – a dose below which no adverse effect should occur. Several researchers have discussed various approaches to setting acceptable daily intakes or exposure limits for developmental and reproductive toxicants. It is thought that an acceptable limit of exposure could be determined using cancer models, but today they are considered inappropriate because of thresholds (Paustenbach 1989).

For a variety of reasons, it is difficult to precisely evaluate toxic responses caused by acute exposures to hazardous substances. Reasons for this include:

- Humans experience a wide range of acute adverse health effects, including irritation, narcosis, asphyxiation, sensitization, blindness, organ system damage, and death. In addition, the severity of many of these effects varies with the intensity and duration of exposure.
- There is a high degree of variation in response among individuals in a typical population.

- For the overwhelming majority of substances encountered in industry, there are not enough data on the toxic responses of humans to permit an accurate or precise assessment of the substance's hazard potential.
- Many releases involve multiple components. There are presently no rules on how these types of releases should be evaluated.
- There are no toxicology testing protocols that exist for studying episodic releases on animals. In general, this has been a neglected area of toxicology research.

There are many useful measures available to use as benchmarks for predicting the likelihood that a release event will result in serious injury or death. Several references review various toxic effects and discuss the use of various established toxicological criteria (Clayson et al. 1985; Foa 1987).

Dangers are not necessarily defined by the presence of a particular chemical, but rather by the amount of that substance one is exposed to, also known as the dose. A dose is usually expressed in milligrams of chemical received per kilogram of body weight per day. For toxic substances other than carcinogens, a threshold dose must be exceeded before a health effect will occur, and for many substances, there is a dosage below which there is no harm. A health effect will occur or at least be detected at the threshold. For carcinogens, it is assumed that there is *no* threshold, and, therefore, any substance that produces cancer is assumed to produce cancer at any concentration. It is vital to establish the link to cancer and to determine if that risk is acceptable. Analyses of cancer risks are much naturally more complex than non-cancer risks (Clayson et al. 1985; Foa 1987).

Not all contaminants or chemicals are created equal in their capacity to cause adverse effects. Toxicity data are derived largely from animal experiments in which the animals (primarily mice and rats) are exposed to increasingly higher concentrations or doses. Responses or effects can vary widely from no observable effect to temporary and reversible effects, to permanent injury to organs, to chronic functional impairment to ultimately, death.

Another health effect is involved when two different chemicals enter the body simultaneously; the result can be intensified or compounded. A *synergistic* effect results when one substance intensifies the damage done by the other. Synergism complicates almost any exposure due to a lack of toxicological information. For just one chemical, it may typically take a toxicological research facility approximately two years of studies to generate valid data. The data produced in that two-year time frame applies only to the effect of that one chemical acting alone. With the addition of another chemical, the original chemical may have a different effect on the body. This fact results in a great many unknowns when dealing with toxic substances, and therefore-increases risks due to a lack of dependable information (Theodore and Dupont 2012).

11.4 THE HAZARD RISK ASSESSMENT PROCESS

As indicated earlier, many practitioners, researchers, and regulators have confused health risk with hazard risk, and vice versa. Although both employ a four-step method of analysis, the procedures are quite different, with each providing different results, information, and conclusions. Both, however, do share a common concern in that they can negatively impact individuals, society, and the environment.

As with health risks, there is a serious lack of information on hazards and the associated implications of these hazards. The unknowns in this risk area may be larger in number and greater in potential consequences than in the health risk area. It is the authors' judgment that hazard risks have unfortunately received something less than the attention they deserve. However, hazard risk analysis (HZRA) details are available, and the traditional approaches successfully applied in the past can be found in this section.

Much has been written about Michael Crichton's powerful earlier science fiction-thriller novel titled *Prey*. The book was not only a bestseller but also the movie rights were sold for $5 million. In it, Crichton provides a frightening scenario in which swarms of nanorobots, equipped with special power generators and unique software, prey on living creatures. To compound the problem, the robots continue to reproduce without any known constraints. This scenario is an example of an accident and represents only one of a nearly infinite number of potential hazards that can arise in the real world. Although the probability of the horror scene portrayed by Crichton, as well as other similar hypothetical scenarios, is extremely low, steps and procedures need to be put into place to reduce, control, and hopefully eliminate low probability hazardous events from actually happening. This section attempts to provide some of this information (Theodore and Dupont 2012).

The Introduction section defined both *chronic* and *acute* problems. When the two terms are applied to emissions, the former usually refers to ordinary, round-the-clock, low-level, everyday emissions, while the latter term deals with short-duration, out-of-the-norm, high-level, accidental emissions. Thus, acute problems normally refer to accidents and/or hazards. The Crichton scenario discussed is an example of an acute problem, and one whose solution should be addressed and treated by a HZRA, rather than the health risk approach provided in Section 11.2.

There are several steps in evaluating the risk of an accident. These are detailed in Figure 11.2, below, and involve the following (Flynn and Theodore 2002):

- Any existing or potential hazard in a system must be identified (Reynolds et al. 2004; Theodore 2006; Theodore and Theodore 2021).

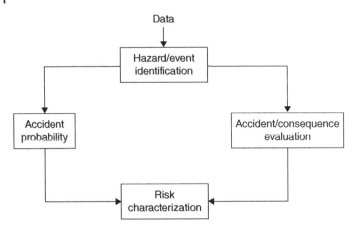

Figure 11.2 Hazard risk assessment flowchart. Source: Compliments of Abhishek Garg, adapted from Flynn and Theodore (2002).

- The probability that the accident will occur also has to be determined.
- The severity of the consequences of the accident must be determined.
- The probability of the accident and the severity of its consequences are combined to characterize the hazard risk.

The reader should note that the analysis and approach to describing potential health problems presented in Section 11.2 – HRA apply to health risks associated with chemicals. There are other classes of environmental health risks that do not pertain to chemicals. For example, health problems can arise immediately soon after a natural hazard, such as a tornado, hurricane, or earthquake, that leave local inhabitants without potable water for extended periods. This class of environmental health risk can be assessed and determined by replacing the steps of toxicity – exposure assessment pictured in Figure 11.1 – with the steps of the probability of occurrence – consequence(s) shown in Figure 11.2. The calculational approach then becomes similar to that provided in Figure 11.2 for HZRA problems.

11.5 HAZARD RISK VERSUS HEALTH RISK

The differences between HRA and HZRA can qualitatively be explained in terms of the words health and hazard. As noted earlier, health problems are generally chronic and related to the general state of well-being. Hazard problems are generally acute and relate to accidents that can either be immediately harmful to life or immediately produce unhealthy results, or both.

The difference between HRA and HZRA is also demonstrated in Figures 11.1 and 11.2. Although both assessment processes involve a four-step procedure, the middle steps, that is, Steps 2 and 3, differ. Consider Step 2. For an HRA, this step is concerned with toxicology and ultimately requires information of a dose–response nature for the health problem in question. For an HZRA, this step is concerned with probability and requires information on the probability of the hazard problem occurring. Also consider Step 3. For an HRA, this step is concerned with exposure and ultimately demands information on a receptor's level and duration of exposure for the health problem in question. For an HZRA, Step 3 involves determining the consequence of the hazard problem being evaluated.

Combining the data/information/calculations generated in Steps 2 and 3 provides risk results for either the HRA or HZRA. The HRA risk is concerned with a health problem, while the HZRA risk is concerned with a hazard problem. The differences can also be illustrated via the simple examples as follows.

11.5.1 HEALTH RISK ASSESSMENT (HRA) EXAMPLE

Suppose the following information is provided for an HRA:

- *The health problem identification:* Ethylene oxide, EtO
- *Toxicology:* Concentration in excess of a given value produces premature death
- *Exposure:* Number of individuals exposed to adverse health (premature death in this example) concentrations
- *Risk:* Number of exposed individuals who will experience adverse health effects (premature death in this example)

If 10 people are exposed to the adverse health effect concentration of EtO from a population of one million, i.e., 10^6, the risk to the population is 10×10^{-6} or 10^{-5}. If this information and results are based on what can occur on an annual basis, then the risk is also based on an annual basis.

11.5.2 HAZARD RISK ASSESSMENT (HRZA) EXAMPLE

Suppose the following is provided for an HZRA:

Hazard problem (event) identification: A biological attack on NYC by terrorists

Probability (of occurrence): One in a million, that is, 10^{-6}

Consequences (of attack): 1 million people will die, that is, 10^6

Assuming the probability is provided on an *annual* basis, the annual risk from this hazard is then calculated to be $(10^{-6})(10^6) = 10^0 = 1$ death. This indicates that the risk of death to the 10×10^6 New York metropolitan area residents due to a biological attack by terrorists is $1.0/10 \times 10^6$ or 10^{-7} or 0.1×10^{-6} on an *annual* basis.

One may therefore conclude that both the HRA and HZRA provide risk information and numbers, but they are related/associated with three different risk problem categories.

11.6 COVID-19 PANDEMIC HAZARD RISK

In response to the COVID-19 pandemic, the Occupational Safety and Health Administration (OSHA) installed Emergency Temporary Standards (EMTS). These aimed to reduce workers' risk of occupational exposure to SARS-CoV-2 during the pandemic. Details of pandemic regulation may vary from state to state, and even community to community, depending on local conditions or outbreaks. Further information is covered in Chapter 8. OSHA requires employers to provide a workplace safe from recognized hazards that are likely to cause physical harm or death (29 U.S.C. §654(a)(1)). To meet this obligation, employers need to assess occupational hazards to which their workers may be exposed. Some OSHA standards, such as those for personal protective equipment (PPE) (29 CFR 1910.132) and respiratory protection (29 CFR 1910.134), include requirements that will help protect workers from exposure to SARS-CoV-2 (OSHA 2021).

Exposure risk depends in part on the type of work, working environment, individual health, infection rates at the local level, and adherence to Centers for Disease Control and Prevention (CDC) guidelines regarding social distancing, and vaccination by the employees and around them (OSHA 2021). The exposure limit for SARS-CoV-2 is considered to be within 6 ft for a total of 15 minutes or more over a 24-hour period. OSHA has a stratified risk of potential exposure on the job into four levels. These categories are provided in Table 11.1.

Finally, it is important to note that the CDC states that, "While airflows within a particular space may help spread disease among people in that space, there is

Table 11.1 OSHA risk levels.

Risk level	Job description examples
Lower exposure risk (caution)	Jobs requiring minimum contact with others, such as remote workers
Medium exposure risk	Workers in outdoor industrial or manufacturing positions with contact with the public in outdoors or well-ventilated areas, such as retail or grocery stores
High exposure risk	Workers in outdoor industrial and manufacturing or contact with the public in poorly ventilated areas, or healthcare workers treating known COVID-19 patients; industrial or manufacturing workers in poorly ventilated areas
Very high exposure risk	Healthcare workers performing aerosol-generating procedures (e.g. intubation, dental procedures, etc.)

no definitive evidence to date that viable virus has been transmitted through an HVAC system to result in disease transmission to people in other spaces served by the same system" (CDC 2021).

11.7 THE UNCERTAINTY FACTOR

Qualitatively, *uncertainty* may be viewed as having two components: variability and lack of knowledge. Uncertainty, whether applied to toxicological values, probability, consequences. risks, and so on, may be described qualitatively or quantitatively. Qualitatively, descriptions include large, huge, monstrous, tiny, very small, and so forth. Quantitative terms describing the uncertainty associated with a value x are normally in the form $x \pm u$, where u provides a measure of the uncertainty (e.g., standard deviation, 95% confidence limit, etc., for a mean of replicate measurements) (Theodore and Theodore 2021; Shaefer and Theodore 2007).

A substantial amount of information on uncertainty and uncertainty analysis is available. Useful references abound, but in general, there are three main sources of uncertainty that have been earmarked by practicing engineers and applied scientists:

- Model uncertainty
- Data uncertainty
- Problem formulation uncertainty

Model uncertainty reflects the weaknesses, deficiencies, and inadequacies present in any model and may be viewed as a measure of the displacement of the model from reality. *Data uncertainty* results from incomplete data measurement, estimation, inference or supposed expert opinion. *Problem formulation uncertainty* arises because the practitioner often cannot identify every health problem or hazard incident that a given population might be exposed to. Naturally, the risk engineer's objective is to be certain that the major contributors to the risk are identified, addressed, and quantified. Uncertainty here arises from not knowing the individual risk contributions from those risk problems that have not been identified or that have explicitly been omitted; one therefore may not be able to accurately predict the overall or combined risk.

Byrd and Cothern have expanded this three-part uncertainty categorization in the following manner (Byrd and Cothern 2000):

- Subjective judgment
- Linguistic imprecision
- Statistical variation

- Sampling error
- Inherent randomness
- Mathematical modeling
- Causality
- Lack of data or information
- Problem formulation

As noted previously, uncertainty analysis is used to estimate the effect of data and model uncertainties on the resulting risk estimate. Sensitivity analysis estimates the effect on calculated outcomes of varying inputs to the models individually or in combination. Importance analysis quantifies and ranks risk estimate contributions from subsystems or components of the complete system, e.g., individual incidents, groups of incidents, sections of a process, etc. (Theodore and Dupont 2012).

To summarize, different assumptions can change any quantitative risk characterization by several orders of magnitude. The uncertainty that arises is related to how well (and often, consistently) input data are obtained, generated, or measured, and the degree to which judgment is involved in developing risk scenarios and selecting input data. Simply put, uncertainty arises from how data/evidence was both measured and interpreted. Despite these limitations and uncertainties, risk characterizations provide the practitioner with assessment capabilities that are critical to improving process safety and that contribute to risk reduction goals (Theodore and Morris 1998).

11.8 APPLICATIONS

Four illustrative examples complement the presentation for this chapter on HRA and HZRA.

Illustrative Example 11.1 PROBABILITY IN RELATION TO RISK
Does probability play a role in the characterization of risk for both health problems and hazards?

Solution
An integral part of describing risk is probability. Therefore, it plays a role in both health risk assessment and hazard risk assessment. The role probability plays in hazard risk assessment is more apparent, as explicitly seen in Figure 11.2.

Illustrative Example 11.2 HEALTH RISK ASSUMPTIONS
Certain assumptions are usually made about an "average" person's attributes when applying health risk assessments to large groups of individuals. Describe these values.

Solution

The values normally employed for humans are

- Average body weight is 70 kg for an adult and 10 kg for a child.
- The average daily drinking water intake is 2 l for an adult and 1 l for a child.
- The average amount of air breathed per day is 20 m^3 for an adult and 10 m^3 for a child.
- The average expected life span is 70 years.
- The average dermal contact area is 1000 cm^2 for an adult and 300 cm^2 for a child.

Illustrative Example 11.3 HEALTH RISK VERSUS HAZARD RISK
Is exposure to dioxins a health problem or a hazard problem?

Solution

It depends. If it is a continuous emission/exposure over an extended period at a low but chronically toxic concentration, it is a health problem. If there is a catastrophic release/exposure for a short duration but at a very high (acutely toxic) concentration, it is considered a hazard problem and should be assessed using hazard risk analysis techniques.

Illustrative Example 11.4 HEALTH RISK ASSESSMENT
New York City health officials have hired you to conduct a future health risk assessment that is concerned with predicting a possible future viral pandemic. The probability of a deadly influenza pandemic occurring is 10^{-2}/year (once every 100 years).

An earlier study conducted by Farber and Associates indicated that:

1. 1000 of the 10 000 exposed individuals located within 1 mile of Times Square (TS) will die.

2. 100 of the 100 000 exposed individuals located between 1 and 5 miles of TS will die.

3. 10 of the 1 000 000 exposed individuals located between 5 and 20 miles of TS will die.

Calculate the health risk associated with this potential pandemic.

Solution

The result is provided in Table 11.2. Note that the local risk near the heart of the city is unacceptably high, i.e., 1×10^{-3}, while the local risk at a distance significantly displaced from the city is low, i.e., 1×10^{-7}. The overall risk, OR, to the total population exposed is given by 9.99×10^{-6}. This is excessive if the standard for protection of the general public from health risks is 1 in a million, i.e., 1×10^{-6}, particularly if the endpoint for predicted effects is death.

Table 11.2 Illustrative Example 11.4 risk calculation.

Location (miles)	Number of exposed individuals	Number of exposed individuals that will die	Annual risk of deaths, deaths × probability of pandemic	Local risk of annual deaths	Overall annual risk
<1	10 000	1000	$(1000)(0.01) = 10$	10/10 000	10/1 111 000
1–5	100 000	100	1.0	1.0/100 000	1.0/1 110 000
5–20	1 000 000	10	0.1	0.1/1 000 000	0.1/111 000
Total	1 110 000	1110	11.1		

11.9 CHAPTER SUMMARY

- Health risk assessment is defined as the process or procedure used to estimate the likelihood that humans or ecological systems will be adversely affected by a chemical or physical agent under a specific set of conditions.
- The health risk evaluation process consists of four steps: health problem identification, dose–response assessment, exposure assessment, and risk characterization.
- In dose–response assessment, effects are evaluated, and these effects vary widely because their capacities to cause adverse effects differ.
- Exposure assessment is the determination of the magnitude, frequency, duration, and routes of exposure to human populations and ecosystems.
- In risk characterization, the toxicology and exposure data are combined to obtain a quantitative or qualitative expression of risk.
- The hazard risk assessment process consists of hazard identification, accident probability, accident or consequence evaluation, and risk characterization.
- The Occupational Safety and Health Administration (OSHA) installed Emergency Temporary Standards (EMTS) to reduce workers' risk of occupational exposure to SARS-CoV-2 during the pandemic.

11.10 PROBLEMS

1 Is an explosion at a gas station a health problem or a hazard problem?

2 Describe the role event trees play in hazard risk analysis.

3 Briefly describe radon and the risks associated with it.

4 Briefly describe a process checklist.

REFERENCES

Burke, G., Singh, B., and Theodore, L. (2000). *Handbook of Environmental Management and Technology*. 2nd. ed. Hoboken, NJ: John Wiley & Sons.

Byrd, D. and Cothern, C. (2000). *Introduction to Risk Analysis*. Rockville, MD: Government Industries.

Canadian Centre for Occupational Health and Safety (CCOHS) (2021). Risk assessment: OSH answers. Government of Canada. https://www.ccohs.ca/oshanswers/hsprograms/risk_assessment.html (accessed 15 November 2021).

Centers for Disease Control and Prevention (2021). *Ventilation in buildings. Centers for Disease Control and Prevention*. https://www.cdc.gov/coronavirus/2019-ncov/community/ventilation.html (accessed 21 November 2021).

Clayson, D.B., Krewski, D., and Munro, I. (1985). *Toxicological Risk Assessment*. Boca Raton, FL: CRC Press, Inc.

Environmental Protection Agency (EPA) (2021). About risk assessment. EPA. https://www.epa.gov/risk/about-risk-assessment#whatisrisk (accessed 16 November 2021).

Environmental Protection Agency (EPA) (1990). *Risk Assessment for Toxic Air Pollutants: A Citizen's Guide. EPA-450/3-90-024*. Washington, DC: Office of Air and Radiation, EPA.

Flynn, A.M. and Theodore, L. (2002). *Health, Safety, and Accident Management in the Chemical Process Industries*. Boca Raton, FL: CRC Press (Originally published by Marcel Dekker.).

Foa, V. (1987). *Occupational and Environmental Chemical Hazards*. Chichester, England: Ellis Horwood Limited.

Merriam-Webster (2015). *Risk Definition & Meaning*. Merriam-Webster.

Occupational Safety and Health Administration (OSHA) (2021). COVID-19 – hazard recognition. U.S. Department of Labor. https://www.osha.gov/coronavirus/hazards (accessed 16 November 2021).

Paustenbach, D. (1989). *The Risk Assessment of Environmental and Human Health Hazards: A Textbook of Case Studies*. Hoboken, NJ: John Wiley & Sons.

Reynolds, J., Jeris, J., and Theodore, L. (2004). *Handbook of Chemical and Environmental Engineering Calculations*. Hoboken, NJ: Wiley-Interscience.

Shaefer, S. and Theodore, L. (2007). *Probability and Statistics Applications for Environmental Science*. Boca Raton, FL: CRC Press/Taylor & Francis Group.

Stander, L. and Theodore, L. (2008). *Environmental Regulatory Calculations Handbook*. Hoboken, NJ: John Wiley & Sons.

Theodore, L. (2006). *Nanotechnology: Basic Calculations for Engineers and Scientists*. Hoboken, NJ: John Wiley & Sons.

Theodore, L. (2008). *Air Pollution Control Equipment Calculations*. Hoboken, NJ: John Wiley & Sons.

Theodore, L. and Dupont, R. (2012). *Environmental Health and Hazard Risk Assessment: Principles and Calculations.* Boca Raton, FL: CRC Press/Taylor & Francis Group.

Theodore, L. and Morris, K. (1998). *Health, Safety and Accident Prevention: Industrial Applications, A Theodore Tutorial.* East Williston, NY: Theodore Tutorials (originally published by US EPA/APTI, RTP, NC).

Theodore, M.K. and Theodore, L. (2021). *Introduction to Environmental Management, 2nd edition.* Boca Raton, FL: CRC Press/Taylor & Francis Group.

Part III

Engineering Considerations

Engineering can be defined as the science concerned with utilizing scientific knowledge and explanations to apply to practical use. The chapters in this final part contain material that one might view as a prerequisite for the detailed engineering calculations that are encountered in practice.

There are six chapters in part III. The chapter numbers and accompanying titles are listed below.

Chapter 12: Introduction to Mathematical Methods
Chapter 13: Probability and Statistical Principles
Chapter 14: Linear Regression
Chapter 15: Ventilation Calculations
Chapter 16: Pandemic Health Data Modeling
Chapter 17: Optimization Procedures

A Guide to Virology for Engineers and Applied Scientists: Epidemiology, Emergency Management, and Optimization, First Edition. Megan M. Reynolds and Louis Theodore.
© 2023 John Wiley & Sons, Inc. Published 2023 by John Wiley & Sons, Inc.

12

Introduction to Mathematical Methods
Contributing Author: Julian Theodore

CHAPTER MENU

Mathematics is generally defined as the relationships among quantities, magnitudes, and properties and of the logical operations by which unknown quantities, magnitudes, and properties may be deduced. Engineers and applied scientists are known for their problem-solving ability. In healthcare, proper interpretation of clinical trial and public health study results is critical in order to identify patterns within massive amounts of data. In problem-solving, considerable importance is attached to the proper analysis of a problem, often requiring the application of mathematical methods. This chapter will serve as an introduction to Part III: Engineering Considerations. While much of the topic is theoretical and not directly applicable to healthcare and public health scenarios, it is an important base from which to begin any discussion of the more difficult topics detailed in the chapters to follow.

The authors have assumed that the reader has some familiarity with topics that can include differentiation, integration, simultaneous linear algebraic equations, nonlinear equations, and the solution of both ordinary differential equations

A Guide to Virology for Engineers and Applied Scientists: Epidemiology, Emergency Management, and Optimization, First Edition. Megan M. Reynolds and Louis Theodore.
© 2023 John Wiley & Sons, Inc. Published 2023 by John Wiley & Sons, Inc.

and partial differential equations. Numerical solution approaches are primarily applied in the material to follow although analytical methods are included, particularly in the applications section.

Many problems usually require an answer to an equation (or equations), and the answer(s) may be approximate or exact. Obviously, an exact answer is preferred, but because of the complexity of some equations, exact solutions may not be attainable. Furthermore, a precise answer may not be necessary. For this instance, one may resort to the aforementioned method that has come to be defined as a *numerical* method. Unlike the exact solution, which is continuous and in closed form, numerical methods provide an inexact (but often reasonably accurate) solution. These numerical methods often involve a stepwise procedure that ultimately leads to an answer and a solution to a particular problem. The method usually requires a large number of calculations and is therefore ideally suited for digital computation.

Finally, it should be noted that numerical methods were taught in the past as a means of providing engineers and applied scientists with ways to solve complicated mathematical expressions that they could not otherwise solve. However, with the advent of computers, these solutions can now be readily obtained. A overview of numerical methods is included to provide the practicing engineer with some insight into what many of the currently used software packages (Excel MathCad, Mathematica, MatLab, etc.) are doing.

12.1 DIFFERENTIATION

Several differentiation methods are available for generating expressions of a derivative. Refer to Table 12.1. Consider the problem of determining the benzene concentration–time gradient dC/dt at $t = 4.0$ seconds (data point 4).

Method 1: This method involves the selection of any three data points and calculating the slope m of the two extreme points. This slope is approximately equal to the slope at the point lying in the middle. The value obtained is the "equivalent" of the derivative at that point 4. Using data points from 3 to 5, one can obtain:

$$\frac{dC}{dt} = m = \frac{C_5 - C_3}{t_5 - t_3} = \frac{1.63 - 2.70}{5.0 - 3.0} = -0.535 \tag{12.1}$$

Method 2: This method involves determining the average of two slopes. Using the same points chosen above, two slopes can be calculated, one for points 3 and 4 and the other for points 4 and 5. Adding the two results and dividing them by 2 will provide an approximation of the derivative at point 4. For the points used in this method, the results are:

$$m_1 = \text{slope}_1 = \frac{C_4 - C_3}{t_4 - t_3} = \frac{2.01 - 2.70}{4.0 - 3.0} = -0.69 \tag{12.2}$$

Table 12.1 Concentration–time data.

Data point	Time, seconds	Concentration of benzene, mg/l
0	0.0	7.46
1	1.0	5.41
2	2.0	3.80
3	3.0	2.70
4	4.0	2.01
5	5.0	1.63
6	6.0	1.34
7	7.0	1.17

Source: L. Perez, homework assignment submitted to
L. Theodore, Manhattan College, Bronx, NY, 2003.

$$m_2 = \text{slope}_2 = \frac{C_5 - C_4}{t_5 - t_4} = \frac{1.63 - 2.01}{5.0 - 4.0} = -0.38 \tag{12.3}$$

$$m_{\text{avg}} = \text{slope}_{\text{avg}} = \frac{m_1 + m_2}{2} = \frac{-0.69 + (-0.38)}{2} = -0.535 \tag{12.4}$$

Method 3: This method consists of selecting any three data points (in this case the same points chosen before) and fitting a curve to it. The equation for the curve is obtained by employing a second-order equation and solving it with the three points. Further details are available in Chapter 14 (Shaefer and Theodore 2007).

$$C = 0.155t^2 - 1.775t + 6.63 \tag{12.5}$$

The derivative of the equation is then calculated and evaluated at any point.

$$\frac{dC}{dt} = 0.31t - 1.775 \tag{12.6}$$

Evaluated at $t = 4.0$ seconds:

$$\frac{dC}{dt} = 0.31(4.0) - 1.775 = -0.535 \tag{12.7}$$

Method 4: This method uses the method of least squares. In this case, all the data points are used to generate a second-order polynomial equation. This equation is then differentiated and evaluated at the point where the value of the derivative is required. For example, Microsoft Excel can be employed to generate the regression equation. Further details are provided in Chapter 14. (Shaefer and Theodore 2007)

$$(C = 0.1626t^2 - 1.9905t + 7.3108 \tag{12.8}$$

Once all the coefficients are known, the equation has only to be differentiated:

$$\frac{dC}{dt} = 0.3252t - 1.9905 \tag{12.9}$$

Evaluate at $t = 4.0$ seconds:

$$\frac{dC}{dt} = 0.3252(4.0) - 1.9905 = -0.6897 \tag{12.10}$$

Methods 5 and 6: These two methods are somewhat similar. They are based on using five data points to generate coefficients. Additional methods are provided in the literature (Lavery 1979; Shaefer and Theodore 2007).

Comparing all four values obtained for the derivative at $t = 4.0$ seconds, one can conclude that the answers are in close proximity to each other. It is important to note that these are approximate values and that they vary depending on the approach and the number of data points used to generate the equations.

Some useful *analytical* derivatives in engineering calculations are provided in the literature (Ketter and Prawler 1969; Reynolds et al. 2004; Prochaska and Theodore 2018; Theodore and Behan 2018).

12.2 INTEGRATION

Numerous engineering applications and science problems often require the solution of integral equations. In a general sense, the integration problem is to evaluate the function on the right-hand side (RHS) of Eq. (12.11):

$$I = \int_a^b f(x)dx \tag{12.11}$$

where I is the value of the integral. There are two key methods employed in their solution: analytical and numerical. If $f(x)$ is a simple function, it may be integrated analytically. For example, if $f(x) = x^2$, then

$$I = \int_a^b x^2 dx = \frac{1}{3}(b^3 - a^3) \tag{12.12}$$

If however, $f(x)$ is a function too complex to integrate analytically {e.g. $\log[\tan(e^{x^3-2})]$}, one may resort to any of the many numerical integration methods available. Two simple numerical integration methods that are commonly employed in engineering are the *trapezoidal rule* and *Simpson's rule*. These are described in the next 2 subsections.

12.2.1 THE TRAPEZOIDAL RULE

To use the trapezoidal rule to evaluate the integral I given by Eq. (12.11) as

$$I = \int_a^b f(x)dx$$

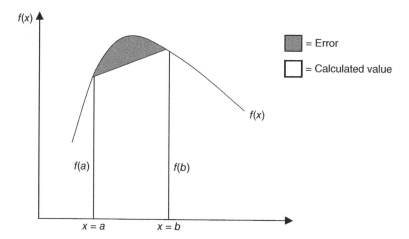

Figure 12.1 Trapezoidal rule analysis and error. Source: Courtesy of Abhishek Garg. Adapted from Bhatty et al. 2017.

One may use the equation

$$I = \frac{h}{2}[y_0 + 2y_1 + 2y_2 + \dots + 2y_{n-1} + y_n] \tag{12.13}$$

where h is the incremental change in x, i.e., Δx, and y_i is the value of $f(x)$ at x_i, i.e., $f(x_i)$. In addition,

$$y_0 = f(x_0) = f(x = a)$$
$$y_n = f(x_n) = f(x = b)$$
$$h = \frac{b/a}{n}$$

This method is known as the trapezoidal rule because it approximates the area under the function $f(x)$ – which is generally curved – with a two-point trapezoidal rule calculation. The error associated with this rule is illustrated in Figure 12.1.

There is an alternative method available for improving the accuracy of this calculation – the interval $(a–b)$ can be subdivided into smaller units. The trapezoidal rule can then be applied repeatedly in turn over each subdivision.

12.2.2 Simpson's Rule

A higher degree interpolating polynomial scheme can be employed for more accurate results. One of the more popular integration approaches is Simpson's rule. For Simpson's 3-point (or one-third) rule, one may use the equation

$$I = \frac{h}{3}\left[y_a + 4y_{\frac{b+a}{2}} + y_b\right] \tag{12.14}$$

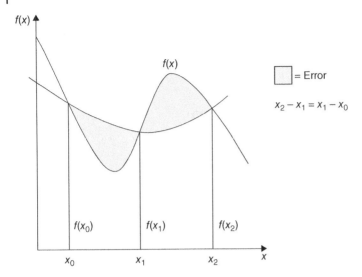

Figure 12.2 Simpson's rule analysis and error. Source: Courtesy of Abhishek Garg. Adapted from Bhatty et al. 2017.

For the general form of Simpson's rule (where n is an even integer), the equation is

$$I = \frac{h}{3}[y_0 + 4y_1 + 4y_2 + \ldots 4y_{n-1} + y_n] \tag{12.15}$$

This method also generates an error, although it is usually smaller than that associated with the trapezoidal rule. A diagrammatic representation of the error for a 3-point-calculation is provided in Figure 12.2.

The reader should note that the trapezoid rule is often the quickest but least accurate way to perform a numerical integration by hand. The results of each numerical integration must be added together to obtain the final answer for smaller step sizes. However, if the step size is decreased, the answer should converge – subject to roundoff errors – to the analytical solution.

Some useful analytical integrals in engineering calculations are provided in the literature (Ketter and Prawler 1969; Reynolds et al. 2004).

12.3 SIMULTANEOUS LINEAR ALGEBRAIC EQUATIONS

The engineer and applied scientist often encounter problems that contain not only more than two or three simultaneous algebraic equations but also those that can be nonlinear. There is, therefore, an obvious need for systematic methods of solving simultaneous linear and simultaneous nonlinear equations. This section addresses linear sets of equations; information on nonlinear sets is available in the literature (Ketter and Prawler 1969).

Consider the following set of n equations:

$$a_{11}x_1 + a_{12}x_2 + \cdots + a_{1n}x_n = y_1$$

$$a_{21}x_1 + a_{22}x_2 + \cdots + a_{2n}x_n = y_2$$

$$\vdots$$

$$a_{n1}x_1 + a_{n2}x_2 + \cdots + a_{nn}x_n = y_n \tag{12.16}$$

where a is the coefficient of the variable x and y is a constant. This set is considered to be linear as long as none of the x terms are nonlinear, e.g., $(x_2)^2$ or $\ln x_1$. Thus, a linear system requires that all terms in x be linear.

The above system of linear algebraic equations may be set in matrix form:

$$\begin{bmatrix} a_{11} & a_{21} & \cdots & a_{1n} \\ a_{12} & a_{22} & \cdots & a_{2n} \\ \cdots & \cdots & \cdots & \cdots \\ a_{n1} & a_{n2} & \cdots & a_{nn} \end{bmatrix} \begin{bmatrix} x_1 \\ x_2 \\ \cdots \\ x_n \end{bmatrix} = \begin{bmatrix} y_1 \\ y_2 \\ \cdots \\ y_n \end{bmatrix} \tag{12.17}$$

However, it is often more convenient to represent Eq. (12.16) in the *augmented matrix* form provided in the following equation:

$$\begin{bmatrix} a_{11} & a_{21} & \cdots & a_{1n} & y_1 \\ a_{12} & a_{22} & \cdots & a_{2n} & y_2 \\ \cdots & \cdots & \cdots & \cdots & \cdots \\ a_{n1} & a_{n2} & \cdots & a_{nn} & y_n \end{bmatrix} \tag{12.18}$$

Methods of solution available for solving these linear sets of equations include (Ketter and Prawler 1969):

- Gauss–Jordan reduction
- Gauss elimination
- Gauss–Seidel approach
- Cramer's rule
- Cholesky's rule

Only methods 1–3 are discussed in this section. Ketter and Prawler provide several excellent Illustrative Examples (Ketter and Prawler 1969).

12.3.1 Gauss–Jordan Reduction

Carnahan and Wilkes provide an example that solved the following two simultaneous equations using the Gauss–Jordan reduction method (Carnahan and Wilkes 1973):

$$3x_1 + 4x_2 = 29 \tag{12.19}$$

$$6x_1 + 10x_2 = 68 \tag{12.20}$$

The four-step procedure is provided below:

1. Divide Eq. (12.19) by the coefficient of x_1

$$x_1 + \frac{4}{3}x_2 = \frac{29}{3} \tag{12.21}$$

2. Subtract a suitable multiple (6, in this case) of Eq. (12.21) from Eq. (12.20), so that x_1 is *eliminated*. Eq. (12.21) remains intact so what remains is

$$\frac{24}{3}x_2 - \frac{30}{3}x_2 = \frac{6(29)}{3} - \frac{68(3)}{3} \tag{12.22}$$

Solving yields

$$-6x_2 = 6(29) - 68(3); \quad -x_2 = 29 - 34 \tag{12.23}$$

3. Divide Eq. (12.23) by the coefficient of x, i.e., solve Eq. (12.23).

$$x_2 = 5 \tag{12.24}$$

4. Subtract a suitable factor of Eq. (12.24) from Eq. (12.22) so that x_2 is eliminated. When $(4/3)x_2 = 20/3$ is subtracted from Eq. (12.22), one obtains

$$x_1 = 3 \tag{12.25}$$

12.3.2 GAUSS ELIMINATION

Gauss Elimination is another method used to solve linear sets of equations. This method utilizes the augmented matrix as described in Eq. (12.18). The goal with Gauss elimination is to rearrange the augmented matrix into a triangle form where all the elements below the diagonal are zero. This is accomplished in much the same way as in the Gauss–Jordan reduction. The procedure is as follows: start with the first equation in the set – this is known as the pivot equation and will not change through the procedure. Once the matrix is in a triangle form, back substitution can be used to solve for the variables.

Gauss elimination is useful for systems that contain fewer than 30 equations. Systems larger than 30 equations become subject to *roundoff errors* where numbers are truncated by computers performing the calculations.

12.3.3 GAUSS–SEIDEL APPROACH

Another approach to solving an equation or series/sets of equations is to make an informed or educated guess. If the first assumed value(s) does not work, the value is updated. By carefully noting the influence of these guesses on each variable, one can obtain these answers or correct the set of values for a system of equations. The reader should note that when this type of iterative procedure is employed, a poor initial guess does not prevent the correct solution from ultimately being obtained.

12.4 NONLINEAR ALGEBRAIC EQUATIONS

The subject of the solution to a nonlinear algebraic equation is considered in this section. Although several algorithms are available in the literature, the presentation will focus on the classic *Newton–Raphson* (NR) method of evaluating the root(s) of a nonlinear algebraic equation. The solution to the equation

$$f(x) = 0 \tag{12.26}$$

is obtained by guessing a value for x, e.g., (x_{new}) that will satisfy this equation. This value is continuously updated (x_{new}) using the equation (the prime represents a derivative)

$$x_{new} = x_{old} - \frac{f(x_{old})}{f'(x_{new})} \tag{12.27}$$

until either little or no change in $(x_{new} - x_{old})$ is obtained. One can also express this operation graphically (see Figure 12.3). Noting that

$$f'(x_{old}) = \frac{df(x)}{dx} \approx \frac{\Delta f(x)}{\Delta x} \tag{12.28}$$

one may rearrange Eq. (12.28) to yield Eq. (12.29) as follows.

$$f'(x_{old}) = \frac{f(x_{old}) - 0}{x_{old} - x_{new}} \tag{12.29}$$

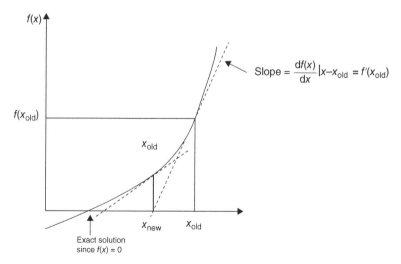

Figure 12.3 Newton–Raphson method for nonlinear equations Source: Courtesy of Abhishek Garg. Adapted from Bhatty et al. 2017.)

The x_{new} then becomes the x_{old} in the next calculation. This method is also referred to by some as *Newton's method of tangents* (NMT) and is a widely used method for improving a first approximation to a root to the aforementioned equation of the form $f(x) = 0$. This development can be rewritten in a subscripted form to (perhaps) better accommodate a computer calculation. Here,

$$f'(x_n) = \frac{f(x_n)}{x_n - x_{n+1}} \tag{12.30}$$

from which

$$x_{n+1} = x_n - \frac{f(x_n)}{f'(x_n)} \tag{12.31}$$

Despite the popularity of this approach, there are two major limitations. First, an analytical expression for the derivative, specifically $f'(x_n)$, is required. In addition to the problem of having to compute an analytical derivative at each iteration, one would expect Newton's method to converge fairly rapidly to a root in the majority of cases. However, as it is common with some numerical methods, it may fail occasionally in certain instances. A possible initial oscillation followed by a displacement away from a root can occur. Note, however, that the method could converge if the initial guess is somewhat closer to the exact root. Thus, the first guess may be critical to the success of this calculation.

12.5 ORDINARY DIFFERENTIAL EQUATIONS

The *Runge–Kutta* (RK) method is one of the most widely used techniques in engineering practice for solving first-order differential equations. For the equation

$$\frac{dy}{dx} = f(x, y) \tag{12.32}$$

the solution takes the form

$$y_{n+1} = y_n + \frac{h}{6}[D_1 + 2D_2 + 2D_3 + D_4] \tag{12.33}$$

where

$$D_1 = hf(x, y)$$
$$D_2 = hf\left(x_n + \frac{h}{2}, y_n + \frac{D_1}{2}\right)$$
$$D_3 = hf\left(x_n + h, y_n + D_2\right)$$
$$D_4 = hf(x_n + h, y_n + D_3) \tag{12.34}$$

The term h represents the increment in x. The term y_n is the solution to the equation at x_n and y_{n-1} is the solution to the equation at x_{n+1} where $x_{n+1} = x_n + h$. Thus, the RK method provides a straightforward means for developing expressions for Δy, namely, $y_{n+1} - y_n$, in terms of the function $f(x, y)$ at various "locations" along the interval in question.

For a simple equation of the form

$$\frac{dC}{dt} = a + bC \tag{12.35}$$

where $t = 0$, $C = C_0$, the RK algorithm given above becomes

$$C_1 = C_0 + \frac{h}{6}[D_1 + 2D_2 + 2D_3 + D_4] \tag{12.36}$$

where

$$D_1 = hf(x, y) = hf(a + bC_0)$$

$$D_2 = hf\left(x_n + \frac{h}{2}, y_n + \frac{D_1}{2}\right) = h\left[a + b\left(C_0 + \frac{D_1}{2}\right)\right]$$

$$D_3 = hf\left(x_n + \frac{h}{2}, y_n + \frac{D_2}{2}\right) = h\left[a + b\left(C_0 + \frac{D_2}{2}\right)\right]$$

$$D_4 = hf\left(x_n + \frac{h}{2}, y_n + \frac{D_3}{2}\right) = h[a + b(C_0 + D_3)] \tag{12.37}$$

The same procedure is repeated to obtain values for C_2 at $t = 2h$, C_3 at $t = 3h$, and so on.

The RK method can also be used if the function in question also contains the independent variable. Consider the following equation:

$$\frac{dC}{dt} = f(C, t) \tag{12.38}$$

For this situation, one once again obtains

$$C_1 = C_0 + \frac{h}{6}[D_1 + 2D_2 + 2D_3 + D_4] \tag{12.39}$$

with

$$D_1 = hf(C, t)$$

$$D_2 = hf\left(C_0 + \frac{D_1}{2}, t_0 + \frac{h}{2}\right)$$

$$D_3 = hf\left(C_0 + \frac{D_2}{2}, t_0 + \frac{h}{2}\right)$$

$$D_4 = hf(C_0 + D_3, t_0 + h) \tag{12.40}$$

If one considers the following equation

$$\frac{dC}{dt} = 5C - e^{-Ct} \tag{12.41}$$

then (for example)

$$D_2 = h \left\{ 5 \left(C_0 + \frac{D_1}{2} \right) - e^{\left[-\left(C_0 + \frac{D_1}{2} \right)\left(t_0 + \frac{h}{2} \right) \right]} \right\} \tag{12.42}$$

Situations may also arise when there is a need to simultaneously solve *more* than one ordinary differential equation (ODE). In a more general case, one could have n dependent variables – $y_1, y_2 \dots y_n$ – with each related to a single *independent variable* x by a system of n simultaneous first-order ODE. The resulting system of equations receives treatment in the literature (Prochaska and Theodore 2018; Theodore and Behan 2018).

Although the RK approach (and other similar methods) has traditionally been employed to solve first-order ODEs, it can also treat higher ODEs. The procedure requires reducing an nth-order ODE to n first-order ODE. Details are available in the literature (Prochaska and Theodore 2018; Theodore and Behan 2018).

The selection of increment size remains a variable for the practicing engineer. Few numerical analysis methods provided in the literature are concerned with error analysis. In general, roundoff and numerical errors are demonstrated in Figure 12.4. In the limit, when the increment →0, one approaches an analytical solution. However, the number of calculations correspondingly increases the error ε, which increases exponentially as the increment →0. Note that selecting

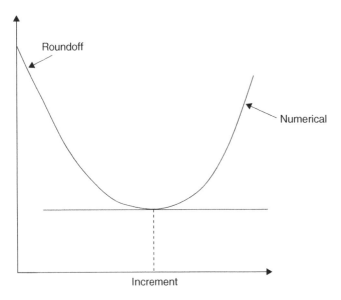

Roundoff

Numerical

Increment

Figure 12.4 Error analysis in numerical calculations. Source: Courtesy of Abhishek Garg. Adapted from Bhatty et al. (2017).

the increment size that will minimize the error is rarely a problem in practice; in addition, computing time is rarely a concern.

12.6 PARTIAL DIFFERENTIAL EQUATIONS

Many practical problems in engineering practice involve at least two independent variables; thus, the dependent variable is defined in terms of (or is a function of) more than one independent variable. The derivatives describing these independent variables are defined as *partial derivatives*. Differential equations containing partial derivatives are referred to as *partial differential equations* (PDEs).

It has been said that "the solution of a partial differential equation is essentially a guessing game" (Prochaska and Theodore 2018; Theodore and Behan 2018). In other words, one cannot always expect to be provided with a formal method that will yield exact solutions for all partial differential equations. Fortunately, numerical methods for solving these equations were developed during the mid to late twentieth century and are routinely used today (L. Theodore, personal notes, East Williston, NY, 1961; J. Reynolds, class notes (with permission), Manhattan College, Bronx, NY, 2001).

The three main PDEs encountered in engineering practice are briefly introduced below, employing T (e.g. the temperature as the dependent variable) with t (time) and x, y, z (position) as the independent variables. Note that any dependent variable, such as pressure or concentration, could have been selected (Prochaska and Theodore 2018; Theodore and Behan 2018).

Parabolic equation:

$$\frac{\partial T}{\partial t} = a\frac{\partial^2 T}{\partial x^2} \tag{12.43}$$

Elliptical equation:

$$\frac{\partial^2 T}{\partial x^2} + \frac{\partial^2 T}{\partial y^2} = 0 \tag{12.44}$$

Hyperbolic equation:

$$\frac{\partial^2 T}{\partial t^2} = a\frac{\partial^2 T}{\partial x^2} \tag{12.45}$$

The preferred numerical method of solution involves finite differencing (Carnahan and Wilkes 1973; Prochaska and Theodore 2018; Theodore and Behan 2018).

12.7 APPLICATIONS

Four illustrative examples complement the presentation for this chapter concerned with mathematical methods.

Illustrative Example 12.1   ANALYTIC QUADRATIC SOLUTION

Outline how to analytically *solve the quadratic equation.*

$$ax^2 + bx + c = 0$$

Solution

The following equation can be solved algebraically to obtain an exact solution to the above equation.

$$x = \frac{-b \pm \sqrt{b^2 - 4ac}}{2a}$$

If a, b, c, are real, and if D is defined as

$$D = b^2 - 4ac$$

then the solutions are

1. real and unequal if $D > 0$
2. real and equal if $D = 0$
3. complex conjugate if $D < 0$

In addition, if x_1, x_2 are the roots, then

$$x_1 + x_2 = -\frac{b}{a}$$

and

$$x_1 x_2 = \frac{c}{a}$$

Illustrative Example 12.2   ANALYTIC THIRD-ORDER ORDER SOLUTION

Provide a simple *solution to the following third-order equation*

$$f(x) = x^3 + 6.6x^2 - 29.05x + 22.64 = 0$$

Solution

One approach is to attempt to reduce the third-order equation to a second-order equation. For example, divide by x, drop $22.64/x$, and solve the resulting quadratic equation.

$$x^2 + 6.6x - 29.05 = 0$$

The solution is

$$x = -9.6$$

This provides a starting value for a trial-and-error solution. One now notes that

$$f(-9.6) = 0.025 \neq 0$$

This provides an excellent starting value in an attempt to obtain a solution to the cubic equation. The solution is approximately $x = -9.8$.

Illustrative Example 12.3 PROBLEMS WITH NO ANALYTICAL SOLUTION
Outline how one can obtain a solution for the following equation that does not have an analytical solution.

$$e^{x^2} - x = 14.2$$

Solution
A convenient way to solve this equation is to assume values of x until the left side of the equation equals the right. However, the equation should be inspected carefully to determine the general effects of a variation in x. Perhaps one term is substantially smaller than the other so that it may be neglected for a first estimation of x. Note that x is probably small compared to e^{x^2}. As a first approximation for x, assume $x = 0$ so what remains is that $e^{x^2} = 14$ for which, $x = 1.63$. But

$$e^{x^2} - x = 14.2 - 1.63 = 12.6 \neq 14.2$$

Inspection of this equation indicates that a somewhat larger value of x should be used. For the second trial, set $x = 2.60$ and proceed as before.

Illustrative Example 12.4 INTEGRAL SOLUTIONS
Provide solutions to the following 8 integrals.

1. $\int a \, dx$
2. $\int af(x) \, dx$
3. $\int (u \pm v \pm w \pm \ldots) \, dx.$
4. $\int (du \pm dv \pm dw)$
5. $\int u \, dv$
6. $\int u^n \, du$
7. $\int \frac{du}{u} \, dx$
8. $\int e^u \, du$

Solution

1. $\int a \, dx = ax$
2. $\int af(x) \, dx = a \int f(x) \, dx$
3. $\int (u \pm v \pm w \pm \ldots) \, dx = \int u \, dx \pm \int v \, dx \pm \int w \, dx \pm \ldots$
4. $\int (du \pm dv \pm dw) = \int du \pm \int dv \pm \int dw$
5. $\int u \, dv = uv - \int v \, du$ [integration by parts]
6. $\int u^n \, du = \frac{u^{n+1}}{n+1}, n \neq -1$
7. $\int \frac{du}{u} \, dx = \ln u$ if $u > 0$ or $\ln(-u)$ if $u < 0 = \ln |u|$
8. $\int e^u \, du = e^u$

12.8 CHAPTER SUMMARY

- Mathematics is generally defined as the relationships among quantities, magnitudes, and properties and of logical operations by which unknown quantities, magnitudes, and properties may be deduced.
- Several differentiation methods are available for generating expressions for a derivative
- There are two key methods employed to evaluate integrals: analytical and numerical. Two simple numerical integration methods that are commonly employed in engineering are the *trapezoidal rule* and *Simpson's rule*.
- Methods of solution available for solving linear sets of equations include:
  Gauss–Jordan reduction
  Gauss elimination
  Gauss–Seidel approach
  Cramer's rule
  Cholesky's rule
- *Newton's method of tangents* (NMT) is a widely used method for improving the first approximation to a root of an equation of the form $f(x) = 0$.
- The *Runge–Kutta* (RK) method is one of the most widely used techniques in engineering practice for solving first-order differential equations.
- Many practical problems in engineering practice involve at least two independent variables; thus, the dependent variable is defined in terms of (or is a function of) more than one independent variable. The derivatives describing these independent variables are defined as *partial* derivatives. Differential equations containing partial derivatives are referred to as *partial differential equations* (PDEs).

12.9 PROBLEMS

1 Provide an analytical solution to the equation

$$x^2 - 6x + 2 = 0$$

2 Verify that the solution(s) provided in Problem (1) are correct.

3 Outline how to analytically solve a cubic equation of the form

$$x^3 + a_1 x^2 + a_2 x + a_3 = 0$$

4 Solve the cubic equation

$$3x^3 + 4x^2 + 6x - 3 = 0$$

REFERENCES

Bhatty, V., Butron, S., and Theodore, L. (2017). *Introduction to Engineering: Fundamentals, Principles, and Calculations.* New York City, NY: Theodore Tutorials and Butte, MT: Montana State University, Department of Environmental Engineering Graduate Division.

Carnahan, B. and Wilkes, J. (1973). *Digital Computing and Numerical Methods.* Hoboken, NJ: John Wiley & Sons.

Ketter, R. and Prawler, S. (1969). *Modern Methods of Engineering Computations.* New York City, NY: McGraw-Hill.

Lavery, F. (1979). *The Perils of Differentiating Engineering Data Numerically.* New York City, NY: Chemical Engineering.

Prochaska, C. and Theodore, L. (2018). *Introduction to Mathematical Methods for Environmental Engineers and Scientists.* Beverly, MA: Scrivener-Wiley.

Reynolds, J., Jeris, J., and Theodore, L. (2004). *Handbook of Chemical and Environmental Engineering Calculations.* Hoboken, NJ: John Wiley & Sons.

Shaefer, S. and Theodore, L. (2007). *Probability and Statistics Applications for Environmental Science.* Boca Raton, FL: CRC Press/Taylor & Francis Group.

Theodre, L. (1961). *Personal notes.* NY: East Williston.

Theodore, L. and Behan, K. (2018). *Introduction to Optimization for Chemical and Environmental Engineers.* Boca Raton, FL: CRC Press/Taylor & Francis Group.

13

Probability and Statistical Principles

The purpose of this chapter is to present the reader with an introduction to probability calculations and provide an overview of statistical methods. Statistics are of great importance in all the life sciences and are fundamental to experimental design, data collection and analysis, and interpretation of results.

Webster defines *probability* as "the quality or state of being probable; likelihood; something probable; math; the number of times something will probably occur over the range of possible occurrences expressed as a ratio," (Merriam-Webster 2021a). and the term *statistics* as "facts or data of numerical kind, assembled, classified, and tabulated to present significant information about a given subject, the science of assembling, classifying, tabulating, and analyzing such facts or data." (Merriam-Webster 2021b). There are many other definitions.

One of the key areas of interest to engineers and applied scientists working in fields related to virology is biostatistics, which is used to analyze epidemiologic and medical studies. Challenges encountered are typically related to interpreting limited data and/or information, and having to make important conclusions based on that interpretation. This can entail any one of several options:

A Guide to Virology for Engineers and Applied Scientists: Epidemiology, Emergency Management, and Optimization, First Edition. Megan M. Reynolds and Louis Theodore.
© 2023 John Wiley & Sons, Inc. Published 2023 by John Wiley & Sons, Inc.

- Obtaining additional data or information.
- Deciding which data or information to use, and which to disregard.
- Generating a mathematical model (generally an equation) to best represent data or information.
- Generating information about unknowns, a process often referred to as inference.

Much of the material in this chapter has been drawn from the work of Theodore and Taylor (1993). Please refer to References for further information.

13.1 PROBABILITY DEFINITIONS AND INTERPRETATIONS

Probabilities are nonnegative numbers associated with the outcomes of the so-called random experiments. A random experiment is an experiment whose outcome is uncertain. The set of possible outcomes of a random experiment is designated the sample space and is usually represented by S. $P(A)$ represents the sum of the probabilities assigned to outcomes constituting the subset A within the sample space S. A population is a collection of objects (or people) having observable or measurable characteristics defined as variates, while a sample is a group of objects drawn from a population (usually random) that are equally likely to be drawn.

Consider, for example, tossing a two-headed (H, T) coin twice. The sample space can be described as

$$S = \{HH, HT, TH, TT\} \tag{13.1}$$

This represents all possible outcomes of the coin landing on either H (heads) of T (tails).

If probability $1/4$ is assigned to each element of S, and A is the event of at least one head (H), then

$$A = \{HH, HT, TH\} \tag{13.2}$$

The sum of the probabilities assigned to the elements of A is $3/4$. Therefore, $P(A) = 3/4$. The description of the sample space is not unique. The sample space S in the case of tossing a coin twice could be described in terms of the number of heads obtained. Then

$$S = \{0, 1, 2\} \tag{13.3}$$

Suppose that probabilities $1/4$, $1/2$, and $1/4$ are assigned to the outcomes 0, 1, and 2, respectively. Then, A, the event of at least one head, would have for its probability,

$$P(A) = P(1,2) = 1/2 + 1/4 = 3/4 \tag{13.4}$$

How probabilities are assigned to the elements of the sample space generally depends on the desired interpretation of the probability of an event. Thus, $P(A)$ can be interpreted as a theoretical relative frequency, which is a number about which the relative frequency of event A tends to cluster as n, the number of times that the random experiment is performed increases indefinitely; this is the objective interpretation of probability. Under this interpretation, to say that $P(A)$ is ¾ in the example above means that if a coin is tossed n times, the proportion of times that one or more heads occur clusters about 75% as n increases indefinitely.

As another example, consider a single valve in a piping system of a medical research laboratory, which can be stuck in an open (O) or closed (C) position. The sample space can be described as follows:

$$S = \{O, C\} \tag{13.5}$$

Suppose that the valve sticks twice as often in the open position as it does in the closed position. Under the theoretical relative frequency interpretation, the probability assigned to element O in S would be ⅔, twice the probability assigned to the element C, i.e., 1/3. If two such valves are observed, the sample space, S, can be described as

$$S = \{OO, OC, CO, CC\} \tag{13.6}$$

Assuming the two valves operate independently, a reasonable assignment of probabilities to the element of S, as just listed, should be 4/9, 2/9, 2/9, and 1/9. If A is the event of at least one valve sticking in the closed position, then

$$A = \{OC, CO, CC\} \tag{13.7}$$

The sum of the probabilities assigned to the element of A is then 5/9. Therefore, $P(A) = 5/9$.

Probability $P(A)$ can also be interpreted subjectively as a measure of the degree of belief on a fractional scale from 0 to 1, that event A occurs. This interpretation is frequently used in medicine, particularly with clinical drug and vaccine trials, to report overall success rates, or adverse events in a study group to extrapolate the data to be inclusive of the predictive measure of the success in the larger target population. This interpretation is also used when, in the absence of the concrete medical data necessary to estimate an unknown probability on the basis of observed relative frequency, the opinion of an expert is sought. For example, if a surgeon states that, "the probability that this surgery will be 75% effective on this patient," this statement is a measure of the doctor's belief that the operation will be successful based on the physician's knowledge of the operation, the overall health of the patient, the clinical data available, and all other factors the surgeon utilized in making that determination.

13.2 INTRODUCTION TO PROBABILITY DISTRIBUTIONS

The probability distribution of a random variable concerns the distribution of probability over the range of the random variables, and the probability distribution function (pdf) represents those random variables together with their associated probabilities.

This section is devoted to providing general properties of the pdf in the case of discrete and continuous random variables, as well as provide an introduction to the cumulative distribution function (cdf). Specific pdfs with extensive application in engineering analysis are considered in the next two sections.

The pdf of a *discrete* random variable, X, is specified by $f(x)$, where $f(x)$ has the following three essential properties:

1. $F(x) = P(X = x)$
2. $F(x) \geq 0$
3. $\Sigma f(x) = 1$

Property 1 indicates that the probability assigned to the outcome corresponding to the number x in the range of X, (i.e. X is specifically designated a value of x).

Property 1 also indicates the pdf of a discrete random variable generates probability by substitution. Properties 2 and 3 restrict the values of $f(x)$ to nonnegative real numbers and numbers whose sum is 1, respectively.

The pdf of a *continuous* random variable X has the following properties:

$$\int_b^a f(x)\mathrm{d}x = P(a < X < b) \tag{13.8}$$

$$F(x) \geq 0 \tag{13.9}$$

$$\int_{-\infty}^{\infty} f(x)\mathrm{d}x = 1 \tag{13.10}$$

Equation (13.8) indicates that the pdf of a continuous random variable generates probability by integration of the pdf over the interval whose probability it requires. When this interval contracts or reduces to a single value, the integral over the interval becomes *zero*. Therefore, the probability associated with any value of a continuous random variable is zero. Consequently, if X is continuous, then

$$P(a < X \leq b) = P(a \leq X < b) \tag{13.11}$$

Equation (13.9) restricts the values of $f(x)$ to nonnegative numbers. Equation (13.10) follows from the fact that

$$P(-\infty < X < \infty) = 1 \tag{13.12}$$

The expression $P(a < X < b)$ can be interpreted geometrically as the area under the pdf curve over the interval (a, b). Integration of the pdf over the interval yields the probability assigned to the interval. For example, the probability that the time in hours between successive failures, x, of a hospital emergency room air-conditioning system is greater than 6 but less than 10 is $P(6 < X < 10)$.

Another function used to describe the probability distribution of a random variable X is the aforementioned cumulative distribution function (cdf). If $f(x)$ specifies the pdf of a random variable X, then $F(x)$ is used to specify the cdf. For both discrete and continuous random variables, the cdf of X is defined by

$$F(x) = P(X \geq x), \quad \text{where} (-\infty < X < \infty) \tag{13.13}$$

Note that the cdf is defined for all real numbers, not just the values assumed by the random variable. It is helpful to think of $F(x)$ as an accumulator of probability as x increases through all real numbers. In the case of a discrete random variable, the cdf is a step function increasing by finite jumps as the values of x in the range of X. In the case of a continuous random variable, the cdf is a continuous function.

The following properties of the cdf of a random variable X can be deduced directly from the definition of $F(x)$:

$$F(a) - F(b) = P(a \leq X \leq b)$$

$$F(+\infty) = 1$$

$$F(-\infty) = 0 \tag{13.14}$$

Also, note that $F(x)$ is a nondecreasing function of X. These properties apply to the cases of both discrete and continuous random variables.

Shaefer and Theodore provide numerous illustrative examples of probability distributions in their CRC Press/Taylor & Francis Group work, *Probability and Statistics Applications for Environmental Science, 2007.*

13.3 DISCRETE PROBABILITY DISTRIBUTIONS

There are numerous discrete probability distributions. This section examines four distributions that engineers and applied scientists involved with virology are likely to encounter in their careers. These are listed here and further explained below (see also Shaefer and Theodore):

- The binomial distribution
- The multinomial distribution
- The hypergeometric distribution
- The Poisson distribution

13.3.1 THE BINOMIAL DISTRIBUTION

Consider n independent performances of a random experiment with mutually exclusive outcomes that can be classified as *success* or *failure*. The words success and failure are to be regarded as labels for two mutually exclusive categories of outcomes of the random experiment. In the development (i.e., text and equations) that follows, they do not necessarily have the ordinary connotation of success or failure.

Assume that p, the probability of success on any performance of a random experiment is constant. Let Q be the probability of failure, so that

$$Q = 1 - p \tag{13.15}$$

The probability distribution of X, the number of successes in n performances of the random experiment, is the binomial distribution with a pdf specified by

$$f(x) = \frac{n!}{x!(n-x)!} p^x Q^{n-x} \quad x = 0.1, \ldots, n \tag{13.16}$$

where $f(x)$ is the probability of x successes in n performances.

The binomial distribution can be used to calculate the reliability of a redundant system. A redundant system consisting of n identical components is a system that fails only if more than r components fail. Familiar examples include single-usage equipment such as missile engines, short-life batteries, flashbulbs, and numerous electrical components which are required to operate for one-time period and are not reused.

Once again, associate success with the failure of a component. Assume that the n components are independent with respect to failure, and that the reliability of each is $1-p$. Then the number of failures X has the binomial pdf in Eq. (13.16) and the reliability of the redundant system is

$$P(X \le r) = \sum_{x=0}^{r} \frac{n!}{x!(n-x)!} p^x Q^{n-x} \tag{13.17}$$

The binomial distribution occurs in problems in which samples are drawn from a *large* population with specified success or failure probabilities and there is a desire to evaluate instances of obtaining a certain number of successes in the sample. It has applications in quality control, reliability studies, hospital environmental management, consumer sampling, and many other cases.

13.3.2 MULTINOMIAL DISTRIBUTION

The multinomial distribution is a generalization of the binomial distribution described in the previous subsection. It is similar, except that instead of two outcomes, i.e., success or failure, there are multiple possible outcomes. A simple example would be that instead of flipping a coin multiple times looking for an outcome, one rolls a pair of dice several times.

In the equation below, the term n is defined by the multinomial expansion

$$(X \leq r) = \sum_{x=0}^{r} \frac{n!}{x_1! x_2! \dots x_r!} p_1^{x_1} p_2^{x_2} \dots p_r^{x_r}, \tag{13.18}$$

where $n = x_1 + x_2 + \dots + x_r$

13.3.3 Hypergeometric Distribution

The hypergeometric distribution applies to situations in which a random sample of r items is drawn *without replacement* from a set of n items. Without replacement means that an item is not returned to the set after it is drawn. Recall that the binomial distribution is frequently applicable in cases where the item is drawn *with replacement*.

Suppose that it is possible to classify each of the n items as a success or a failure. Again, the words success and failure do not have the usual connotation. They are merely labels for two mutually exclusive categories into which n items have been classified. Thus, each element of the population may be dichotomized as belonging to one of these two disjointed classes.

In the development that follows, set a as the number of items in the category labeled success. Then $n-a$ will be the number of items in the category labeled failure. Let X denote the number of successes in a random sample of r items drawn without replacement from the set of n items. Then the random variable X has a hypergeometric distribution whose pdf is specified as follows:

$$f(x) = \frac{\{a!/[x!(a-x)!]\}\{(n-a!)/[(r-x)!(n-a-r+x)!]\}}{n!/[r!(n-r)!]}$$

$$[x = 0, 1, \dots, \min(a, r)] \tag{13.19}$$

The term $f(x)$ is defined as the probability of x number of successes within a random sample of n items drawn without replacement. Here, the terms a are classified as successes and $(n-a)$ as failures. The term $\min(a, r)$ represents the smaller of the two numbers of a and r, e.g., $\min(a, r) = r$ if $r \leq a$.

The hypergeometric distribution is applicable in situations similar to those when the binomial distribution is used, except that samples are taken from a small population. Examples arise in sampling from small numbers of chemical samples as well as medical and hospital environmental samples, and from manufacturing lots.

13.3.4 POISSON DISTRIBUTION

The pdf of the Poisson distribution can be derived by taking the limit of the binomial pdf as $n \to \infty$, $P \to 0$, with $nP = \mu$ remaining constant. The Poisson pdf is then given by

$$f(x) = \frac{e^{-\mu} \mu^x}{x!} (x = 0, 1, 2, \dots) \tag{13.20}$$

Here, $f(x)$ is the probability of x occurrences of an event that occurs on the average μ times per unit of space or time. Both the mean and the variance of a random variable X having a Poisson distribution are μ.

The Poisson distribution can be used to estimate probabilities obtained from the binomial pdf given earlier for large n and small P. In general, good approximations will result when n exceeds 100 and nP is less than 10.

If λ is the failure rate (per unit of time) of each component of a system, then λt is the average number of failures rate (per unit of time) of each component of a system. The probability of x failures in a specified unit of time is obtained by substituting $\mu = \lambda t$ in Eq. (13.19) to obtain

$$f(x) = \frac{e^{-\lambda t} \lambda t^x}{x!} (x = 0, 1, 2, \dots) \tag{13.21}$$

In addition to the applications cited earlier, the Poisson distribution is widely utilized in medicine, particularly when studying virology. The distribution can be used when considering how a virus infects and multiplies in the body. The *Multiplicity of Infection (MOI)* is defined as the number of viral particles that enter a particular cell. In Eq. (13.20), x refers to the MOI and λ is the number of particles entering a cell. $f(x)$ can be used to obtain the probability that a target cell will be infected by x virions. With an MOI of 1, the probability that a solitary particle of virus will infect a single cell within a population of cells is $f(1)$, which is equal to 36.8%. The probability of the cell being infected with two viral particles is $f(2)$ is equal to 18.4%, and so on. This relation also often occurs in microbiology (Reynolds 2021).

The Poisson distribution is a distribution which also arises in many other different applications. For instance, it provides probabilities of specified numbers of COVID-19 testing procedures at a given testing site, of a given number of defects per 1000 NK95 masks, of various numbers of bacterial colonies per unit volume, etc.

13.4 CONTINUOUS PROBABILITY DISTRIBUTIONS

Any data collected from healthcare studies require careful evaluation to be of any use. Epidemiology relies on statistical measures for the analysis of results. Continuous probability distributions are one of many statistical analysis tools used to quantify and summarize data. This section reviews measures of central tendency and scatter, and examines four of the most common continuous distributions utilized in healthcare and epidemiology:

- The normal distribution
- The lognormal distribution
- The exponential distribution
- The Weibull distribution

13.4.1 MEASURES OF CENTRAL TENDENCY AND SCATTER

One basic way of summarizing data is by the computation of a central value. The most commonly used central value statistic is the arithmetic average, or the *mean*. This statistic is particularly useful when applied to a set of data having a fairly symmetrical distribution. The mean is an efficient statistic because it summarizes all the data in the set and each piece of data is taken into account in its computation. However, the arithmetic mean is not a perfect measure of the true central value of a given data set because arithmetic means can overemphasize the importance of one or two extreme data points.

When a distribution of data is asymmetrical, it is sometimes desirable to compute a different measure of central value. This second measure, known as the *median*, is simply the middle value of a distribution, or the number above which half the data lie and below which the other half lie. If n data points are listed in their order of magnitude, the median is the $[(n+1)/2]$th value, if n is an odd number. If the number of data is even, then the numerical value of the median is the value midway between the two data points nearest to the middle. The median, being a positional value, is less influenced by extreme values in a distribution than the mean. However, the median alone is usually not a good measure of central tendency. To obtain the median, the data provided must first be arranged in order of magnitude, such as 8, 10, 13, 15, 18, and 22. Here, the median is 14, or the value halfway between 13 and 15, because this data set has an even number of measurements (Shaefer and Theodore 2007).

Generally, the mean falls near the "middle" of the distribution. The mean may be thought of as the "center of gravity" of the distribution. The mean has another important property. If each measurement is subtracted from the mean, one obtains n "discrepancies" or differences. Some of these are positive and some are negative, but the algebraic sum of all the differences is equal to zero.

The mean of a set of measurements provides some information about the location of the "middle" or "center of gravity" of the set of measurements, but it gives no information about the scatter (or dispersion or amount of concentration) of the measurements. For example, the five measurements, 14.0, 24.5, 25.0. 25.5, and

36.0, have the same mean as the five measurements, 24.0, 24.5, 25.0, 25.5, and 26.0, but the two sets of measurements have different amounts of scatter. One simple indication of the scatter of a set of measurements is the *range*, i.e., the largest measurement minus the smallest. In the two sets of measurements mentioned above, the ranges are 22 and 2, respectively. One would find the range to be convenient with very small sample sizes. It is difficult, however, to compare a range for one sample size with that for a different sample size. For this and other reasons, the range, in spite of its simplicity, convenience, and importance, is used only in rather restricted situations. Finally, the *mode* is the most frequently occurring value.

One clearly needs a measure of scatter, which can be used in samples of any size and in some sense makes use of all the measurements in the sample. There are several measures of scatter that can be used for this purpose, and the most common of these is the standard deviation, which is discussed later in the chapter. The standard deviation may be thought of as the "natural" measure of scatter. Calculation details for not only the standard deviation but also the variance are provided in the literature (Shaefer and Theodore 2007).

13.4.2 THE NORMAL DISTRIBUTION

The normal distribution is widely used in epidemiologic studies. Also referred to as a Gaussian or Bell curve distribution, classic normal distributions are defined by the fact that the measures of central tendency (i.e. mean, median, and mode) are all equivalent. Other variables, although not normally distributed per se, sometimes approximate a normal distribution after an appropriate transformation such as taking the logarithm (see next subsection) or the square root of the original variable. The area under the curve of the normal distribution allows for the determination of measures of spread, including the standard deviation and confidence interval (see also Section 13.5), both of which are regularly used in epidemiologic studies, especially in clinical trials (CDC 2012).

The normal distribution has the advantages of being tractable mathematically, and more straightforward than other methods. Consequently, many of the techniques of statistical inference have been derived under the assumption of underlying normal variants. A recent example of the use of Gaussian curves in virology is in predicting growth models for COVID-19 cases in various hospitals, towns, and countries. Although it is a relatively simplistic method, utilization in real-world practical matters is well established.

Actual (i.e., experimental) data have shown many physical variables to be normally distributed. There are numerous examples of industrial physical measurements on living organisms, molecular velocities in an ideal gas, scores on an intelligence test, the average annual influenza rates in a given locality, etc (Bhatty et al. 2017; CDC 2012).

The initial presentation in this subsection on *normal distributions* will focus on the time-to-failure rate. This can be used to measure the time it will take for any one component out of a population to fail. An example includes the estimated failure rate of HEPA filters utilized in hospitals. When time-to-failure has a normal distribution, its pdf is given by

$$f(t) = \frac{1}{\sqrt{2\pi}\sigma} \exp\left[-\frac{1}{2}\left(\frac{t-\mu}{\sigma}\right)^2\right] \quad (-\infty < t < \infty) \tag{13.22}$$

where μ is the mean value of T and σ is its standard deviation, which is a measure of the spread of the data. The graph of $f(t)$ is the familiar bell-shaped curve (see also Figure 13.2). The reliability function corresponding to the normally distributed failure time is given by

$$R(t) = \frac{1}{\sqrt{2\pi}\sigma} \exp \int_t^\infty \left[-\frac{1}{2}\left(\frac{t-\mu}{\sigma}\right)^2\right] dt \tag{13.23}$$

If T is normally distributed with mean μ and standard deviation σ, then the random variable $(T-\mu)/\sigma$ is normally distributed with a mean of 0 and a standard deviation of 1. The term $(T-\mu)/\sigma$ is called a *standard normal curve* that is represented by Z, not to be confused with the failure rate $Z(t)$ to be introduced later in this section.

Table 13.1 is a tabulation of areas under a standard normal curve to the right of Z_0. Probabilities about a standard normal variable Z can be determined from these tables. For example, $P(Z > 1.54) = 0.062$ is obtained directly from the table as the area to the right of 1.54. As presented in Figure 13.1, the symmetry of the standard normal curve about zero implies that the area to the right of zero is 0.5. If X is not normally distributed, then $\overline{X}$, the mean of a sample n observations on X, is *approximately* normally distributed with mean μ and standard deviation $\sigma/\sqrt{n}$, provided the sample size n is large (>30). This result is based on an important theorem in probability called the *central limit theorem* (CLT) which is widely used in healthcare and epidemiology. The CLT states that when a sample size is sufficiently large, the average of any random sample will fall in a Gaussian curve distribution with the central value defined as the population average. The random variable's *standard error* is then equivalent to the standard deviation of the entire population, divided by the square root of the sample size. In a standard normal distribution, 68% of the population falls within one standard deviation of the mean, while approximately 95% falls within two standard deviations (CDC 2012).

Normal distributions can have differently sized modal peaks, depending on the measure of dispersion. The curves in Figure 13.2 are all normal, meaning they all have central tendency, but have different shapes, based on their levels of dispersion. The standard deviation measures this level of spread. When data is presented in clinical trials, any resulting mean should include the associated standard deviation.

Table 13.1 The standard normal distribution.

z	0.00	0.01	0.02	0.03	0.04	0.05	0.06	0.07	0.08	0.09
0	0.5	0.496	0.492	0.488	0.484	0.48	0.476	0.472	0.468	0.464
0.1	0.46	0.456	0.452	0.448	0.444	0.44	0.436	0.433	0.429	0.425
0.2	0.421	0.417	0.413	0.409	0.405	0.401	0.397	0.394	0.39	0.386
0.3	0.382	0.378	0.374	0.371	0.367	0.363	0.359	0.356	0.352	0.348
0.4	0.345	0.341	0.337	0.334	0.33	0.326	0.323	0.319	0.316	0.312
0.5	0.309	0.305	0.302	0.298	0.295	0.291	0.288	0.284	0.281	0.278
0.6	0.274	0.271	0.268	0.264	0.261	0.258	0.255	0.251	0.248	0.245
0.7	0.242	0.239	0.236	0.233	0.23	0.227	0.224	0.221	0.218	0.215
0.8	0.212	0.209	0.206	0.203	0.2	0.198	0.195	0.192	0.189	0.187
0.9	0.184	0.181	0.179	0.176	0.174	0.171	0.169	0.166	0.164	0.161
1.0	0.159	0.156	0.154	0.152	0.149	0.147	0.145	0.142	0.14	0.138
1.1	0.136	0.133	0.131	0.129	0.127	0.125	0.123	0.121	0.119	0.117
1.2	0.115	0.113	0.111	0.109	0.107	0.106	0.104	0.102	0.1	0.099
1.3	0.097	0.095	0.093	0.092	0.09	0.089	0.087	0.085	0.084	0.082
1.4	0.081	0.079	0.078	0.076	0.075	0.074	0.072	0.071	0.069	0.068
1.5	0.067	0.066	0.064	0.063	0.062	0.061	0.059	0.058	0.057	0.056
1.6	0.055	0.054	0.053	0.052	0.051	0.049	0.048	0.047	0.046	0.046
1.7	0.045	0.044	0.043	0.042	0.041	0.04	0.039	0.038	0.038	0.037
1.8	0.036	0.035	0.034	0.034	0.033	0.032	0.031	0.031	0.03	0.029
1.9	0.029	0.028	0.027	0.027	0.026	0.026	0.025	0.024	0.024	0.023
2.0	0.023	0.022	0.022	0.021	0.021	0.02	0.02	0.019	0.019	0.018
2.1	0.018	0.017	0.017	0.017	0.016	0.016	0.015	0.015	0.015	0.014
2.2	0.014	0.014	0.013	0.013	0.013	0.012	0.012	0.012	0.011	0.011
2.3	0.011	0.01	0.01	0.01	0.01	0.009	0.009	0.009	0.009	0.008
2.4	0.008	0.008	0.008	0.008	0.007	0.007	0.007	0.007	0.007	0.006
2.5	0.006	0.006	0.006	0.006	0.006	0.005	0.005	0.005	0.005	0.005
2.6	0.005	0.005	0.005	0.005	0.005	0.004	0.004	0.004	0.004	0.004
2.7	0.003	0.003	0.003	0.003	0.003	0.003	0.003	0.003	0.003	0.003
2.8	0.003	0.002	0.002	0.002	0.002	0.002	0.002	0.002	0.002	0.002
2.9	0.002	0.002	0.002	0.002	0.002	0.002	0.002	0.002	0.002	0.002

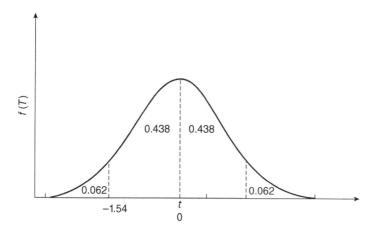

Figure 13.1 Areas under a standard normal curve.

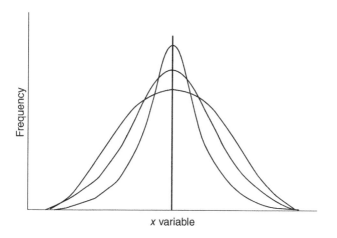

Figure 13.2 Various normal curves with different standard deviations. Source: CDC (2012)/U.S. Department of Health and Human Services/public domain.

It is important to note, that in epidemiology, there are many other frequency distributions that are considered non-normal. A curve may be skewed positively or negatively or may have multiple peaks, such as in bimodal curves. This is also addressed in Chapter 16. Figure 13.3 shows three separate distributions with curve *A* representing a positive skewed modal peak (skewed to the right), curve *B* a standard normal curve, and curve *C* as a negatively skewed peak (skewed to the left).

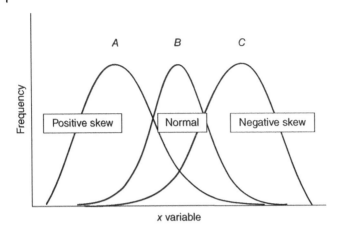

Figure 13.3 Differently skewed curves. Source: CDC (2012)/U.S. Department of Health and Human Services/public domain..

13.4.3 The Lognormal Distribution

A nonnegative random variable X has a lognormal distribution whenever $\ln X$, that is, the natural logarithm of X, has a normal distribution. The pdf of a random variable X having a lognormal distribution is specified by

$$f(x) = \left\{ \frac{1}{\sqrt{2\pi}\beta} x^{-1} \exp\left[-\frac{(\ln x - \alpha)^2}{2\beta^2} \right] \right\} \quad \text{for } x > 0$$

$$f(x) = 0; \quad \text{elsewhere} \tag{13.24}$$

The mean and variance of a random variable X having a lognormal distribution are given by

$$\mu = e^{\alpha + (\beta^2/2)} \tag{13.25}$$

$$\sigma^2 = e^{\alpha + (\beta^2/2)}(e^{\beta^2} - 1) \tag{13.26}$$

Probabilities concerning random variables having a lognormal distribution can be calculated from the previously employed table of the normal distribution (see Table 13.1). If X has a lognormal distribution with parameters α and β, then the $\ln X$ has a normal distribution with $\mu = \alpha$ and $\sigma = \beta$. Probabilities concerning X can therefore be converted into equivalent probabilities concerning $\ln X$.

Estimates of the parameters α and β in the pdf of a random variable X having a lognormal distribution with mean α and standard deviation β. Therefore, the mean and standard deviation of the natural logarithms of the sample observations of X furnish estimates of α and β.

The lognormal distribution has been employed as an appropriate model in a wide variety of situations including engineering, biology, and economics. Additional applications include the distributions of personal incomes, inheritances, bank account deposits, and the distribution of organism growth subject to many small impurities. The lognormal distribution has also been used to represent particle size distribution in gaseous emissions, bacterial cell growth, and chemical products from many industrial processes (Theodore 2008).

13.4.4 THE EXPONENTIAL DISTRIBUTION

The exponential distribution is important in that it represents the distribution of the time required for a single event from a Poisson process to occur. In particular, in sampling from a Poisson distribution, i.e., $f = \mu^x/x!$ with parameter, μ, the probability that no events occur during $(0, t)$ is $e^{-\lambda t}$. Consequently, the probability that an event will occur during $(0, t)$ is (Shaefer and Theodore 2007).

$$F(t) = 1 - e^{-\lambda t} \tag{13.27}$$

This represents the cumulative distribution function (cdf) of t. One can therefore show that the pdf is

$$f(t) = e^{-\lambda t} \tag{13.28}$$

Note that the parameter $1/\lambda$ (sometimes denoted as μ) is the expected value. Normally, the reciprocal of this value is specified and represents the expected value of $f(t)$. Because the exponential function appears in the expression for both the pdf and the cdf, the distribution is justifiably called the exponential distribution (Shaefer and Theodore 2007).

Alternatively, the cumulative exponential distribution can be obtained from the pdf (with x replacing t):

$$F(x) = \int_0^x \lambda e^{-\lambda x} dx = 1 - e^{-\lambda x} \tag{13.29}$$

All that remains is a simple evaluation of the negative exponent in Eq. (13.29).

In statistical and reliability engineering applications, a failure density or hazard function is common used in such techniques. One often encounters a random variable's conditional failure density, $g(x)$, where $g(x)dx$ is the probability that a "product" will fail during $(x, x + dx)$ under the condition that it had not failed before time x. Consequently

$$g(x) = \frac{f(x)}{1 - F(x)} \tag{13.30}$$

If the probability density function $f(x)$ is exponential, with parameter λ, it follows from Eqs. (13.29) and (13.30) that

$$g(x) = \frac{\lambda e^{-\lambda x}}{1 - (1 - \lambda e^{-\lambda x})} = \frac{\lambda e^{-\lambda x}}{e^{-\lambda x}}$$

$$= \lambda \tag{13.31}$$

Equation (13.31) indicates that the failure probability is constant, irrespective of time. It implies that the probability of failure that a component whose time-to-failure distribution is exponential during the first hour of its life is the same as the probability that it will fail during an instant in the thousandth hour, presuming that it has survived up to that instant. It is for this reason that the parameter λ is usually referred to in life-test applications as the *failure rate*. This definition generally has meaning only with an exponential distribution.

The natural association with life testing makes the exponential distribution appealing as a representative of the life distribution of complex systems. The exponential distribution is as prominent in reliability analysis as the normal distribution is in other branches of statistics (Shaefer and Theodore 2007).

13.4.5 THE WEIBULL DISTRIBUTION

Unlike the failure rate of the exponential distribution, equipment frequently exhibits three stages:

- A break-in stage with a declining failure rate
- A useful-life stage is characterized by a constant failure rate
- A wear-out period characterized by an increasing failure rate

Many medical device components follow this path. A failure curve exhibiting these three phases (see Figure 13.4) is called a "bathtub curve." *Weibull* introduced the distribution, which bears his name principally on empirical grounds, to represent certain life-test data. The Weibull distribution provides a mathematical model of all three stages of the bathtub curve. This is further discussed as follows.

An assumption about the failure rate $Z(t)$ that reflects all three stages of the bathtub stage is

$$Z(t) = \alpha \beta t^{\beta - 1} \ (t > 0) \tag{13.32}$$

where α and β are constants. For $\beta < 1$, the failure rate $Z(t)$ decreases with time. For $\beta = 1$, the failure rate is constant and equal to α. For $\beta > 1$, the failure rate increases with time.

Translating the assumption about failure rate into a corresponding assumption about the pdf of T, time-to-failure, one obtains

$$f(t) = \alpha \beta t^{\beta - 1} \exp\left(\int_0^t \alpha \beta t^{\beta - 1} dt\right) = \alpha \beta t^{\beta - 1} \exp(-\alpha t^\beta); \quad t > 0, \alpha > 0, \beta > 0) \tag{13.33}$$

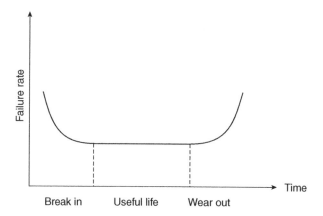

Figure 13.4 Bathtub curve. Source: Bhatty et al. (2017).

Equation (13.33) defines the pdf of the Weibull distribution. The exponential distribution discussed in the preceding subsection, whose pdf is given in Eq. (13.28), is a special case of the Weibull distribution with $\beta < 1$.

The Weibull distribution's ability to calculate failure rate (and the probability distribution of time-to-failure) makes it applicable in a wide variety of areas, such as describing failure-time distributions in medical devices and other equipment, in order to plan for replacement. Another important example of the utility of this distribution data is in determining the lifecycle of a pacemaker. These electric devices are limited by their battery life, and they must be surgically replaced before they begin to fail. (Theodore and Dupont 2012). One of the authors is currently developing an improvement to the Weibull model (Shaefer and Theodore 2007).

13.5 CONTEMPORARY STATISTICS

This section examines seven topics that Shaefer and Theodore (2007) have classified as contemporary statistics. These measures are widely used in most epidemiologic studies that utilize one or more of these measures. Hypothesis testing is the basis of analytic epidemiology, as discussed later in the section. The following subject areas are reviewed:

- Confidence intervals for means
- Confidence intervals for proportions
- Hypothesis testing
- Hypothesis test for means and proportions
- The F distribution
- Analysis of variance (ANOVA)
- Nonparametric tests

Although some of the material is beyond the scope of an introductory statistics chapter, these seven topics are briefly discussed. Additional details and numerous illustrative examples are provided by Shaefer and Theodore, including the design of experiments (DOE) (Shaefer and Theodore 2007).

13.5.1 CONFIDENCE INTERVALS FOR MEANS

As described in an earlier section the sample mean $\overline{X}$ constitutes a so-called point estimate of the mean μ of the population from which the same was selected at random. Instead of a *point* estimate, an *interval* estimate of μ may be required along with an indication of the confidence that can be associated with the interval estimate. Such an interval estimate is called a *confidence interval*, and the associated confidence is indicated by a *confidence coefficient*. The length of the confidence interval varies directly with the confidence coefficient for fixed values of the sample size n; the larger the value of n, the shorter the confidence interval. Thus, for fixed values of the confidence coefficient, the limits that contain a parameter with a probability of 95% (usually chosen as the conventional cutoff for medical study design) are defined as 95% (or any other percentage) confidence limits for the parameter; the interval between the confidence limits is the aforementioned confidence interval (Bhatty et al. 2017).

When considering the results of a clinical trial or other medical study, precision and accuracy are both critical. Precision is defined as the consistency and reliability of the resulting estimate. The more precise an estimate, the less random variation is reflected in it. Accuracy reflects the validity of the result, indicating how close it is to the *true* value. A high accuracy reflects a lack of bias or systemic error in the study.

13.5.2 CONFIDENCE INTERVALS FOR PROPORTIONS

Consider a random variable X that has a binomial distribution. For large n, the random variable X/n is approximately normally distributed with mean p and standard deviation equal to the square root of pQ/n. If n is the size of a random sample from a population in which p is the proportion of elements classified as a *success* because of the possession of a specified characteristic, this serves as the basis for constructing a confidence interval for P, the corresponding population proportion. The term n is *large* means that np and $n(1-p)$ are both greater than 5.

13.5.3 HYPOTHESIS TESTING

In hypothesis testing, one makes a statement (hypothesis) about an unknown population parameter and then uses statistical methods to determine (test) whether the observed sample data support that statement. Note that a statistical hypothesis is an assumption made about some parameter, i.e., about a statistical measure of a population. One may also define hypothesis as a statement about one or more parameters. A hypothesis can also be tested to verify its credibility. Hypothesis testing forms the foundation of *analytic epidemiology,* as presented below. The hypothesis is predetermined during the design of an experiment. After the study is concluded, the hypothesis is then analyzed for statistical significance. From this result, the hypothesis is either accepted or rejected. Clinical trial research uses hypothesis testing in order to determine whether a medication or treatment is an improvement over the standard treatment, as was discussed in Part I, Chapter 6.

13.5.4 HYPOTHESIS TEST FOR MEANS AND PROPORTIONS

It is important to note that testing a statistical hypothesis concerning a population involves the setting of a rule of deciding, based on a random sample from the population, whether to reject the hypothesis being tested; this is the so-called *null* hypothesis. The test is formulated in terms of a test statistic, i.e., some function of the sample observations. In the case of a large sample ($n > 30$), testing a hypothesis concerning the population mean μ involves the use of the following test statistic:

$$Z = \frac{\overline{X} - \mu_0}{\sigma/\sqrt{n}} \tag{13.34}$$

where $\overline{X}$ = sample mean
μ_0 = value of μ specified by null hypothesis H_0
σ = population standard deviation
n = sample size
When σ is unknown, the sample standard deviations may be employed to estimate it. This test statistic is approximately distributed as a standard normal variable when the sample originates from a normal population whose standard deviation σ is known. In clinical trial design, the gold standard for demonstrating the effectiveness of two or more treatments is a randomized, controlled, double-blind study.

The conventional level of significance is 0.95. The *p* value is then set to $p < 0.05$, to test for statistical significance. If the result of the trial shows a *p* value greater than 0.05, the null hypothesis is rejected, and the results are determined to be not statistically significant. For example, if a trial was designed to show that a new vaccine was more effective than the current vaccine, and the resulting p value was 0.07, the trial would have been a failure. In this situation, the pharmaceutical company would not be able to claim that their vaccine was more effective than that which was currently on the market. (CDC 2012).

13.5.5 THE F DISTRIBUTION

Variance *provides a* measure of variability. Comparison of variability, therefore, involves comparison of variances. Suppose that σ_1^2 and σ_2^2 represent the unknown variances of two independent normal populations. The null hypothesis $H_0: \sigma_1^2 = \sigma_2^2$ asserts that the two populations have the same variance and are therefore characterized by the same variability. When the null hypothesis is true, the test statistic s_1^2/s_2^2 has an *F distribution* with parameters $n_1 - 1$ and $n_2 - 1$ called the *degree of freedom*. The term s_1^2 is the sample variance of a random sample of n_1 observations from the normal population having a variance σ_1^2, while s_2^2 is the sample variance of a random sample of n_2 observations from a normal population having a variance σ_2^2. When statistic s_1^2/s_2^2 is used as a test statistic, the critical region depends on the alternative hypothesis H_1 and α, the tolerated probability of a type I error. (See Shaefer and Theodore 2007 for additional details.)

13.5.6 ANALYSIS OF VARIANCE (ANOVA)

Analysis of variance (ANOVA) is a statistical technique featuring the splitting of a measure of total variation into components measuring variation attributable to one or more factors or combinations of factors. ANOVA is widely used in healthcare as a parametric statistical test. Other parametric tests include t-tests, described above, and linear correlation, which is the subject of Chapter 14.

The essence of ANOVA is to provide information on the reason for the displacement of all sample data from different groups from the overall mean via the sum of squares method. This total is then split into two components – one measuring the displacement *within* each group from the mean and another measuring the displacement *among* (or *between*) the means of each group with the overall mean. The former provides a measure of the error associated with data, measurements, etc, and the latter provides a measure of the variation between groups. If the latter is larger than the former, there is evidence for contradicting the null hypothesis that there is no difference between the groups; in effect, variations should not be attributed to error.

The simplest application of analysis of variance involves data classified in categories (levels) of one factor, as in the following example. Suppose that k antivirals used for the treatment of COVID-19 are to be compared with respect to remaining virus levels after use. Let X_{ij} denote the viral load level of the jth patient in a random sample of n_i patients receiving the ith antiviral ($i = 1, \ldots, k$). Let n, $\overline{X}_i$, and $\overline{X}$ denote, respectively, the total sample size, the mean of the observations in the ith sample, and the mean of the observation in all k samples (Shaefer and Theodore 2007) Then,

$$n = \sum_{i=1}^{k} m_i = km \tag{13.35}$$

$$X_i = \frac{\sum_{j=1}^{m_i} X_{ij}}{m_i} \tag{13.36}$$

$$\overline{X} = \frac{\sum_{j=1}^{m_i} \sum_{j=1}^{m_i} X_{ij}}{n} = \frac{\sum_{i=1}^{k} m_i \overline{X}_i}{n} \tag{13.37}$$

The term $\sum_{i=1}^{k} \sum_{j=1}^{m_j} (X_{ij} - \overline{X})^2$, a measure of the total variation of the observations, can be algebraically split into components (derivation not provided) as follows:

$$\sum_{i=1}^{k} \sum_{j=1}^{m_j} (X_{ij} - \overline{X})^2 = \sum_{i=1}^{k} \sum_{j=1}^{m_j} [(X_{ij} - \overline{X}_i) + (\overline{X}_i - \overline{X})]^2$$

$$= \sum_{i=1}^{k} \sum_{j=1}^{m_j} (X_{ij} - \overline{X}_i)^2 + \sum_{i=1}^{k} m_i (\overline{X}_i - \overline{X})^2 \tag{13.38}$$

In essay form,

Total sum of squares (TSS) = Residual sum of squares (RSS)

+ *Among* group sum of squares (GSS)

$$\tag{13.39}$$

The RSS can then be written as

$$\text{RSS} = \sum_{j=1}^{m_i} \sum_{i=1}^{k} (X_{ij})^2 - \sum_{i=1}^{k} \left(T_{i.}^2 / m_i \right) \tag{13.40}$$

$$\text{GSS} = \sum_{i=1}^{k} \left(T_{i.}^2 / m_i \right) - \left(T_{..}^2 / n \right) \tag{13.41}$$

where t_i is the total sum of the j's in group i and $T_{..}$ is the grand total. Thus, the measure of total variation has been expressed in terms of two components, the first measuring variation *within* the k samples, and the second measuring the variation *among* the k samples. In the terminology of analysis of variance, the *within* group sum of squares is also called the aforementioned RSS or the *error* sum of squares. The analysis of variance statistic for testing the null hypothesis of no significant difference among the antivirals with respect to reduction in viral load features comparison of the *among* group sum of squares with the *within* group sum of squares. If the mean *among* group sum of squares, i.e., the *among* group sum of squares divided by $k - 1$ (the degrees of freedom), is large relative to the mean within group sum of squares, i.e., the *within* group sum of squares divided by $n - k$, there is evidence for contradicting the null hypothesis. Additional details and illustrative examples are provided in the literature (Shaefer and Theodore 2007).

13.5.7 NONPARAMETRIC TESTS

Classical tests of statistical hypotheses are often based on the assumption that the populations sampled are normal. In addition, these tests are frequently confined to statements regarding a finite number of unknown parameters on which the specification of the pdf of the random variable under consideration depends. Efforts to eliminate the necessity of the restrictive assumptions about the population sampled have resulted in statistical methods called *nonparametric* methods, directing attention to the fact that these methods are not limited to inferences concerning population parameters. These methods are also referred to as distribution-free, emphasizing their applicability in cases in which little is known about the functional form of the pdf of the random variable observed. Nonparametric tests of statistical hypotheses are tests whose validity generally require only the assumption of a continuity of the cumulative distribution function (cdf) of the random variable involved. These approaches are highly useful in applications such as hospital length of stay assessments and quality improvement measures. (Bhatty et al. 2017).

13.6 APPLICATIONS

The following four illustrative examples are intended to portray fundamental statistical concepts in both a quantitative and qualitative manner.

Illustrative Example 13.1 DISTRIBUTION CALCULATION
Testing of effluent sewage streams from individual buildings or entire communities has been shown to be effective in detecting the presence of SARS-CoV-2. This method of detection is so effective that the CDC launched the National Wastewater Surveillance System (NWSS) to work with health departments in building new capabilities for tracking infections in municipal water systems.

The concentration of viral particles in a SARS-CoV-2 positive effluent stream is known to be normally distributed with mean $\mu = 400 \times 10^3$ RNA copies/ml and standard deviation $\sigma = 8 \times 10^3$ RNAcopies/ml. Calculate the probability that the virus concentration, C, is between 392×10^3 and 416×10^3 RNA copies/ml.

Solution

As noted in the chapter, if C is normally distributed with a mean μ and a standard deviation σ, then the random variable $(C - \mu)/\sigma$ is also normally distributed with a mean of 0 and a standard deviation of 1, with the term $(C - \mu)/\sigma$ referred to as a *standard normal variable*, with its pdf being a *standard normal curve*. The probabilities about a standard normal variable Z can be determined from Table 13.1 presented earlier in the chapter.

Since C is normally distributed with a mean $\mu = 400 \times 10^3$ and a standard deviation $\sigma = 8 \times 10^3$ RNA copies/ml. It follows that:

$$P(392 \times 10^3 < C < 416 \times 10^3)$$

$$= P\left\{ \frac{392 \times 10^3 - 416 \times 10^3}{8 \times 10^3} < \left[\frac{(C - 400 \times 10^3)}{8 \times 10^3} \right] < \frac{416 \times 10^3 - 400 \times 10^3}{8 \times 10^3} \right\}$$

$$= P\left\{ -1 < \left[\frac{(C - 400 \times 10^3)}{8 \times 10^3} \right] < 2 \right\} = P(-1 < Z < 2)$$

From tabulated normal probability tables,

$$P(392 \times 10^3 < C < 416 \times 10^3) = 0.341 + 0.477 = 0.818 = 81.8\%$$

Illustrative Example 13.2   PROBABILITY CALCULATION

A procuring agent for a regional hospital system is asked to sample a lot of 100 oxygen canisters for use in their hospital's emergency department. The sampling procedure calls for the inspection of 20 canisters. If there are any defective canisters among those 20, the entire lot is rejected; otherwise, it is accepted. Assume a 4% overall defective rate.

Suppose that, at the most, two defective canisters were allowed in the sample. What kind of protection would the hospital system have?

Solution

For this case, one needs to calculate the probability of accepting the lot (with four defective oxygen canisters) if 0, 1, or 2 defectives is allowed in the sample. Therefore,

$$P(X \le 2) = P(X = 0) + P(X = 1) + P(X = 2)$$

$$= \frac{20!}{(0!)(20!)}(0.04)^0(0.96)^{20} + \frac{20!}{(1!)(19!)}(0.04)^1(0.96)^{19}$$

$$+ \frac{20!}{(2!)(18!)}(0.04)^2(0.96)^{18}$$

$$= 0.442 + (20)(0.04)(0.4604) + (10)(19)(0.0016)(0.48)$$

$$= 0.442 + 0.368 + 0.146$$

$$= 0.956 = 95.6\%$$

Illustrative Example 13.3   Poisson Distribution

The probability that two fully vaccinated SARS-CoV-2 infected individuals will die of COVID-19 has been shown by one study to be 0.001. Calculate the probability that exactly three people will die from a sample population of 2000 patients.

Solution

For this example,

$$\lambda - (2000)(0.001) = 2$$

The Poisson equation applies,

$$f(x) = \frac{e^{-\lambda t} \lambda t^x}{x!}$$

Set x equal to 3, so that

$$f(x = 3) = \frac{e^{-2} 2^3}{3!}$$

$$= 0.18 = 18\%$$

There is an 18% probability that exactly three people will die within the sample population of 2000 COVID-19 patients.

Illustrative Example 13.4   ANOVA

After several outbreaks of COVID-19 among personnel at a regional hospital, a study is conducted to compare case numbers in various departments. Cases from five different departments are studied over three separate time frames. The following shows the number of cases for each department (group) during each time frame.

Emergency Department (Group 1)	Intensive Care Unit (Group 2)	Department of Surgery (Group 3)	Department of Medicine (Group 4)	Cardiac Care Unit (Group 5)
3	5	7	6	4
2	8	8	8	9
4	8	6	7	5

Based on the data above, calculate the following parameters:

1. *The among group of squares (GSS)*
2. *The RSS*
3. *The total sum of squares (TSS)*

(Refer to Eqs. (13.38) through (13.41) in Section 13.5.6 – Analysis of Variance.)

Solution

The data listed in above are used to calculate the total value T, and the mean, M, for each group.

Emergency Department (Group 1)	Intensive Care Unit (Group 2)	Department of Surgery (Group 3)	Cardiac Care Unit (Group 4)	Medical Floor (Group 5)
3	5	7	6	4
2	8	8	8	9
4	8	6	7	5
$T_1 = 9$	$T_2 = 21$	$T_3 = 21$	$T_4 = 21$	$T_5 = 18$
$M_1 = 3$	$M_2 = 7$	$M_3 = 7$	$M_4 = 7$	$M_5 = 6$

1. The among group of squares (GSS) is given by:

$$\text{GSS} = \sum_{i=1}^{5} \frac{T_i^2}{m_i} - \frac{T_{..}^2}{n}$$

The GSS calculation can now be completed:

$$\text{GSS} = \frac{(9)^2 + (21)^2 + (21)^2 + (21)^2 + (9)^2}{3} - \frac{90^2}{15}$$

$$= \frac{1728}{3} - \frac{8100}{15}$$

$$= 36$$

2. The RSS is given by:

$$\text{RSS} = \sum_{j=1}^{3} \sum_{i=1}^{5} x_{ij}^2 - \sum_{i=1}^{k} \frac{T_i^2}{n}$$

The RSS calculation can now be completed:

$$\text{RSS} = 602 - \frac{(9)^2 + (21)^2 + (21)^2 + (21)^2 + (9)^2}{3}$$

$$= 602 - 576$$

$$= 26$$

3. The TSS is given by:

$$\text{TSS} = \sum_{j=1}^{3} \sum_{i=1}^{5} x_{ij}^2 - \sum \frac{T_{..}^2}{n}$$

The TSS calculation can now be completed:

$$\text{TSS} = 602 - 540$$

$$= 62$$

13.7 CHAPTER SUMMARY

- An understanding of statistics is essential across all scientific fields. The emphasis of this chapter is biostatistics with an overview of statistical methods and an introduction to probability calculations.
- Probabilities are nonnegative numbers associated with the outcomes of the so-called random experiments.
- A random experiment is an experiment whose outcome is uncertain.
- A population is a collection of objects (or people) having observable or measurable characteristics defined as variates, while a sample is a group of objects drawn from a population (usually random) that are equally likely to be drawn.
- There are numerous discrete probability distributions, including the binomial distribution, the multinomial distribution, the hypergeometric distribution, and the Poisson distribution.
- There are numerous continuous probability distributions, including the exponential distribution, the Weibull distribution, the normal distribution, and the lognormal distribution.
- Knowledge of different experimental measures and mathematical modeling is key to understanding how to read and interpret experiment results. Parameters such as confidence interval, statistical significance, various distributions, and hypothesis testing are frequently encountered.
- A further depth of understanding is beneficial for experimental design, data collection, and analysis.
- Analysis of variance (ANOVA) is a statistical technique featuring the splitting of a measure of total variation into components measuring variation attributable to one or more factors or combinations of factors. The essence of ANOVA is to provide information on the reason for the displacement of all sample data from different groups from the overall mean via the sum of squares method.

13.8 PROBLEMS

1 Why is the general subject of probability and statistics important in healthcare?

2 Discuss the relationships between the hypergeometric and binomial distributions.

3 Describe how the pdf and cdf for a random discrete variable differ from that of a random continuous variable.

4 A pharmaceutical company recently investigated ten antiviral compounds in preclinical trials. Assume the success rate of each compound reaching the market to be 0.5. Using the binomial theorem, find the probability that five of the compounds are successful. Assume $n = 10$, $p = 0.5$, and $x = 5$.

REFERENCES

Bhatty, V., Butron, S., and Theodore, L. (2017). *Introduction to Engineering. Course 2: Introduction to Engineering Principles*. New York: Theodore Tutorials and Montana State University, Department of Environmental Engineering Graduate Division.

Centers for Disease Control and Prevention (CDC) (2012). Principles of epidemiology. Centers for Disease Control and Prevention. https://www.cdc.gov/csels/dsepd/ss1978/lesson2/section4.html (accessed 27 January 2022).

Merriam-Webster (2021a). Probability. https://www.merriam-webster.com/dictionary/probability (accessed 12 November 2021).

Merriam-Webster (2021b). Statistics. https://www.merriam-webster.com/dictionary/statistics (accessed 12 November 2021).

Reynolds, M. (2021). *Personal notes*, Feiburg, Germany.

Shaefer, S. and Theodore, L. (2007). *Probability and Statistics Applications for Environmental Science*. Boca Raton, FL: CRC Press/Taylor & Francis Group.

Theodore, L. (2008). *Air Pollution Control Equipment Calculations*. Hoboken, NJ: John Wiley & Sons.

Theodore, L. and Dupont, R. (2012). *Environmental Health and Hazard Risk Assessment: Principles and Calculations*. Boca Raton, FL: CRC Press/Taylor & Francis Group.

Theodore, L. and Taylor, F. (1993). *Probability and Statistics, A Theodore Tutorial*. East Williston, NY: Theodore Tutorials (originally published by USEPA/APTI, RTP, NC).

14

Linear Regression

CHAPTER MENU

As noted in the previous chapter, experimental and statistical data are often better understood when presented in the form of graphs or mathematical equations. The preferred method of obtaining the mathematical relationships between variables is to plot the data as straight lines and use the slope–intercept method to obtain coefficients and exponents; therefore, it best suits the reader to be aware of the methods of obtaining straight lines on various types of graph paper, and of determining the equations of such lines. One usually strives to plot data as straight lines because of the simplicity of the curve, and ease of interpolation and extrapolation. Graphical methods prove invaluable in straightforward investigations as well as in streamlining more complex, multivariate data analysis. This chapter will focus on regression analysis while introducing graphical methods which can be applied to a wide variety of applications in the applied sciences and engineering fields.

One or more of the many types of graphical representations may be employed for the following purposes:

- as an aid in visualizing a process for the representation of quantitative data
- for the representation of quantitative data

A Guide to Virology for Engineers and Applied Scientists: Epidemiology, Emergency Management, and Optimization, First Edition. Megan M. Reynolds and Louis Theodore.

- for the representation of a theoretical equation
- for the representation of an empirical equation
- for the comparison of experimental data with a theoretical expression
- for the comparison of experimental data with an experimental expression

Graphical or tabular presentation of data is usually used if no theoretical equations can be developed to fit the data. This type of presentation of data is one method of reporting experimental results. For example, virus droplet concentrations might be tabulated at various distances. This data may also be presented graphically. One should note that graphs are inherently less accurate than numerical tabulations. However, they are useful for visualizing variations in data and for interpolation and extrapolation.

The relation between two quantities y, the dependent variable, and x, the independent variable, is commonly obtained as a tabulation of values of y for several of different values of x. The relation between y and x may not be easy to visualize by studying the tabulated results and is often best seen by plotting y vs x. If the conditions are such that y is known to be a function of x only, the functional relationship will be indicated by the fact that the points may be represented graphically by a smooth curve. Deviations of the points from a smooth curve can indicate the reliability of the data. If y is a function of the two variables x_1 and x_2, a series of results of y in terms of x_1 may be obtained in terms of x_2. When plotted, the data will be represented by a family of curves, each curve representing the relation between x_1 and y for a definite constant value of x_2.

The bulk of the material for this chapter has been drawn from the literature. (Theodore and Taylor 1993; Shaefer and Theodore 2007)

14.1 RECTANGULAR COORDINATES

Rectangular (sometimes referred to as Cartesian) coordinates are most generally used to represent an equation of the form

$$y = mx + b \tag{14.1}$$

where

y = variable represented on the ordinate
x = variable represented on the abscissa
m = slope of the line
b = y-intercept at $x = 0$

All graphs are created by drawing a line, or lines, about one or more axes. For the purposes of this section, the presentation will be primarily concerned with graphs built around two axes – the x-axis and the y-axis. The graphical data may be presented as coordinates of the x- and y-axis in the form (x, y), or they may be presented as the abscissa (distance from the y axis or the x coordinate) and the ordinate (distance from the x-axis or the y coordinate) (Bhatty et al. 2017).

14.2 LOGARITHMIC COORDINATES

Plotting (on rectangular paper) numbers from one to ten on the ordinate versus the logarithm of the number on the abscissa easily constructs a logarithmic scale. If the points representing the numbers on the ordinate are projected to the resultant curve and then upward, the scale formed by the vertical lines will be a logarithmic scale.

The utility of the logarithmic scale lies in the fact that one can use actual numbers on the scale instead of the logarithms. When two logarithmic scales are placed perpendicular to each other and lines are drawn vertically and horizontally to represent major divisions, full logarithmic graphs result, i.e., a log–log graph. The term log–log or full logarithmic is used to distinguish between another logarithmic kind yet to be discussed, e.g., semilogarithmic (to be discussed shortly). Equations of the general form

$$y = bx^m \tag{14.2}$$

where

y = a variable
m = a constant
x = a variable
b = a constant

will plot as straight lines on (full) logarithmic paper. A form of the equation analogous to the slope-intercept equation for a straight line is obtained by plotting the equation in logarithmic form, i.e.,

$$\log y = \log b + m \log x \tag{14.3}$$

If $\log y$ vs $\log x$ were plotted on rectangular coordinates, a straight line of slope m and y-intercept $\log b$ would be obtained. An expression for the slope is obtained by differentiating Eq. (14.3).

$$\frac{d(\log y)}{d(\log x)} = m \tag{14.4}$$

When $x = 1$, $m \log x = 0$, and the equation reduces to

$$\log y = \log b \tag{14.5}$$

Hence, $\log b$ is the y-intercept.

Graphs prepared with one logarithmic scale and one arithmetic scale are termed semilogarithmic. Normally, the logarithmic scale is the *ordinate*, and the arithmetic scale is the *abscissa* on semilogarithmic paper. Thus, a semilogarithmic graph employs a standard scale for one axis and a logarithmic scale for the other axis, i.e., the graph uses scales of different mathematical proportions. The reason for this use of scales is that a logarithmic (exponential) plot would quickly exceed the physical boundaries of the graph if one were to plot values of some base having exponents of 2, 3, 4, 5, 6, 7, and 8 against an ordinary numeric scale.

Semilogarithmic paper can be used to represent – as straight lines – equations of the form

$$y = ne^{mx} \tag{14.6}$$

or

$$y = n10^{mx} \tag{14.7}$$

where

y = dependent variable
x = independent variable
n = a constant
m = a constant
e = the natural logarithm

Here, again, a form of the equation analogous to the slope–intercept equation can be obtained by placing the above equation in logarithmic form. For example, Equation (14.6) may be written as

$$\log y = b + mx \tag{14.8}$$

where

$$b = \log n$$

An expression for slope is obtained from the differential form of Eq. (14.8)

$$\frac{d(\log y)}{dx} = m \tag{14.9}$$

so that the y-intercept is log n or b. Expressions for the slope and intercept of Eq. (14.6) are similar to those of Eq. (14.7) except for the fact that natural logarithms are employed.

In addition to the types of coordinates previously discussed, the practicing engineer and applied scientist may also have need for probability versus arithmetic, probability versus logarithmic, logarithmic versus reciprocals, and other special types of graphs. Nonstandard scales, if desired, may be constructed in a

Table 14.1 Method for plotting various equation functions.

$y = a + bx$	→	Plot y vs x
$y = ax^n$	→	Plot log y vs log x or y vs x on logarithmic coordinates
$y = c + ax^n$	→	First obtain c as an intercept on a plot of y vs x, then plot log $(y - c)$ vs x on logarithmic coordinates
$y = ae^{bx}$	→	Plot log y vs x or y vs x on semilogarithmic coordinates
$y = ab^x$	→	Plot log y vs x or y vs x on semilogarithmic coordinates
$y = a + \dfrac{b}{x}$	→	Plot y vs $1/x$
$y = \dfrac{x}{a + bx}$	→	Plot x/y vs x or $1/y$ vs $1/x$

similar manner to the logarithmic scale. For example, logarithmic-probability (log-normal) graphics find application in viral transmission studies to depict fine particle size distribution. (Theodore and Ricci 2010). For more details on viral transmission, the reader is referred to Part I, Chapter 6.

14.3 METHODS OF PLOTTING DATA

The simplest procedure to employ in plotting equations of various forms is provided in the Table 14.1. Various additional forms are available in the literature provided in the reference section. Details on statistical methods for calculating the coefficients in the equations specified in Figure 14.1 are provided in a later section.(See also Chapter 13). (Theodore and Taylor 1993; Shaefer and Theodore 2007; Theodore 2008).

14.4 SCATTER DIAGRAMS

Scatter diagrams are useful in determining how two variables are related. The graph resulting from plotting the data depicts how closely changes to one variable affect another, and thus how well they are correlated. As noted in the introduction, engineers and applied scientists often encounter applications that require the need to develop a mathematical relationship between data for two or more

(a)
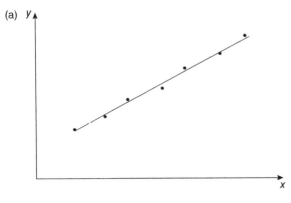

Figure 14.1 Scatter diagrams: (a) linear relationship, (b) parabolic relationship, and (c) dual-linear relationship. Source: Bhatty et al. (2017).

(b)
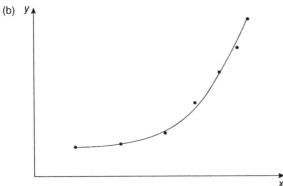

(c)
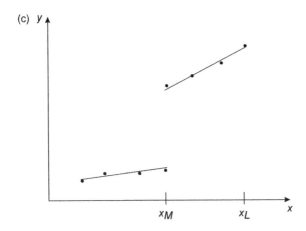

variables. For example, if y (a dependent variable) is a function of, or depends on x (an independent variable), i.e.,

$$y = f(x) \tag{14.10}$$

one may require expressing this (x, y) in equation form. (This process is referred to as regression analysis, and the regression method most often employed is the method of least squares, a topic to be addressed in Section 14.6.) The result, referred to as a scatter diagram, could take on any form. Three such plots are provided in Figure 14.1a–c.

The first plot (a) suggests a linear relationship between x and y, i.e.,

$$y = a_0 + a_1 x \tag{14.11}$$

The second graph (b) appears to be best represented by a second-order (parabolic) relationship, i.e.,

$$y = a_0 + a_1 x + a_2 x^2 \tag{14.12}$$

The third plot (c) suggests a linear model that applies over two different ranges should represent the data

$$y = a_0 + a_1 x; \quad x_0 < x < x_M \tag{14.13}$$

and

$$y = a_0' + a_1' x; \quad x_M < x < x_L \tag{14.14}$$

This multi-equation model has uses in a wide variety of applications in healthcare, epidemiology, and in pharmaceutical drug development. In any event, a scatter diagram and individual judgement can suggest an appropriate model at an early stage in the analysis (Theodore 2008).

There are no clear rules to follow for determining the type of mathematical equation which will best represent a data set. The following are some suggested approaches (Theodore 2014):

- Plot the data and look for any obvious patterns.
- Examine the first few value differences for any indications of a polynomial.
- Look for symmetry across the entire range of values.
- Look for asymptotes.
- Look for periodicity since harmonic behavior may be evident.
- Look for logarithmic or exponential tendencies by plotting on a log or semilog graph.
- Consider dividing the curve into sections and fitting these separately, with individual curves that do not necessarily match at the endpoints.
- Look for maximums and minimums.
- Look for inflections.

In any event, if one wishes to draw a line on a graph to represent data as well as possible, considering experiment scatter, one must first postulate a model representing the situation.

14.5 CURVE FITTING

Fitting a polynomial *exactly* to a set of experimental data is often not the best method for developing mathematical models representing data sets for several reasons. If a large number of data points are available, it is necessary to select the least possible number of data points to evaluate the constants of the proposed model when other data points are not considered. The resulting model will fit the selected points *exactly*, but it may oscillate tremendously between points and *not* represent the true smooth function at all (refer to Figure 14.2).

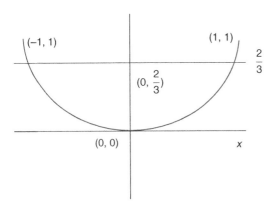

Figure 14.2 Parabola representation. Source: Courtesy of Abhishek Garg. Adapted from Theodore and Taylor.

Generally, experimental data are subject to a host of errors, many of which are difficult to quantify. If the data are subject to random errors, the data will differ from the actual values, so it is risky to fit a polynomial exactly to data that are not exact.

The method of least squares to be discussed in the next section overcomes the objections discussed above. It may be used to fit an equation, as well as possible to all the available data. The resulting equation may not fit any point exactly, but it will give a better representation of the true smooth function. The method is based on the principle of least squares, which effectively states that "the best or most probable value obtainable from a set of measurements of equal precision is that value for which the sum of the squares of the errors is a minimum" (Shaefer and Theodore 2007).

In fitting experimental data with an approximating formula or model, there are two possible approaches. One is to have the approximating function pass through the observed points. This function will then reproduce the original observed data *exactly*. The other approach is to have the approximating function retain some properties of the data, such as the shape of the curve, and to have it pass as close as possible but not necessarily through the original data points. This is usually acceptable when dealing with experimental data which is subject to errors due to a host of reasons. Once the formula is established, it may be used to obtain other values that may be desired. The method of least squares, developed by French mathematician Adrien-Marie Legendre in the nineteenth century, has been used in statistics for many years, and has become a useful technique for practitioners, engineers, and scientists (Theodore and Taylor 1993).

As stated above, it is not a necessary condition that the approximating equation take on the values of the given function at the specified data points. A natural consequence of this presumption is that an nth-order polynomial approximating equation requires exactly $n+1$ data points for evaluation. In dealing with experimental data or in developing other types of approximating equations, it is often desirable to include in the consideration a larger number of points than that which would be dictated by the direct application of the aforementioned approximation. For such cases, it is desirable to specify at the outset, not the necessity that the equation pass through a given point or set of points, but the necessity that the sum of errors at these points (or the sum of the squares of the errors between the assumed and actual values) be minimized, as originally dictated by Legendre. In such a case, a best-fitting relationship is realized. Thus, the method of least squares is based on the aforementioned assumption that the sum of the squares of the deviations between the assumed functions and the given data, i.e., the sum of the squares of the errors must be a minimum (Theodore and Taylor 1993).

As noted above, instead of requiring that an equation pass through the points y_1, $y_2 \ldots, y_n$, a weaker approximation may be used in which a suitable approximating function is made to pass as close as possible, in some sense, to the points. Care must be exercised since the model chosen may not be a good one, and Legendre's technique will yield the best of that form in the least-squares sense only.

Consider the three points on the parabola in Figure 14.2. The best straight line found by the least-squares method is the one that passes through the point $(0, \frac{2}{3})$ and is parallel to the x-axis. Obviously, extreme care must be used in choosing an appropriate form for the approximating mathematical model.

Finally, experimental data (and data in general) in many applications may be presented using a table, a graph, or a mathematical model as described earlier. Tabular presentation permits the retention of all significant figures of the original numerical data. Therefore, it is the most numerically accurate way of reporting data. However, it is often difficult to interpolate between or to extrapolate beyond the data points.

14.6 METHOD OF LEAST SQUARES

There are various categories of models that may be employed as the solution to many societal problems. These models are generally mathematical or physical in nature and the type of model often depends on the system under analysis. The mathematical approach often employed for these models is addressed in the development to follow (Shaefer and Theodore 2007).

Some of the models often employed in epidemiology are as follows:

$$y = a_0 + a_1 x; \text{Linear} \tag{14.15}$$

$$y = a_0 + a_1 x + a_2 x^2; \text{Parabolic} \tag{14.16}$$

$$y = a_0 + a_1 x + a_2 x^2 + a_3 x^3; \text{Cubic} \tag{14.17}$$

$$y = a_0 + a_1 x + a_2 x^2 + a_3 x^3 + a_4 x^4; \text{Quadratic} \tag{14.18}$$

Procedures to evaluate the regression coefficients, a_0, a_1, a_2, etc., are provided as follows. As noted in Section 14.5, the reader should note that the analysis is based on the method of least squares. This technique provides numerical values for the regression coefficients a_i such that the sum of the square of the difference (error) between the actual y and y_e predicted by the equation or model is minimized. This is shown in Figure 14.3 (Bhatty et al. 2017).

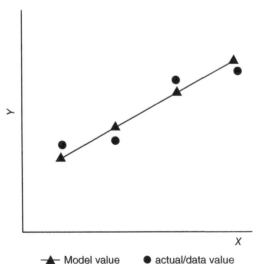

Figure 14.3 Error difference: actual and modeled predicted values. Source: Courtesy of Abhishek Garg.

─▲─ Model value   ● actual/data value

In Figure 14.3, the dots (experimental value of *y*) and triangles (equation or model value of *y*, i.e., y_e) represent the data and model values, respectively. On examining the figures one can immediately conclude that the error $(y - y_e)$ squared and summed for the points is relatively low. Also, note that a dashed line represents the error. The line that ultimately produces a minimum of the sum of the individual errors squared, i.e., has its smallest possible value, is the regression model (based on the method of least squares) (Shaefer and Theodore 2007).

To evaluate a_0 and a_1 for a linear model (see Eq. (14.11)), one employs the following least-squares algorithm for *n* data points of *y* and *x*.

$$a_0 n + a_1 \sum x = \sum y \tag{14.19}$$

$$a_0 \sum x + a_1 \sum x^2 = \sum xy \tag{14.20}$$

All the quantities given, except a_0 and a_1, can easily be calculated from the data. Because there are two equations and two unknowns, this set of equations can be solved for a_0 and a_1. For this case,

$$a_1 = \frac{n \sum xy - \sum x \sum y}{n \sum x^2 - \left(\sum x\right)^2} \tag{14.21}$$

Dividing the numerator and denominator by *n*, and defining

$$\bar{x} = \sum \frac{x}{n} \tag{14.22}$$

$$\bar{y} = \sum \frac{y}{n} \tag{14.23}$$

leads to

$$a_1 = \frac{\sum xy - \frac{\sum x \sum y}{n}}{\sum x^2 - \frac{(\sum x)^2}{n}} \tag{14.24}$$

$$= \frac{\sum xy - n\bar{x}\bar{y}}{\sum x^2 - \bar{x}^2} \tag{14.25}$$

Using this value of a_1 produces the following equation for a_0:

$$a_0 = \bar{y} - a_1 \bar{x} \tag{14.26}$$

If the model (or line of regression) is forced to fit through the origin, then the calculated value of $y_e = 0$ at $x = 0$. For this case, the line of regression takes the form

$$y_e = a_1 x; a_0 = 0 \tag{14.27}$$

with

$$a_1 = \frac{\sum xy}{\sum x^2} \tag{14.28}$$

A cubic model takes the form of Eq. (14.29)

$$y = a_0 + a_1 x + a_2 x^2 + a_3 x^3 \tag{14.29}$$

For n pairs of x–y values, the constants, a_0, a_1, a_2, a_3 can be obtained by the method of least squares so that $\sum(y - y_e)^2$ again has the smallest possible value, i.e., is minimized. The coefficients a_0, a_1, a_2, a_3 are the solution of the following system of four linear equations:

$$a_0 n + a_1 \sum x + a_2 \sum x^2 + a_3 \sum x^3 = \sum y \tag{14.30}$$

$$a_0 \sum x + a_1 \sum x^2 + a_2 \sum x^3 + a_3 \sum x^4 = \sum xy \tag{14.31}$$

$$a_0 \sum x^2 + a_1 \sum x^3 + a_2 \sum x^4 + a_3 \sum x^5 = \sum x^2 y \tag{14.32}$$

$$a_0 \sum x^3 + a_1 \sum x^4 + a_2 \sum x^5 + a_3 \sum x^6 = \sum x^3 y \tag{14.33}$$

Because there are four equations and four unknowns, this set of equations can be solved for a_0, a_1, a_2, a_3. This development can be extended to other regression equations, e.g., exponential, hyperbola, higher-order models, etc.

The correlation coefficient provides information on how well the model, or line of regression, fits the data. It is denoted by r and is given by

$$r = \frac{\sum xy - \frac{\sum x \sum y}{n}}{\sqrt{\left(\sum x^2 - \frac{(\sum x)^2}{n}\right)\left(\sum y^2 - \frac{(\sum y)^2}{n}\right)}} \tag{14.34}$$

or

$$r = \frac{n \sum xy - \sum x \sum y}{\sqrt{n\left(\sum x^2 - (\sum x)^2\right)\left(\sum y^2 - (\sum y)^2\right)}} \tag{14.35}$$

or

$$r = \frac{\sum xy - n\overline{x}\,\overline{y}}{\sqrt{\left(\sum x^2 - n\overline{x}^2\right)\left(\sum y^2 - n\overline{y}^2\right)}} \tag{14.36}$$

This equation can also be shown to take the form

$$r = \pm\left[\frac{\sum(\overline{y} - y_e)^2}{\sum(y - \overline{y})^2}\right]^{0.5} \tag{14.37}$$

The correlation coefficient satisfies the following six properties:

- If all points of scatter diagram lie on a line, then $r = +1$ or -1. In addition, $r^2 = 1$. The square of the correlation coefficient is defined as the coefficient of determination.
- If no linear relationship exists between the x's and y's, then $r = 0$. Furthermore, $r^2 = 0$. It can be concluded that r is always between -1 and $+1$, and r^2 is always between 0 and 1.

- Values of r close to $+1$ or -1 are indicative of a strong linear relationship.
- Values of r close to 0 are indicative of a weak linear relationship.
- The correlation coefficient is positive or negative depending on whether the linear relationship has a positive or negative slope. Thus, positive values of r indicate that y increases as x increases; negative values indicate that y decreases as x increases.
- If $r = 0$, it only indicates a lack of linear correlation; x and y might be strongly correlated by some nonlinear relation, as discussed earlier. Thus, r can only measure the strength of linear correlations; if the data are nonlinear, one should attempt to linearize before computing r.

Another measure of the model's fit to the data is the standard error of the estimate, or s_e. It is given as

$$s_e = \pm \left[\frac{\sum (y - y_e)^2}{n} \right]^{0.5} \tag{14.38}$$

Finally, the equation of the least-squares line of best fit for the given observations in x and y can be written in terms of the correlation coefficient as follows:

$$y - \bar{y} = r \frac{s_y}{s_x} (x - \bar{x}) \tag{14.39}$$

where,

$\bar{x}$ and $\bar{y}$ = sample means

s_x and s_y = sample standard deviations

r = sample correlation coefficient obtained from the n pairs of observations on x and y, respectively

It should be noted that the correlation coefficient only provides information on how well a model fits the data. It is emphasized that r provides no information on how good the model is or, to reword this, whether this is the correct or best model to describe the functional relationship of the data. In addition, r values for nonlinear regression models are not a valid way to determine the best fit. Although simulations and software often calculate the r-squared in nonlinear regression, researchers agree that this is an inadequate measure and, thus, an invalid, statistic. Unfortunately, r-squared results published using nonlinear models are still used often in the literature, particularly in the biochemicals and pharmaceutical fields (Theodore 1964; Spiess and Neumeyer 2010).

No discussion involving regression analysis would be complete without mention of analysis of variance, often referred to as ANOVA. As discussed in Chapter 13, ANOVA consists of splitting a measure of total variation into components measuring variation attributable to one or more factors or combinations of factors. This technique is widely used in public health and applied sciences and is especially valuable for its adaptability. ANOVA can differentiate between the variability *within* groups versus the variability *between* different groups, where there are two or more groups. As an example, ANOVA could be used to determine whether the variations in efficacy of three new COVID-19 vaccines are statistically significant or not. The pharmaceutical industry gold standard for significance level is 5%, meaning the *p* value is less than 0.05 (Bhatty et al. 2017; Theodore and Taylor 1993; Shaefer and Theodore 2007).

Theodore has defined ANOVA as a statistical technique that allows the practitioner to answer the question: *Is this the best and/or correct model?* (Theodore 2014) The method quantifies the displacement of *experimental* data from a proposed model in terms of the error associated with both the measurement of the data and the error associated with employing a poor or incorrect model. In effect, if the displacement can be attributed to experimental error, the model could be assumed to be valid.

14.7 APPLICATIONS

Four illustrative examples complement the presentation for the chapter concerning regression analysis.

Illustrative Example 14.1 METHOD OF LEAST SQUARES
Concisely explain the method of least squares.

Solution
For the purpose of this question, the method of least squares is the technique of fitting a line to a set of *n* points in such a way so the $\sum(y - y_e)^2$ has its smallest possible value (is minimized), where the sum is calculated for the given *n* pairs of *x* and *y*.

Illustrative Example 14.2 LINEAR REGRESSION
A city in the northwestern region of New York State was hard hit by COVID-19 in early 2020. Per capita weekly death rates per one thousand residents for the first five weeks of their local epidemic is provided by the following data:

Week(x)	Deaths/1000 residents (y)
1	2.3
2	4.1
3	6.2
4	8.0
5	10.4

At the end of week 5, the Center for Disease Control (CDC) was asked to provide an estimate of the potential death rate at the end of week 10. As a member of CDC's statistical analysis team, you were asked to provide this estimate.

Solution

First, plot the data available. The resulting scatter diagram, in Figure 14.4, immediately suggests that the death rate (y)–time (x) can be represented in a linear form, $y = mx + b$.

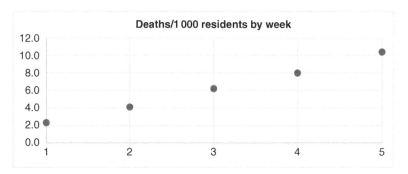

Figure 14.4 Death rates: experimental data. Source: Reynolds, M. 2022.

Using the same data as above, it is possible to use linear regression to provide the following result: $y = 2.01x + 0$.

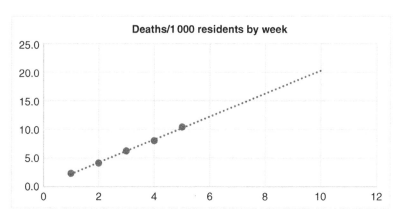

Figure 14.5 Death rates: predicted data. Source: Reynolds, M. 2022.

From the chart in Figure 14.5, it is evident that at week 10 there will be slightly more than 20 deaths per 1000 residents. (By also substituting week 10 into the derived linear equation, $y = 2*10 + 0 = 20$. Using Microsoft Excel, the exact number at week 10 is 20.1.)

Illustrative Example 14.3 LOGARITHMIC REGRESSION

Refer to Illustrative Example 14.2, above. Two weeks later, the CDC modified the initial death-rate calculations after discovering that a few death reports had been accidently duplicated due to reporting from both hospitals and the medical examiner. The new, lower, rates are included in the modified data below. In addition, data from weeks 6 to 7 are now available.

Week (x)	Deaths/1000 residents (y)
1	2.1
2	4.1
3	5.4
4	6.8
5	8.0
6	8.7
7	9.1

Recalculate the results.

Solution

The data can be plotted as was done in Illustrative Example 2. The scatter diagram with the modified data, below, now suggests that the death rate (*y*)–time (*x*) is no longer linear. One option is to select an equation in the logarithmic form, $y = a*\ln x + b$.

Using the new data, as shown in Figure 14.6, it is possible to again use data regression to provide the following result:

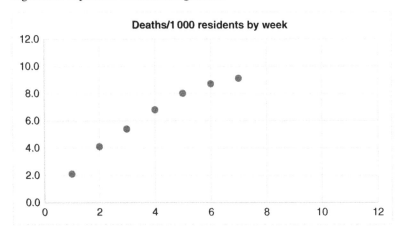

Deaths/1 000 residents by week

Figure 14.6 Modified death rate data. Source: Reynolds, M. 2022.

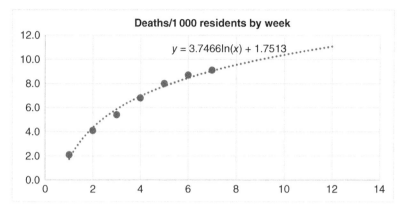

Figure 14.7 Predicted data. Source: Reynolds, M. 2022.

Using Figure 14.7, the resulting regression shows that, at week 10, there will be 10.4 deaths per 1000 people. The corresponding logarithmic regression equation, $y = 3.75*\ln(x) + 1.75$, is also shown in the chart. Solving for $x = 10$, yields

$$y = 3.75^* \ln(10) + 1.75 \qquad (14.40)$$

$$= 10.4$$

Illustrative Example 14.4 LINEAR VERSUS NON-LINEAR
Explain the differences between linear and nonlinear.

Solution
Equations are identified by the type of line or curve they produce when plotted on a graph. A linear equation, as the name implies, will result in a plot of a straight line on linear coordinate graph paper. With the linear equation, each incremental change of x (or y) will result in an incremental change of a fixed ratio in y (or x). The ratio of change between y and x is known as the slope of the line. A linear equation is also referred to as a first-degree equation.

A nonlinear equation will result in a line having one or more curves when plotted and the ratio of change between x and y is *not* constant. Nonlinear equations come in a variety of forms and may be quickly identified since they may have at least one term containing a nonlinear term. The equation $a = \pi r^2$ is representative of a nonlinear equation.

14.8 CHAPTER SUMMARY

- The preferred method of obtaining the mathematical relations between variables is to plot the data as straight lines and use the slope–intercept method to obtain coefficients and exponents.

- Rectangular (sometimes referred to as Cartesian) coordinates are most generally used to represent an equation of the form $y = mx + b$.
- The utility of the logarithmic scale lies in the fact that one can use actual numbers on the scale instead of the logarithms.
- Engineers and applied scientists often encounter applications that require the need to develop a mathematical relationship between data for two or more variables.
- Procedures to evaluate regression coefficients are based on the method of least squares.

14.9 PROBLEMS

1. Discuss the differences between regression analysis and analysis of variance. Which is more important in terms of applications?
2. Develop an original problem in linear regression that would be suitable as an illustrative example in a book.
3. Use any suitable method to linearize the following equation $ae^y = \ln x + bx$.
4. Briefly discuss intercepts.

REFERENCES

Bhatty, V., Butron, S., and Theodore, L. (2017). *Introduction to Engineering. Course 2: Introduction to Engineering Principles*. New York: Theodore Tutorials and Butte, MT: Montana State University, Department of Environmental Engineering Graduate Division.

Reynolds, M. (2022). Personal notes. Merano, Italy.

Shaefer, S. and Theodore, L. (2007). *Probability and Statistics Applications for Environmental Science*. Boca Raton, FL: CRC Press/Taylor & Francis Group.

Spiess, A. and Neumeyer, N. (2010). An evaluation of R^2 as an inadequate measure for nonlinear models in pharmacological and biochemical research: a Monte Carlo approach. *BMC Pharmacology* 10: 6.

Theodore, L. (1964). Personal notes. East Williston, NY.

Theodore, L. (2008). *Air Pollution Control Equipment Calculations*. Hoboken, NJ: John Wiley & Sons.

Theodore, L. (2014). *Chemical Engineering: The Essential Reference*. New York City, NY: McGraw Hill.

Theodore, L. and Ricci, F. (2010). *Mass Transfer Operations for the Practicing Engineer*. Hoboken, NJ: John Wiley & Sons.

Theodore, L. and Taylor, F. (1993). *Probability and Statistics, A Theodore Tutorial*. East Williston, NY: Theodore Tutorials, originally published by USEPA/APTI, RTP, NC.

15

Ventilation

CHAPTER MENU

Ventilation is one of the most important techniques available to control viruses and other contaminant levels in the workplace. (The reader should note that reference to the workplace in this chapter refers to hospitals, traditional workplace, home, laboratories, etc.) As will be discussed later in the chapter, the control of a potentially infectious disease can be achieved in two ways: by dilution of the contaminant concentration before it reaches an individuals' breathing zone by mixing it with uncontaminated air (called ventilation or dilution ventilation) or by the removal of the virus or contaminant at or near its source or point of generation to prevent its release into the environment. The latter approach is called local exhaust ventilation (LEV).

Ventilation systems can be designed for commercial, residential, or industrial applications. Industrial ventilation systems, however, require unique and specific

A Guide to Virology for Engineers and Applied Scientists: Epidemiology, Emergency Management, and Optimization, First Edition. Megan M. Reynolds and Louis Theodore.
© 2023 John Wiley & Sons, Inc. Published 2023 by John Wiley & Sons, Inc.

design features for the ancillary equipment that must be taken into account in order for the design to be successful. The technology for the design and operation of industrial ventilation systems were not well developed in the past, but significant progress has been made in recent years.

A local exhaust is generally preferred over a dilution ventilation system for virus-control purposes because a local exhaust system removes the contaminants directly from the source whereas dilution ventilation merely mixes the contaminant with uncontaminated air to reduce the contaminant concentration. Dilution ventilation may be acceptable when the contaminant concentration has a low toxicity and the rate of contaminant emission is constant and sufficiently low that the quantity of required dilution air is not prohibitively large. However, dilution ventilation generally is not practical when the acceptable concentration of the contaminant is less than 100 ppm. Additional details are provided later in the chapter.

In determining the quantity of dilution air required, one must consider the mixing characteristics of the work area in addition to the quantity (mass or volume) of the contaminant to be diluted. Thus, the amount of air required in a dilution ventilation system is much higher than the amount required in a local exhaust system. In addition, if the replacement air requires heating or cooling to maintain an acceptable workplace temperature, then the operating cost of a dilution ventilation system may greatly exceed the cost of a local exhaust system.

The reader is referred to the work of Heinsohn for an excellent treatment of ventilation (Heinsohn 1991). In addition, it should be noted that several of the illustrative examples and problems to follow were excerpted and/or adapted from publications resulting from National Science Foundation (NSF) sponsored faculty workshops and two subsequent publications (Ganesan et al. 1996; Dupont et al. 1998).

15.1 INTRODUCTION TO INDUSTRIAL VENTILATION SYSTEMS

When designing manufacturing sites or other industrial construction, ventilation must be considered an integral part of the overall industrial process. This approach is essential to ensure a safe and efficient design of the ventilation system within the overall process. Naturally, much of this applies to hospitals and other medical facilities. The application areas for ventilation systems (both GVA and LEV), are as follows:

- Control of contaminants to acceptable levels
- Control of heat and humidity for comfort
- Prevention of fires and explosions

For general, as well as industrial, applications, all these factors are important considerations at the design stage. In general, the controlling design criteria will be to achieve acceptable contaminant levels in the location in question.

The following are four steps to be considered in solving *industrial* ventilation problems.

1. *Process modifications.* This should always be the first approach to establish whether the process can be modified, and the contaminant can be controlled to an acceptable level. The result of this process modification approach is that there is no need for a ventilation system.
2. *Local exhaust ventilation.* With the exception of process modifications, LEV systems will almost always be the most cost-effective technology for controlling contaminant levels in the workplace. Contaminants are controlled at the source with low volumes of gas.
3. *General work area* (i.e., general or dilution ventilation). The concept of general or dilution ventilation is based on supplying or exhausting large volumes of air to an occupied space or the complete processing/manufacturing complex. The ventilation air can be supplied and exhausted by natural ventilation or mechanical ventilation. A subsequent section will describe the features of general ventilation for various facilities.
4. *Personal protective equipment (PPE).* For some specific applications, it may be necessary to use personal protective equipment to reduce the exposure for workers. In general, this is only applicable if process modification or ventilation is not feasible. Personal protective equipment can be considered for nonroutine maintenance work or for processes or systems that do not have feasible engineering control solutions. (See also Part I, Chapter 5.)

15.2 COMPONENTS OF VENTILATION SYSTEMS

The function of a ventilation system is to prevent contaminants from entering the air in the workplace. There are several ways to accomplish this goal. Figure 15.1, below, provides a line diagram of the basic components of a ventilation system. All ventilation systems contain some, if not all, of the components shown in Figure 15.1. Note that either the control device and/or fan can be located in the room/workplace area (Theodore 2008).The major components of a ventilation system include the following:

- Exhaust hood
- Contaminant control device
- Exhaust vent or stack
- Fan

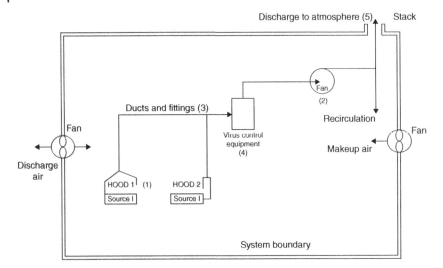

Figure 15.1 Components of an industrial ventilation system. Source: Courtesy of Abhishek Garg. (Adapted from Theodore 2014.)

- Valves and fittings
- Ductwork

The latter three components receive treatment in the next section. Several types of hoods are available. One must select the appropriate hood for a specific operation to effectively remove viruses or contaminants from a work area and transport them into the ductwork. The ductwork (see Section 15.3) must be sized such that the contaminant is transported without being deposited within the duct; adequate velocity must be maintained in the duct to accomplish this (Theodore 2008; Theodore 2014). Selecting a control device that is appropriate for contaminant removal is important to meet certain pollution control removal efficiency requirements. The exhaust fan is the workhorse of the ventilation system. The fan must provide the volumetric flow at the required static pressure and must be capable of handling contaminated air characteristics such as dustiness, corrosivity, and moisture in the air stream. Properly venting the exhaust out of the building is equally necessary to avoid virus or contaminant recirculation into the air intake or the building through other openings. Such problems can be minimized by properly locating the vent pipe concerning the aerodynamic characteristics of the building. In addition, all or a portion of the cleaned air may be recirculated to the workplace Primary (outside) air may be added to the workplace and is referred to as makeup air; the temperature and humidity of the makeup air may have to be controlled. It also may be necessary to exhaust a portion of the room air.

15.3 FANS, VALVES AND FITTINGS, AND DUCTWORK

The three other key components of the ventilation system pictured in Figure 15.1, are (1) Fans, and (2) Valves and Fittings and (3) Ductwork. Each are briefly described below (Theodore 2008).

15.3.1 FANS

If a pressure difference is required between two points in a system, a prime mover such as fan is often used to provide the necessary pressure and/or flow impetus. Engineers and applied scientists are often called on to specify prime movers more frequently than any other piece of processing equipment. In a general sense, these primary movers are to a process plant what the engine is to one's automobile, or what the heart is to a human.

Moving fluid through the various pieces of equipment at a facility, plant or hospital (including piping and duct work) requires mechanical energy not only to impart an initial velocity to the fluid but also, more importantly, to overcome pressure losses that occur throughout the flow path of the moving fluid. This energy may be imparted to the moving stream in at least one of three modes:

- *an increase in stream velocity*, where the additional energy takes the form of an increase in the kinetic energy as the bulk stream velocity increases.
- *an increase in stream pressure*, where the internal energy (mainly potential energy, but usually some kinetic as well) of the stream increases; this pressure increase may also cause a stream temperature rise, which represents an internal energy increase (Theodore 2008).
- *an increase in stream's height, or some combination of the three*, which may be relatively small for some operations, where the bulk fluid experiences an increase in potential energy in Earth's gravitational field.

The devices that convert electrical energy into mechanical energy that can be applied to various ventilator streams are fans, which move low-pressure gases; Except for special applications, centrifugal units are normally employed. These units are normally rated in terms of the following four characteristics:

- *Capacity* – the quantity of fluid discharged per unit time (the mass flow rate).
- *Pressure increase* – often reported for pumps as head: head can be expressed as the energy supplied to the fluid per unit mass and is obtained by dividing the increase in pressure (the pressure change) by the fluid density.
- *Power* – the energy consumed by the mover per unit time.
- *Efficiency* – the energy supplied to the fluid divided by the energy supplied to the unit.

Finally, the net effect of most fans is to increase the pressure of the fluid.

The reader should note that the terms fans and blowers are often used inter-changeably and are not differentiated in the following discussion; thus, any statements about fans apply equally to blowers. Strictly speaking, however, fans are used for low pressure (drop) operations, generally below 2.0 lb/in.^{2} (psi). Blowers are generally employed when generating pressure heads ranging from 2.0 to 14.7 psi. Higher-pressure operations require compressors.

Fans are usually classified as centrifugal (see above) or axial-flow type. In centrifugal fans, the gas is introduced into the center of a revolving wheel (the eye) and discharges at right angles to the rotating blades. In axial-flow fans, the gas moves directly (forward) through the axis of rotation of the fan blades. Both types are used, but, it is the centrifugal fan that is employed at most facilities.

The fan selection procedure for the practitioner requires in part, an examination of the fan curve and the system curve. A fan curve, relating to static pressure with flow rate, is provided in Figure 15.2 denoted as h_c. Note that each type of fan has its own characteristic curve. Also note that fans are usually tested in the factory or laboratory with open inlets and long smooth straight discharge ducts. Since these conditions are seldom duplicated in the field, the actual operation often results in lower efficiency and reduced performance. A system curve is also shown in Figure 15.2. This curve must be calculated or provided prior to the purchase of a fan to obtain the best estimate of the pressure drop across the system through which the fan must deliver the gas. The curve should approach a straight line with an approximate slope of 1.8 on log–log or ln–ln coordinates. The system pressure (drop), defined as the *resistance* associated with ducts, fittings,

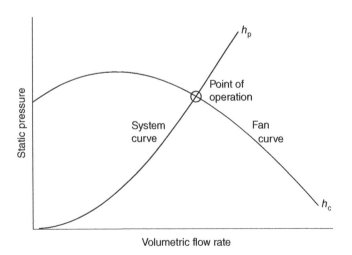

Figure 15.2 System and fan characteristics.

equipment, contractions, expansions, etc., is a measure of h_p, the head loss of a system (Farag 1995; Abulencia and Theodore 2009).

Numerous process and equipment variables are classified as part of a fan specification. These include flow rate (*acfm*), temperature, density, gas stream characteristics, static pressure that needs to be developed, motor type, drive type, materials of construction, fan location, and noise controls (Farag 1995; Abulencia and Theodore 2009).

15.3.2 VALVES AND FITTINGS

Tubing and other conduits are used for the transportation of gases, liquids, and slurries. These are often connected and may also contain various valves and fittings, including expansion and contraction joints. There are four basic types of connecting conduits:

- Threaded
- Bell-and-spigot
- Flanged
- Welded

Extensive information on these connections classes is available in the literature (Farag 1995; Abulencia and Theodore 2009; Theodore 2014).

Because of the diversity of types of systems, fluids, and the environments in which valves must operate, a vast array of valve types have been developed. Examples of the common types are the globe valve, gate valve, ball valve, plug valve, pinch valve, butterfly valve, and check valve. Each type of valve has been designed to meet specific needs. Some valves are capable of throttling flow, other valve types can only stop flow, others work well in corrosive systems, and others handle high-pressure fluids.

Valves have two main functions in a pipeline or duct: to control the amount of flow or to stop the flow completely. Of the many different types of valves employed in practice, the most commonly used are the gate valve and the globe valve. The gate valve contains a disk that slides at right angles to the flow direction. This type of valve is used primarily for on/off control of a liquid flow. Because small lateral adjustments of the disk can cause significant changes in the flow cross-sectional area, this type of valve is not suitable for accurate adjustment of flow rates. As the fluid passes through the gate valve only a small amount of turbulence is generated; the direction of flow is not altered and the flow cross-sectional area inside the valve is only slightly less than that of the pipe or duct. As a result, the valve causes only a minor pressure drop. Problems with abrasion and erosion for the disk arise when the valve is used in positions other than fully open or fully closed. Unlike the gate valve, the globe valve – so called because of the spherical shape of the valve body – is designed for more sensitive flow control. In this type of valve, the fluid passes through the valve in a somewhat tortuous or circuitous route. In one form, the seal is a horizontal ring into which a plug with a slightly beveled edge is inserted when the stem is closed. Good control of flow is achieved with this type of valve, but at the expense of a higher pressure loss than with a gate valve.

A fitting is a piece of equipment whose function is one or a combination of the following (Farag 1995; Abulencia and Theodore 2009; Theodore 2014):

- The joining of two pieces of straight pipe (e.g., couplings and unions)
- The changing of pipeline direction (e.g., elbows and Ts)
- The changing of pipeline diameter (e.g., reducers and bushings)
- The terminating of a pipeline (e.g., plugs and caps)
- The joining of two streams (e.g., Ts and Ys)

Each of these is briefly discussed. A *coupling* is a short piece of pipe threaded on the inside and used to connect straight sections of pipe with no change in direction or size. When a coupling is opened, a considerable amount of piping must usually be dismantled. A *union* is also used to connect two straight sections but differs from a coupling in that it can be opened relatively easily without disturbing the rest of the pipeline. An *elbow* is an angle fitting used to change flow direction, usually by 90°, although 45° elbows are also available. In addition, a T component (shaped like the letter T) can be used to change flow direction; this fitting is more often used to combine two streams into one; i.e., when two branches are to be connected at the same point. A *reducer* is a coupling for two sections of different diameters. A *bushing* is also a connector for pipes or ducts of different diameters, but unlike the reducer-coupling, is threaded on both the inside and outside. The larger pipe screws onto the outside of the bushing and the smaller pipe screws into the inside of the bushing. *Plugs*, which are threaded on the outside, and *caps*, which are threaded on the inside, are used to terminate a pipeline. Finally, a Y component (shaped like the letter Y) is similar to the T and is used to combine two streams (Theodore 2014).

15.4 SELECTING VENTILATION SYSTEMS

Exposure to viruses in a home, hospital, or workplace can be reduced by proper ventilation. As noted above, ventilation can be provided either by dilution ventilation or by a local exhaust system. In dilution ventilation, air is brought into the work area to dilute the contaminant sufficiently to minimize its concentration and subsequently reduce worker exposure. In a local exhaust system, the contaminant itself is removed from the source through hoods.

It should be noted that a local exhaust is generally preferred over a dilution system for health/hazard control. In determining the quantity of dilution air required, one must also consider mixing characteristics of the area in question, in addition to the quantity (mass or volume) of contaminant to be diluted. Thus, the amount of air required in a dilution ventilation system is much higher than the amount required in a local exhaust system. In addition, if the replacement air requires heating or cooling to maintain an acceptable workplace temperature, then the operating cost of a dilution ventilation system may further exceed the cost of a local exhaust system.

As mentioned earlier the major components of any industrial ventilation system generally include the following:

- Exhaust hoods
- Ductwork
- Contaminant control devices
- Exhaust fans
- Exhaust vents or stacks

Several types of hoods are available. One must select the appropriate hood for a specific operation to effectively remove viruses or contaminants from an area and transport them into the ductwork. The ductwork must be sized such that the contaminant is transported without being deposited within the duct; adequate velocity must be maintained in the duct to accomplish this. Selecting a control device that is optimal for the removal of the viruses or contaminants is important to ensure meeting any relevant control removal efficiency requirements.

The exhaust fan is the workhorse of the ventilation system. The fan must provide the necessary volumetric flow at the required static pressure and must be capable of handling contaminated air characteristics such as dustiness, corrosiveness, and moisture in the air streams (Farag 1995; Abulencia and Theodore 2009; This also applies to viruses).

Properly venting the exhaust out of a building is often necessary to avoid viruses or contaminant recirculation into the air intake or into a building through other openings. As noted earlier, such problems can be minimized by properly locating the vent pipe in relation to the aerodynamic characteristics of the building. In addition, all or a portion of the cleaned air may be recirculated. Once again, primary (outside) air may be added, and is referred to as makeup air; the temperature and humidity of the makeup air may have to be controlled. It also may be necessary to exhaust a portion of the room air.

When selecting a ventilation system, one should never overlook the most obvious corrective step – modify the process to generate smaller amounts of contaminant and take steps to prevent contaminants entering the workplace. General practices are not set in stone, operational costs and contaminant exposure can often be reduced simultaneously by modifying the process or choosing different raw (feed) materials.

Once one has eliminated all opportunities to reduce the generation of viruses or contaminants, a ventilation system may have to be designed. To a great extent the choice of a ventilation system is dictated by how the emissions are generated. If there are many sources of emission and they are distributed throughout the process, the emissions are called fugitive and must be controlled one at a time. If the emissions are generated from well-defined points in a process, they are referred to as *point-source emissions* and can be controlled more easily.

Ventilation *hood* is a generic phrase given to uniquely configured air inlets through which contaminated air is withdrawn. The word denotes the function to be performed rather than any particular geometrical configuration.

Design of ventilation systems requires making decisions in several stages:

- Identify the viruses/contaminants and understand their effect on health.
- Select the maximum exposure limits to be used as design criteria and standards to judge the performance of the system.
- Design the "hood" and select the volumetric flow rates of exhaust air, recirculated air, and makeup air.
- Design the duct system, select the fans, and compute the operating costs.
- Select the gas cleaning device (if applicable) to remove contaminants before discharge to the outside environment or recirculation to the workplace.
- Conduct laboratory experiments on the ventilation system, test the full-scale system in the field and sample the air in the vicinity of the source and the patient or worker to ensure that levels remain below maximum exposure limits.

Selecting the maximum exposure limits is often the principal concern.

15.5 KEY PROCESS EQUATIONS

For industrial ventilation applications, gas flow in the ductwork can be treated as one-dimensional if the proper averages are used for the variables across the section flow area. A simplified equation can be written for any two points in a fluid system based on one-dimensional compressible flow with only one entrance and one exit in the control volume. This equation is as follows:

$$\rho_1 A_1 v_1 = \rho_2 A_2 v_2 \tag{15.1}$$

where ρ is the specific density of the fluid (lb/ft^3); A, the cross-sectional area of the duct (ft^2) and v, the average gas velocity (fpm).

For most ventilation applications, the absolute pressure remains relatively constant and hence the density of the gas remains relatively incompressible. For this case $\rho_1 = \rho_2 =$ constant and the above equation becomes, $v_1 A_1 = v_2 A_2$ or $q_1 = q_2$, where $q = Av$, the volumetric flow rate (ft^3/min in English units). These equations must be modified if there is more than one inlet area, A_1 or outlet area A_2 or if steady state conditions do not exist.

Numerous process and equipment equations need to be specified as part of a ventilation system. These include flow rate (acfm), temperature, density, gas stream characteristics, static pressure that needs to be provided.

Some key Equations are provided below. Gas (or air) horsepower (GHP) and brake horsepower (BHP) are the two terms of interest. These may be calculated from Eqs. (15.2) and (15.3):

$$\text{GHP} = 0.0001575 q_a \Delta P = \frac{q_a \Delta P}{6356} \tag{15.2}$$

$$\text{BHP} = \frac{0.0001575 q_a \Delta P}{\eta_f} \tag{15.3}$$

where

$q_a =$ volumetric flow rate, (acfm) actual cubic feet per minute
$\Delta P = SP =$ static pressure head developed, in H_2O
$\eta_f =$ fractional fan efficiency

15.5.1 Regarding Friction Losses

These frictional losses can take several forms. An important engineering problem is the calculation of these losses. It can be shown that the fluid can flow in either of two modes – laminar or turbulent. For laminar flow, an equation is available from basic theory to calculate friction loss in a pipe. In practice, however, fluids (particularly gases) are rarely moving in laminar flow (Theodore 1971; Theodore 2006; Bird et al. 2002).

One can theoretically derive the frictional loss term, h_f, for laminar flow. The equation takes the form

$$h_f = \frac{32 \mu v L}{\rho g_c D^2} \tag{15.4}$$

for a fluid flowing through a straight cylinder of diameter D and length L. A friction factor, f, that is dimensionless, may now be defined as (for laminar flow)

$$f = \frac{16}{Re} \tag{15.5}$$

so that Eq. (15.4) takes the form

$$h_f = \frac{4 f L v^2}{2 g_c D} \tag{15.6}$$

Although the above equation describes friction loss or the pressure drop across a conduit of length L, it can also be used to provide the pressure drop due to friction per unit length of conduit; for example, $\Delta P/L$ by simply dividing the above equation by L.

It should also be noted that another friction factor term exists in the literature that differs from that provided in Eq. (15.5). In this other case, f_D is defined as

$$f_D = \frac{64}{Re} \tag{15.7}$$

In essence, then

$$f_D = 4f \tag{15.8}$$

The term f is defined as the *Fanning friction factor* while f_D is defined as the *Darcy* or *Moody friction factor*. Care should be taken as to which of the friction factors is being used in an application. In general, chemical engineers employ the Fanning friction factor; other engineers prefer the Darcy/Moody friction factor.

If the cross-section of a conduit enlarges gradually so that the flowing fluid velocity does not undergo any disturbances, energy losses are minor and may be neglected. However, if the change is sudden, as in a rapid expansion, additional friction losses can result. For such sudden enlargement/expansion situations, the loss can be estimated by

$$h_{f,e} = \frac{v_1^2 - v_2^2}{2g_c}; \quad [e = \text{sudden expansion (SE)}] \tag{15.9}$$

where $h_{f,e}$ is the loss in head, v_2 is the velocity at the larger cross-section, and v_1, is the velocity at the smaller cross-section. When the cross-section of the pipe is reduced suddenly (as with a contraction), the loss may be expressed by

$$h_{f,e} = \frac{Kv_2^2}{2g_c}; \quad [e = \text{sudden contraction (SC)}] \tag{15.10}$$

where v_2, is the velocity at the smaller cross-section and K is a dimensionless loss coefficient that is a function of the ratio of the two cross-sectional areas. Both of these calculations receive treatment in the literature (Farag 1995; Abulencia and Theodore 2009).

15.6 VENTILATION MODELS

Indoor air pollution is rapidly becoming a major health issue in the United States. Indoor pollutant levels are quite often higher than outdoors, particularly where buildings are tightly constructed to save energy. Since most people spend nearly 90% of their time indoors, exposure to unhealthy concentrations of indoor air pollutants is often inevitable. The degree of risk associated with exposure to indoor pollutants depends on how well buildings are ventilated and the type, mixture, and amounts of pollutants in the building.

As noted earlier, ventilation is the field of applied science concerned with controlling airborne contaminants and viruses to produce healthy conditions for workers, doctors, patients, etc. and a clean environment. However, ventilation cannot completely prevent contaminants from entering the workplace. More to the point, and within the realm of achievement, is the goal of controlling contaminant exposure within prescribed limits. To accomplish this goal, one must be able to predict – hopefully accurately – the concentration of a species in a control volume.

Regarding development of a model, one can make the assumption that the ventilator workplace or room is the control volume. Apply the aforementioned law of conservation of mass (See Part I, Chapter 6) to the species (e.g., a virus) to develop the governing question:

$$\{\text{Rate of mass in}\} - \{\text{Rate of mass out}\}$$

$$\pm \{\text{Rate of mass generated or destroyed in the room}\}$$

$$= \{\text{Rate of mass accumulated in the room}\} \tag{15.11}$$

Employing the notation (in SI units) specified below one may obtain a new equation.

$V =$ volume of the room, m^3

$q_0 =$ volumetric flow rate of ventilation air, m^3/s

$c_0 =$ concentration of the chemical in ventilation air, $gmol/m^3$

$c =$ concentration of the chemical leaving the ventilated room, $gmol/m^3$

$c_1 =$ concentration of the chemical initially present in the ventilated room, $gmol/m^3$

$r =$ rate of disappearance of the chemical in the room due to reaction and/or other effects, $gmol/m^3 \cdot s$

Each term in eq. 15.11 is evaluated below:

$$\{\text{Rate of mass in}\} = qc_{in} \tag{15.12}$$

$$\{\text{Rate of mass out}\} = qc \tag{15.13}$$

$$\{\text{Rate of mass generated}\} = rV \tag{15.14}$$

$$\{\text{Rate of mass accumulated}\} = V\frac{dc}{dt} \tag{15.15}$$

Substituting into Eq. (15.11) leads to:

$$qc_{in} - qc + rV = V\frac{dc}{dt} \tag{15.16}$$

Rearrangement leads to:

$$\frac{q}{V}(c_{in} - c) + r = \frac{dc}{dt} \tag{15.17}$$

The term V/q represents the average time the chemical remains in the room, or its residence time, and is commonly designated as θ. The reciprocal of θ is defined as the space velocity (Theodore 2012).

Equation (15.7) may be written as:

$$\frac{dc}{dt} = \frac{c_{in} - c}{\theta} + r \tag{15.18}$$

These last three equations represent a model that can be employed to describe the concentration of a specific contaminant in a ventilated control volume. The following is also assumed in the development:

- The contents of the control volume are unchanged during the ventilation process.
- The contents are perfectly mixed.
- The (total) mass density of the system is constant.

Finally, there are other forms of the above equation. Several equivalent equations may also be derived on a mass basis containing any of several conversion variables. (Theodore 2012) These equations can be used, provided they are dimensionally consistent. The choice is generally one of convenience.

15.7 MODEL LIMITATIONS

There are numerous deficiencies or inadequacies in the understanding of how to not only design a ventilation system to satisfy standards of performance but also to accurately predict concentration variation in a ventilated control volume. Some of the later problems include:

- Inability to predict the concentration
- Effects of drafts and wakes in the control volume
- New or unusual contaminant sources
- Concept of the capture velocity
- Concept of source term(s)

All of these concerns are deficiencies that arise during the motion of contaminants from their source of origin to the outlet from the ventilation system.

Regarding the effects of drafts and wakes in the control volume, theoretical consideration of ventilation systems is often based on the assumption that surrounding the source is a quiescent body of air. Nothing could be further from the truth. Every *workplace* contains equipment and workers performing tasks that produce room air currents of unique and unpredictable character, i.e., *drafts*. Spurious room air currents can also be produced.

Other factors affecting the motion of contaminants are walls, partitions, equipment surfaces, and other fixtures. In some cases, impediments may be eliminated, but for the most part they may have to be incorporated in the design of the ventilation system from the beginning. An even more vexing circumstance is trying to correct unwanted room air currents produced by equipment or new procedures put in place after the ventilation system has been installed.

Theodore has provided several nonideal approaches that can be divided into three main categories (Theodore 2012):

- Imperfect mixing
- Bypassing/short circuiting
- Both imperfect mixing and bypassing/short circuiting

Pockets of stagnant fluid or *dead spots* can exist. Here, the discharge stream is lower because the remainder of the fluid resides in the control volume for a shorter period of time. The insertion of baffles in the control volume can enhance the mixing process (Fogler 2006). Fogler has classified these as

1. Segregated flow in which the elements of fluid do not mix but follow separate paths through the system so that they have different residence times, and
2. Micromixing in which the adjacent elements of fluid are partially mixed.

Unfortunately, much of this information is not available for a real system, and other approaches must be employed. The important fact for the practicing engineer or applied scientist is to remember that deviations from ideal behavior negatively impacts the performance of a ventilation system. Every attempt should be made to minimize or eliminate this effect (Theodore et al. 2009; Theodore 2011; Theodore 2012).

Finally, Theodore has also noted that if the enthalpy of reaction for any of the species, including viruses, is not negligible, or, if it is required to supply or remove thermal energy in the form of heat from the control volume, an energy balance around the system must be obtained in order to determine the temperature and its variation with control volume (Theodore 2012).

15.8 INFECTION CONTROL IMPLICATIONS

The COVID-19 pandemic has highlighted the crucial need for proper ventilation in order to minimize the viral transmission of SARS-CoV-2. The highly contagious nature of the virus indicates that there is a vast difference in risk of infection between indoor ventilation and natural (outdoor) ventilation, as particles (viruses) spread between people more readily indoors than outdoors. The CDC recommends contagion mitigation strategies to reduce viral concentration. Since COVID-19 is dose-dependent, lower concentrations decrease the potential of inhaling high doses and therefore can reduce the chance of infection (or reduce the severity if it does occur) (CDC 2021).

In 2007, the World Health Organization (WHO) proposed guidelines for proper ventilation in rooms for patients infected with airborne diseases. In hospitals, the requirements for designating a room as under *airborne precaution* includes ventilation with at least 12 air changes per hour (ACH) and a controlled direction of airflow to contain airborne infections. A properly (mechanically) ventilated room should meet the following minimal requirements to ensure proper air handling and airflow direction (Atkinson 2009):

- A negative pressure differential of >2.5 Pa (0.01 in. water gauge).
- An airflow differential >125 cfm (56 l/s) of exhaust versus supply.
- Clean-to-dirty airflow (inlet) with an exhaust to the outside, (or a high efficiency particulate air (HEPA) filter if room air is recirculated).
- Ability to fully seal off the room, permitting a maximum 0.046 m^2 leakage.
- >12 ACH for a new building, and >6 ACH in existing buildings (e.g., equivalent to 40 l/s for a $4 \times 2 \times 3$ m^3 room).

Studies of several infectious diseases have demonstrated the relationship between the transmission of airborne illnesses and poor ventilation, some of which are listed below (Atkinson 2009; CDC 2021).

- Varicella-Zoster virus (Chickenpox)
- Measles
- Smallpox
- Pulmonary Tuberculosis (TB)
- Influenza
- Corona viruses

Lack of ventilation or low ventilation rates are associated with increased infection rates or outbreaks of airborne diseases. The importance of high ventilation rates for infection control cannot be overstated in preventing the spread of COVID-19. However, this is only one of several prevention methods necessary for decreasing risk, including vaccination, frequent testing, personal protection equipment, physical distancing, and frequent handwashing (CDC 2021).

The above precautions apply only to airborne infection, as proper ventilation does not protect against direct contact or normal, close-range droplet transmission. However, viruses such as those causing SARS and COVID-19, which are principally transmitted by droplets, may be considered as "opportunistic airborne" diseases under certain circumstances. This occurs when fine aerosols are generated that can remain in the air for long periods. Such aerosol-generating behaviors include singing, shouting, and coughing, as well as performing high-risk medical procedures, such as intubation, cardiopulmonary resuscitation, and bronchoscopy (Atkinson 2009; CDC 2021; Vance et al. 2021).

According to the Wells–Riley equation, the probability of infection through infectious droplets is inversely correlated to the ventilation rate. As presented below, the parameters used in the Wells–Riley equation include ventilation rate, generation of droplet nuclei from the source (quanta/minute), and duration of exposure (Atkinson 2009).

$$P = \frac{D}{S} = 1 - \exp\left(-\frac{Ipqt}{Q}\right) \qquad (15.19)$$

where:

P = probability of infection for susceptible persons
D = number of disease cases
S = number of susceptible persons
I = number of infectors
p = breathing rate per person (m³/h)
q = quantum generation rate by an infected person (quanta/s)
t = total exposure time (seconds)
Q = outdoor air supply rate (m³/s).

Based on this model, in situations of high quanta production (e.g., high-risk, aerosol-generating procedures), the estimated probability of infection with 15 minutes of exposure in a room with 12 ACH would be below 5% (Atkinson 2009).

15.9 APPLICATIONS

Four illustrative examples complement the presentation for this chapter on ventilation.

Illustrative Example 15.1 Application of Probability Equations
A sample of five ventilators is removed and tested "with replacement" from a government storage facility that is known to contain 95% working ventilators. (The term "with replacement" indicates that each ventilator tested is returned to the lot before the next one is drawn.) What is the probability that the number of defective fans in the sample is at most 2?

Solution

This random process consists of drawing a ventilator at random with replacement from a lot. The random experiment is performed five times because a sample of five ventilators is drawn with replacement from the lot. Therefore n = 5. *Also note that the performances are independent because each ventilator is replaced before the next is drawn.*

Therefore, the composition of the lot is exactly the same before each drawing. (For additional details, the reader is referred to Chapter 13: Probability and Biostatistics.)

For this problem, associate *success, p*, with drawing a *defective* ventilator. (Associate *failure, q*, with drawing a *working* ventilator.) Because 95% of the lot is working, $q = 0.95$, while $p = 0.05$. Substitute these values (n, p and q) in the binomial equation, as described in Chapter 13.

$$f(x) = \frac{5!}{x!(5-x)!}(0.05)^x(0.95)^{5-x}; x = 0, 1, \ldots, 5$$

For less than three defectives, x can assume values of 0, 1, and 2. One may now substitute the appropriate values of X to obtain the required probabilities.

$$P(\text{exactly 0 defectives}) = P(X = 0) = \frac{5!}{0!5!}(0.05)^0(0.95)^5$$

$$= 0.774 = 77.4\%$$

$$P(\text{exactly 1 defectives}) = P(X = 1) = \frac{5!}{1!4!}(0.05)^1(0.95)^4$$

$$= 0.0204 = 2.14\%$$

$$P(\text{exactly 2 defectives}) = P(X = 2) = \frac{5!}{2!3!}(0.05)^2(0.95)^3$$

$$= 0.0214 = 2.14\%$$

Therefore,

$$P(X < 0) = P(X = 0) + P(X = 1) + P(X = 2)$$

$$= 0.774 + 0.204 + 0.0214$$

$$= 0.999 = 99.9\%$$

Illustrative Example 15.2 STANDARD DEVIATION
The temperature of an emergency room in a hospital during the winter months is normally distributed with a mean 66°F and a standard deviation 3.0°F. Calculate the probability that the temperature is between 65 and 72°F.

Solution

Normalizing the temperature T gives

$$Z_1 = \frac{65 - 66}{3.0} = -0.333$$

$$Z_2 = \frac{72 - 66}{3.0} = 2.0$$

Therefore,

$$P(-0.333 < Z < 2.0) = P(0.0 < Z < 2.0) - P(-0.333 < Z < 0.0$$

$$= 0.4722 - (-0.1293)$$

$$= 0.6015 = 60.15\%$$

Illustrative Example 15.3   HALF-LIFE CALCULATIONS

Determine the time, in hours, required for the concentration of a virus in a hospital room to be reduced to one-half of its initial value. Assume the first-order reaction velocity constant for the virus is 0.07/h (Theodore 2012).

Solution

The decay (disappearance) of a species can often be described as a first-order function:

$$\frac{dC}{dT} = -kC$$

where:

C = concentration at time t
t = time
k = first-order reaction rate constant

The integrated form of this equation is,

$$\ln \frac{C_0}{C} = kt$$

where C_0 is the concentration at time zero. When half of the initial material has decayed (reacted), C_0/C is equal to 2; the corresponding time is given by the following expression:

$$t_{1/2} = \frac{\ln(2)}{k}$$

$$= \frac{0.69}{0.07/h}$$

$$= 9.86 \text{ h}$$

Illustrative Example 15.4 Vᴇɴᴛ Cᴀʟᴄᴜʟᴀᴛɪᴏɴꜱ

1. *Develop a solution to the equation describing the concentration in a room as a function of time if there are no "reaction" effects, that is $r = 0$ (see Section 15.5).*
2. *Develop a solution to the equation describing the concentration in a room as a function of time if $r = -k$. Note that the minus sign indicates that the virus is disappearing.*
3. *Develop a solution to the equation describing the concentration in a room as a function of time if $r = -kc$. Once again, note that the minus sign indicates the virus is disappearing.*

Solution

1. If it is assumed that the virus does not react, $r = 0$. Substituting this into Eq. (15.18) (in Section 15.5) for r yields:

$$\frac{dc}{dt} = \frac{C_{in} - c}{\theta} + 0; \quad \frac{dc}{dt} = \frac{C_{in} - c}{\theta}$$

Separating variables and integrating yields:

$$\frac{dc}{c_{in}} = \frac{dt}{t}$$

$$\int_{c_0}^{c} \frac{dc}{c_{in} - c} = \int_0^t \frac{dt}{\theta}$$

$$-\ln\left(\frac{c_{in} - c}{c_{in} - c_0}\right) = \frac{t}{\theta}$$

$$\left(\frac{c_{in} - c}{c_{in} - c_0}\right) = e^{-t/\theta}$$

Therefore, $c = c_{in} - (c_{in} - c_0)e^{-\frac{t}{\theta}} = c_{in} + (c_0 - c_{in})e^{-t/\theta}$

2. If it is assumed that the virus reacts according to a zero-order rate law, $r = -k$. Substituting this into Eq. (15.18) for r yields:

$$\frac{dc}{dt} = \frac{C_{in} - c}{\theta} - k = \frac{C_{in}}{\theta} - k - \frac{c}{\theta} = \left(\frac{c_{in} - k\theta}{\theta}\right) - \frac{c}{\theta} = (c_{in} - k\theta - c)\left(\frac{1}{\theta}\right)$$

Separating variables and integrating yields:

$$\frac{dc}{[(c_{in} - k\theta) - c]} = \frac{dt}{\theta}$$

$$\int_{c_0}^{c} \frac{dc}{[(c_{in} - k\theta) - c]} = \int_0^t \frac{dt}{\theta}$$

$$-\ln\left[\frac{(c_{in} - k\theta) - c}{(c_{in} - k\theta) - c_0}\right] = \frac{dt}{\theta}$$

$$\frac{(c_{in} - k\theta) - c}{(c_{in} - k\theta) - c_0} = e^{-(t/\theta)}$$

Therefore, $c = c_0 e^{-t/\theta} + (c_{in} - k\theta)(1 - e^{-t/\theta})$

3. If it is assumed that the virus reacts according to a first-order rate law, $r = -kc$. Substituting this into Eq. (15.18) for r yields:

$$\frac{dc}{dt} = \frac{C_{in} - c}{\theta} - kc = \frac{C_{in}}{\theta} - \frac{c}{\theta} - kc = \frac{C_{in}}{\theta} = -c\left(\frac{1 + k\theta}{\theta}\right)$$

$$= [c_{in} - (1 + k\theta)c]\left(\frac{1}{\theta}\right)$$

Separating variables and integrating yields:

$$\frac{dc}{[(c_{in} - (1 + k\theta)\,c]} = \frac{dt}{\theta}$$

$$\int_{c_0}^{c} \frac{dc}{[(c_{in} - (1 + k\theta)\,c]} = \int_{0}^{t} \frac{dt}{\theta}$$

$$-\left(\frac{1}{1 + k\theta}\right)\ln\left[\frac{c_{in} - (1 + k\theta)c}{c_{in}(1 + k\theta)c_0}\right] = \frac{t}{\theta}$$

$$\left(\frac{c_{in} - (1 + k\theta)c}{c_{in}(1 + k\theta)c_0}\right) = e^{-(t/\theta)(1+k\theta)}$$

Therefore, $c = c_0 e^{-(t/\theta)(1+k\theta)} + \left(\frac{c_{in}}{1+k\theta}\right)[1 - e^{-(t/\theta)(1+k\theta)}]$

15.10 CHAPTER SUMMARY

- The following steps should be considered when solving ventilation problems: process modifications, local exhaust ventilation, process building or system volume (general or dilution ventilation), and personal protective equipment.
- The major components of a ventilation system include exhaust hoods, exhaust fan ductwork, environmental control devices, exhaust vents or stacks.
- Several types of hoods are available. One must select the appropriate hood for a specific operation to effectively remove viruses or contaminants from a work area and transport them into the ductwork. The ductwork must be sized such that the contaminant is transported without being deposited within the duct; adequate velocity must be maintained in the duct to accomplish this.
- Selecting a control device that is optimal for removal of any viruses or contaminants is important to ensure meeting relevant pollution control removal efficiency requirements.
- There are numerous deficiencies or inadequacies in understanding of how to not only design an industrial ventilation system to satisfy standards of performance

but also to accurately predict concentration variation in a ventilator control volume.

- Problems in ventilator calculations can include the inability to predict the concentration, effects of drafts and wakes in the control volume, new or unusual virus contaminant sources, concept of the capture velocity, concept of source, etc.

15.11 PROBLEMS

1 Describe the following types of fittings: reducers, bushings, and expanders.

2 Refer to Illustrative Example 15.3. Calculate the time it takes for 99% of the virus to disappear.

3 Refer to Illustrative Example 15.4. Assume the following two scenarios:
  a. if $r = 0$, qualitatively discuss the effect on the final equation if the volumetric flow rate, v_0 varies sinusoidally.
  b. if $r = 0$, qualitatively discuss the effect on the final equation if the inlet concentration, c_0 varies sinusoidally.

4 Refer to Illustrative Example 15.4. If the room volume is $142 \, m^3$, the flow rate of the $10 \, ng/m^3$ ventilation air is $12.1 \, m^3/min$, and viruses are being generated at a steady rate of $30 \, ng/min$, calculate how long it would take for the concentration to reach $20.7 \, ng/m^3$. The initial virus concentration in the room is $85 \, ng/m^3$.

REFERENCES

Abulencia, J.P. and Theodore, L. (2009). *Fluid Flow for the Practicing Chemical Engineer*. Hoboken, NJ: John Wiley & Sons.

Atkinson, J. (2009). *Natural Ventilation for Infection Control in Health-care Settings*. World Health Organization.

Bird, R., Stewart, W., and Lightfoot, E. (2002). *Transport Phenomena*. Hoboken, NJ: John Wiley & Sons.

Centers for Disease Control and Prevention (CDC) (2021). Ventilation in buildings. CDC. https://www.cdc.gov/coronavirus/2019-ncov/community/ventilation.html (accessed 20 November 2021).

Dupont, R., Baxter, T., and Theodore, L. (1998). *Environmental Management: Problems and Solutions*. Hoboken, NJ: John Wiley & Sons.

Farag, I. (1995). *Fluid Flow, A Theodore Tutorial*. East Williston, NY: Theodore Tutorials (originally published by USEPA/APTI, RTP, NC).

Fogler, S. (2006). *Elements of Chemical Reaction Engineering*. Upper Saddle River, NJ: Prentice Hall PTR.

Ganesan, K., Theodore, L., and Dupont, R.R. (1996). *Air Toxics: Problems and Solutions*. New York, NY: Routledge.

Heinsohn, R.J. (1991). *Industrial ventilation: Engineering principles*. Hoboken, NJ: John Wiley & Sons.

Theodore, L. (1971). *Transport Phenomena for Engineers*. Scranton, PA: International Textbook Co.

Theodore, L. (2006). *Nanotechnology: Basic Calculations for Engineers and Scientists*. Hoboken, NJ: John Wiley & Sons.

Theodore, L. (2008). *Air Pollution Control Equipment Calculations*. Hoboken, NJ: John Wiley & Sons.

Theodore, L. (2011). *Heat Transfer Applications for the Practicing Chemical Engineer*. Hoboken, NJ: John Wiley & Sons.

Theodore, L. (2012). *Chemical Reactor Analysis and Applications for the Practicing Engineer*. Hoboken, NJ: John Wiley & Sons.

Theodore, L. (2014). *Chemical Engineering: The Essential Reference*. New York City, NY: McGraw Hill.

Theodore, L., Ricci, F., and VanVliet, T. (2009). *Thermodynamics for the Practicing Engineer*. Hoboken, NJ: John Wiley & Sons.

Vance, D., Shah, P., and Sataloff, R.T. (2021). *COVID-19: impact on the musician and returning to singing; a literature review. Journal of Voice: Official Journal of the Voice Foundation* S0892-1997(21)00003-5. https://doi.org/10.1016/j.jvoice.2020.12.042.

16

Pandemic Health Data Modeling

"Mathematical models generally provide mathematicians with opportunities to perform useless exercises while placing great emphasis solely on analytical or (on rare occasions) numerical solutions. Making mathematical models optimally useful is simply not in their dictionary. These models, which are often expressed in analytical form, also provide the engineer and practicing scientist with frustrating experiences in attempting to relate any analysis to real-world happenings"

(L. Theodore, New Hyde Park, NY, 1971)

Life, as we knew it, changed forever with the arrival of COVID-19. The pandemic caused one of the worst health and economic disasters in history. Furthermore, the effects on the healthcare, economic, and judicial systems are certain to have far-reaching effects in both the near term and the future. Tens of millions of Americans lost their jobs and over one million died. Across the planet, over 6 million people lost their lives. One could rightly claim that the death toll was unfathomable, with approximately 80% of the death toll being adults over the age of 65 (Center on Budget and Policy Priorities 2022; CDC 2022).

A Guide to Virology for Engineers and Applied Scientists: Epidemiology, Emergency Management, and Optimization, First Edition. Megan M. Reynolds and Louis Theodore.
© 2023 John Wiley & Sons, Inc. Published 2023 by John Wiley & Sons, Inc.

A coordinated effort by the technical community is now developing in order to help fulfill societal needs while protecting health. Viral illnesses such as measles and polio, once conquered through the use of vaccines and other therapies, are reemerging. As societies change, the challenges facing public health will continue to become more complex, and public health officials and agencies will play an important role in improving the lives of individuals, families, and entire communities. Some of these challenges include collecting, analyzing, and reporting accurate data on health status and virus indicators. Also essential is addressing the impact of chronic diseases, such as the still poorly understood "long-COVID" conditions, have on both health costs and quality of life. Finally, a mission of the healthcare community is to protect communities from health issues, safety concerns, and security threats. These include activities involving emergency planning and response (EP&R), which is covered in detail in Part II, Chapter 9: Emergency Planning and Response.

As discussed in Chapter 9, some of the many common-sense reasons to plan ahead are provided in the following (Krikorian 1982):

- Pandemics will happen; it is only a question of time.
- When pandemics occur, the minimization of loss and the protection of people and the environment can be achieved through the proper implementation of an appropriate emergency response plan.
- Minimizing health problems caused by a pandemic requires planned procedures, understood responsibility, designated authority, accepted accountability, and trained and experienced people. With a fully implemented plan, these goals can almost always be achieved.
- If an emergency such as a pandemic occurs, it may be too late to plan. Lack of preplanning can turn a health crisis into a monumental disaster.

Five specific pandemic-related Emergency Planning and Response (EP&R) objectives also include:

- Putting science into action
- Detecting and responding to new and emerging health threats
- Tackling virus health problems causing death and disabilities
- Putting science and advanced technology into action to prevent diseases and viruses
- Monitoring virus status to identify and solve future health problems

Key among these EP&R objectives and the basis for this chapter is the specific need to better define data using mathematical models and other applicable equations to represent risk factors and other causes of morbidity and mortality. This will better enable the technical community to predict virus activity in the future and respond accordingly. Thus, one key activity will include the development of the aforementioned pandemic health data models.

More must be done in preparation for future pandemics. For decades, infectious disease specialists and scientists have warned that the influenza virus is capable of mutating into a dangerous variation, with effects similar to the 1918 pandemic, which killed at least 50 million worldwide. Likewise, novel pathogens, such as the SARS-CoV-2 – the virus that causes COVID-19 are certain to emerge. (CDC 2018b) More details on influenza and its ability to mutate are available in Part I, Chapter 3: Pandemics, Epidemics, and Outbreaks.

This chapter presents several mathematical models that may be employed to predict a host of health effects associated with virus and related conditions. These predictions can hopefully not only provide planning strategies but also help "flatten health data curves" earlier in a pandemic in order to reduce the adverse effects associated with these diseases. This would help prevent the overcrowding of hospitals that was evident in multiple countries during the course of the COVID-19 pandemic, and threatened breakdowns of healthcare systems. The proposed mathematical models in this chapter can provide valuable medical-related information that could help guide public health engineers, scientists, officials, and others in the healthcare sector through these problems.

16.1   COVID-19: A RUDE AWAKENING

The actual death toll from the COVID-19 pandemic is estimated to be much higher than official accounts (Wise 2022). Across the world, lockdowns and mask mandates created political turmoil and produced emotions and sentiments (even at the time of the preparation of this chapter) that negatively impacted decisions and policies. Despite unprecedented heroic conduct by healthcare workers, hospital systems in many areas were overburdened and patients lost their lives as a result of personnel and equipment (i.e., ventilators) shortages.

This pandemic, however, is not unique; pandemics have been a regular occurrence throughout history. Thanks to modern science, vaccines and drugs, as well as personal protection equipment and ventilators, can be employed to keep patients alive. With appropriate planning and preparation, this can provide meaningful pandemic relief. It should be noted that no country had properly prepared for a pandemic on the scale of COVID-19. Preparedness planning was insufficient, and most countries' leaders were caught unaware when this highly infectious disease emerged. A year into the pandemic, the risk of death was somewhat mitigated with the arrival of highly effective vaccines. This may have been due in part to the decade-long search for a coronavirus vaccine to combat SARS and MERS, which significantly aided the process of developing the COVID-19 vaccine program. Financial commitments and scientific coordination by several countries, and particularly the United States early in the pandemic, made the development of these vaccines possible. Further details on viral vaccines are available in Part I, Chapter 4: Virus Prevention, Diagnosis, and Treatment.

As discussed in Part II, Chapter 9, emergencies such as pandemics, industrial accidents, and natural disasters have occurred in the past and will continue to occur in the future. The COVID-19 pandemic revealed many deficiencies in international emergency preparation. Countries acted much too slowly and too late or made bad decisions that were not based on science, and which in turn accelerated outbreaks. Some laid blame on the World Health Organization (WHO) because it hesitated in declaring COVID-19 a pandemic. By the time of the March 11, 2020 official pandemic declaration, the virus had already spread to 114 countries, although 80% of cases were concentrated in only four countries. While the bureaucracy of national and international agencies and governments certainly complicated initial responses, there was also a fear among nations' leaders at the time that informing the public with the information could cause panic and disruptions among their communities – especially when little was known regarding the novel SARS-CoV-2 virus. In retrospect, the response came too late, as the virus has been proven to be highly contagious as well as pathogenic (WHO 2020).

A main concern among the technical and healthcare communities is whether the world will learn from COVID-19 and be better prepared for the future. Hopefully, it will motivate the technical and public health communities to develop the necessary strategies to prepare adequately for the next pandemic or healthcare crisis. Responsible individuals – particularly engineers and applied scientists can recognize and learn from the COVID-19 pandemic and respond accordingly to the next crisis, whether it be one associated with health from naturally occurring pathogens or those potentially arising from terrorist activities.

The technical and healthcare communities around the world must accept the fact that another pandemic or health-related problems or any other nightmare scenario will strike sometime in the future, and the technical community needs to better prepare *now*!

16.2 EARLIER WORK

One should use the simplest model possible that can still effectively represent the relationship being explored. The modeling described in this chapter focuses on the use of "simple" analytical models that could effectively capture the essential form of pandemic infection rates. Much more complicated, process focused models have been reported on in the literature but the aim of this chapter is to explore relatively simple analytical models that can be widely used and adapted across disciplines, and that capture the complexity of pandemic spread as a tool to support decision makers in dealing with outbreak cases in a timely manner. (Dupont, R. letter to Theodore L., 2022)

> According to R.N. Thompson, in her discussion of epidemiological models applied to COVID-19, "…The spread of communicable diseases throughout a population can be theoretically modelled by numerous methods, each with their own mathematical frameworks, assumptions, and inherent advantages and disadvantages. One…approach is *compartmental modelling*, in which individuals are identified by their infection or symptom status. One particularly useful, yet simple, epidemiological compartmental model developed is the *Susceptible-Infectious-Removed (SIR) model*, which consists of three ordinary differential equations describing the time-rate of change of the subsets of the population who are susceptible to the disease, infectious, or removed (i.e., recovered, in isolation, etc.)" (Thompson 2020)

Thompson discusses the role models can play in guiding COVID-19 scientific advice to decision-makers. The general framework underlying the study is the aforementioned compartmental modeling, in which individuals are categorized according to their infection(s) or symptom status and was defined as the aforementioned SIR model. (Thompson 2020)

$$\frac{ds}{dt} = -\beta SI \tag{16.1}$$

$$\frac{dI}{dt} = \beta SI - \mu I \tag{16.2}$$

$$\frac{dR}{dt} = -\mu I \tag{16.3}$$

The parameter β set the infection rate, and the average infectious period is $1/\mu$ days. The basic reproduction number, $R0 = \beta n/\mu$, represents the expected number of individuals that a single infectious host will infect if introduced into a population of N susceptible hosts. The SIR model can be solved numerically (Thompson 2020).

The above model can capture characteristic behaviors observed in infection trends (e.g., a rapidly increasing infection rate, followed by a peak infection and subsequent decay of case counts). The model can be further extended by incorporating additional effects, including the introduction of *stochasticity* (i.e., randomness), the inclusion of asymptomatic carriers, and the ability to divide the population into age- or gender-based increments (Davies et al. 2016).

With such models, one can attempt to simulate the impact of proactive measures, such as *non-pharmaceutical interventions (NPI)*, e.g., as school closures, social distancing, lockdown, etc., by assigning a percentage reduction in disease spread attributable to each of these measures (Davies et al. 2016). However, one must consider how accurately the impacts of such measures can be predicted a priori, as required for a truly predictive model.

The material to follow represents the authors' relatively simplistic approach to describing public health data. A number of researchers have reported on the complex problem of modeling and forecasting transmission of COVID-19 using much more complicated, process-based models. (Bertozzi et al., 2020) Some have been based on stochastic networks to assess the impact of mitigation measures on disease spread upon school reopenings, (McGee et al., 2021) while others used US population demographics to estimate ICU bed occupancy (Moghadas et al., 2020), and still others explored the spread of COVID-19 across the US using a kernel-modulated, multi-fractal scaling model. (Geng et al., 2021)

16.3 PLANNING FOR PANDEMICS

Because the COVID-19 pandemic has affected virtually everyone, it is important to assist government, businesses, workplaces, schools, etc. This assistance can vary from configuring solutions to preparing for crisis response and recovery. This would naturally be an integral and strategic part of any response efforts. Perhaps the key to any planning strategy will be the availability and proper use and maintenance of medical equipment such as mechanical ventilators. This type of life support equipment replaces an individual breathing functioning capability when unable to breathe on their own. As described earlier in Parts I and II, the ventilator forces atmospheric air under positive pressure into the patient's lungs and subsequently removes carbon dioxide. Equipment settings normally include:

- quantity of air
- air pressure
- airflow discharge
- monitor location
- oxygen concentration

Blood tests and chest X-rays are occasionally available with the unit.

Regarding models, COVID-19 data is extensive and continues to grow with time, and it is important to quickly release information and provide solutions that are specific to reopening and recovery. This information can provide the technical community, residents, and local officials what is needed to protect not only their communities but also prevent against the spread of infections in order to make intelligent and informed decisions. The development of the aforementioned emergency planning and responses should include an attempt to create a mathematical model that can yield health data updates throughout a pandemic crisis.

As previously discussed, the wrong response can turn an epidemic into a pandemic as easily as no response. The proper response requires an understanding of all the data that is available and a plan that can provide the practitioner with the information needed to respond quickly and properly during a health crisis.(Theodore and Dupont 2012) The potential for pandemics, based on history and knowledge of the region, should also be considered early in any study. Thus, successful pandemic planning begins with a thorough understanding of the infectious disease and/or the potential disaster being planned for. In addition, the impacts on public health and the environment must be estimated at an early stage.

The application of the above has recently come under increased scrutiny in the public health field with the arrival of the aforementioned COVID-19 pandemic. It has become more important than ever to put science and engineering into practice to help prevent the effects of such pandemics. One area that has received significant attention is a representation of infectious disease data in equation form for analysis and predictive purposes. This EP&R activity offers the potential to protect the community from injury, better preparing for emergencies such as pandemics and allowing the equitable allocation of emergency funds, medical personnel, and other resources. There is a definite need to include this form of preparedness activity for future pandemics. This action provides an opportunity to restore and protect human health from infectious disease.

The next section attempts to provide the technical and public health communities with the tools necessary to generate mathematical models to describe pandemics health and/or disease data.

16.4 GENERATING MATHEMATICAL MODELS

As noted in both Chapter 12: Introduction to Mathematical Methods and Chapter 14: Linear Regression, engineering and medical data are often better understood when presented in the form of graphs or mathematical equations. The preferred method of engineers for obtaining the mathematical relations between variables is to attempt to plot the data as *straight* lines and use the slope (m)-intercept (*b*) method to obtain coefficients and exponents. Therefore, it behooves the engineer, medical practitioner, and applied scientist to be aware of the methods of obtaining straight lines on various types of graph paper, and of determining the equations of such lines. One usually strives to plot data as straight lines because of the simplicity of the curve and ease of both interpolation and extrapolation. Graphical methods proved invaluable in the past in the analysis of data and relatively simple processes; much of the basic physical and chemical data are still best presented graphically. Some of the information to follow is a repeat of that presented earlier in Chapters 12 and 14 but is included for the convenience of the reader.

One or more of the many types of graphical representations may be employed for the following purposes (Bhatty et al. 2017):

- As an aid in visualizing the representation of quantitative data
- For the representation of quantitative data
- For the representation of a theoretical equation
- For the representation of an empirical equation
- For the comparison of experimental data with a theoretical equation
- For the comparison of experimental data with an experimental equation
- For predictive analysis

As noted in Chapter 14: Linear Regression, the relation between two quantities y, the dependent variable, and x, the independent variable, is commonly obtained as a tabulation of values of y for a number of different values of x. The relation between y and x may not be easy to visualize by studying the tabulated results and is often best seen by plotting y vs x. If the conditions are such that y is known to be a function of x only, the functional relation will be indicated by the fact that the points may be represented graphically by a smooth curve. Deviations of the points from a smooth curve may indicate a problem with the reliability of the data. If y is a function of the two variables x_1 and x_2, a series of results of y in terms of x_1 may be obtained in terms of x_2. When plotted, the data will be represented by a family of curves, each curve depicting the relationship between x_1 and y for a definite constant value of x_2. If another variable x_3 is involved, one may have separate graphs for values of x_3, each showing a family of curves of y vs x_1. One may expand this method of representing data to relate more than three independent variables.

Rectangular (sometimes referred to as Cartesian) coordinates are most generally used to represent an equation of the form

$$y = mx + b \qquad (16.4)$$

where

y = variable represented in the ordinate
x = variable represented on the abscissa
m = slope of the line
$b = y -$ intercept at $x = 0$

All graphs are created by drawing a line, or lines, about one or more axes. Information on other coordinates, e.g., logarithmic, semi-logarithmic, etc. are available in the literature (Shaefer and Theodore 2007). In addition to these types of coordinates, one may also have a need for probability versus arithmetic, probability versus logarithmic, logarithmic versus reciprocals, and other special types of graphs. Nonstandard scales, if desired, may be constructed similarly to the logarithmic scale. For example, logarithmic-log probability (log normal) graphs find application in air pollution control studies (Theodore 2009).

As noted above, one often encounters applications that require the development of a mathematical relationship between data for two or more variables. For example, if y (a dependent variable) is a function of, or depends on x (an independent variable), i.e.,

$$y = f(x) \tag{16.5}$$

One may then express this (x, y) function in equation form. The process of obtaining the coefficients in the equation is referred to as regression analysis, and the regression method most often employed is the method of least squares, a topic earlier addressed in Chapter 14: Linear Regression. This type of study usually begins by attempting to determine the mathematical form of this relationship. The approach, referred to as a scatter diagram, could take on any form. Three such plots are provided in Figure 14.1 in Chapter 14 but are presented once again for the reader's convenience in Figure 16.1.

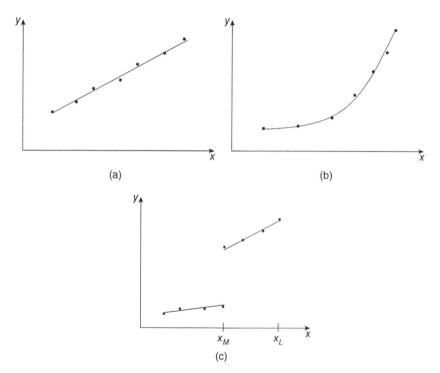

Figure 16.1 Scatter diagrams: (a) linear relationship, (b) parabolic relationship, and (c) dual-linear relationship. *Source*: Courtesy of Bhatty et al. (2017).

The first plot (a) suggests a linear relationship between x and y, i.e.,

$$y = a_0 + a_1 x \tag{16.6}$$

The second graph (b) appears to be best represented by a second order (parabolic) relationship, i.e.,

$$y = a_0 + a_1 x + a_2 x^2 \tag{16.7}$$

The third plot (c) suggests a linear model that applies over two different ranges i.e.,

$$y = a_0 + a_1 x; \qquad x_0 < x < x_M \tag{16.8}$$

and

$$y = a_0' + a_1' x; \qquad x_M < x < x_L \tag{16.9}$$

This multi-equation model finds application in representing adsorption equilibria, multiparticle size distributions, quantum energy relationships, etc. (Theodore 2009) There are numerous additional mathematical models. Several are presented in Table 16.1. Suggested plotting procedures to express these equations and their accompanying slope and intersept are included in the table.

The scatter diagram approach allows the practitioner to generate mathematical models that provide the functional relationship between the aforementioned dependent variable y and the independent variable x. Their application – particularly equations (8), (9), and (10) in Table 16.1 to describe various forms of virus and infectious disease data is demonstrated in the illustrative example in the applications section to follow. In any event a scatter diagram and individual judgment can suggest an appropriate mathematical model at an early stage in an analysis. For a more in-depth explanation, the reader is referred to Chapter 14: Linear Regression (Bhatty et al. 2017) (Shaefer and Theodore 2007).

It is important to note that setting up a mathematical model or equation requires a thorough understanding of the phenomenon. Without this understanding, the resulting expression may not describe the situation in question. Therefore, it is often necessary to check the expression using information separate from the model. If the data agree, one can infer that the expression is adequate to describe the phenomenon but only within the *range* of data used. One may also infer that the proposed description of the phenomenon was adequate. This latter inference is *not* necessarily correct. More data in a different range of some of the variable(s) may not agree at all with the expression.

There are various categories of models that may be employed to the solution of many societal problems. These models are generally mathematical or physical in nature. In addition, the type of model required depends on the system under analysis.

Table 16.1 Mathematical Models: (Theodore, J, and Theodore, L. 2021)

Describing Equation	Plotting Procedure	Intercept (b)/ Slope(m)	Graphical Result
(1) $y = a_0 + a_1 x$	Plot y vs x	$b = a_0, m = a$	
(2) $y = a_0 + a_1 x + a_2 x^2$	Plot y vs x	$b = a_0$	
(3) $y = a_0 + a_1 x + a_2 x^2 + a_3 x^3$	Plot y vs x	$b = a_0$	
(4) $y = a_0 + a_1/x$	Plot y vs $1/x$	$b = a_0, m = a_1$	
(5) $y = 1/(a_0 + a_1 x)$	Plot $1/y$ vs x	$b = a_0, m = a_1$	
(6) $y = x/(a_0 + a_1 x)$	Plot $1/y$ vs $1/x$	$b = a_0, m = a_1$	
(7) $y = a_0 x^{a_1}$	Plot $\log y$ vs $\log x$	$b = \log a_0,$ $m = a_1$	
(8) $y = (a_0 x)^{1/x}$	Plot $x \log y$ vs $\log x$	$b = \log a_0$	
(9) $y = (x)^{a_0/x}$	Plot $x \log y$ vs $\log x$	$m = a_0$	
(10) $y = (a_0 x)^{a_1/x}$	Plot $x \log y$ vs $\log x$	$b = a_1, \log a_0, = a_1$	

(continued)

Table 16.1 (Continued)

Describing Equation	Plotting Procedure	Intercept (b)/ Slope(m)	Graphical Result
$(11)\ y = a_0 + e^{a_1 x}$	Plot $x \log y$ vs x	$b = \log a_0$, $m = a_1$	
$(12)\ y = (a_0)(a_1)^x$	Plot $\log y$ vs x	$b = \log a_0$, $m = \log a_1$	

Many phenomena may be described by rigorous mathematical expressions; others may be described by appropriate expression; and still others are best described by a graphical or tabular presentation. The analysis of most phenomena, including the aforementioned pandemic, may be divided into four basic steps:

- Consideration of the phenomenon
- Mathematical description of the phenomenon
- Solution of mathematical relationships to give a final expression to describe the phenomenon
- Verification of the mathematical model (or expression) with experimental data

In many cases, the model must be simplified so that the resultant mathematical equation is solvable. Unfortunately, this simplification may go so far that the derived expression no longer accurately describes the actual phenomenon in question.

Finally, experimental data (and data in general) in many applications may be presented (as described above) using a table, a graph, or an equation or a mathematical model. Tabular presentation permits the retention of all significant figures of the original numerical data. Therefore, it is the most numerically accurate way of reporting data. However, it is often difficult to interpolate data points or extrapolate beyond the data points (Bhatty et al. 2017).

16.5 PANDEMIC HEALTH DATA MODELS

United States COVID-19 data over the 1/28/20–3/21/21 period is available for the number of individuals infected as a function of time (CDC 2022). This information is presented in Table 16.2 with y representing the number of new infected cases every week, and x based on the week associated with the last seven-day

Table 16.2 New Infection Cases (per 10^3) vs. Week Number

New Cases	Week Number	New Cases	Week Number	New Cases	Week Number
0.0	1 (1/28/20)	36.7	26	173.9	51
0.0	2	47.3	27	161.9	52
0.0	3	53.6	28	195.0	53
0.0	4	62.5	29	214.2	54
0.0	5	66.8	30	216.4	55
0.0	6	66.2	31	206.4	56
0.0	7	60.5	32	182.8	57
0.1	8	53.8	33	216.4	58
0.9	9	52.5	34	250.4	59
5.7	10	46.5	35	217.6	60
15.3	11	42.2	36	175.2	61
26.3	12	42.7	37	154.0	62
31.2	13	40.8	38	130.3	63
28.9	14	34.8	39	104.1	64
27.6	15	40.3	40	80.9	65
29.1	16	42.9	41	64.1	66
27.0	17	43.2	42	67.1	67
24.6	18	45.5	43	59.3	68
23.7	19	52.2	44	54.5	69
23.1	20	57.7	45	53.7	70
21.3	21	70.3	46	57.4	71 (3/24/21)
20.6	22	80.9	47		
21.0	23	110.7	48		
21.2	24	130.1	49		
27.2	25	162.2	50		

average. An y-x plot employing the data in the table, was prepared by Theodore and Theodore (Theodore, L. and Theodore, J., personal notes 2021). The result appears in Figure 16.2.

Based on this scatter diagram plot, it appears that Equation (10) in Table 16.1 might be a viable option to describe this data. For this equation,

$$y = (a_0 x)^{a_1/x} \tag{16.10}$$

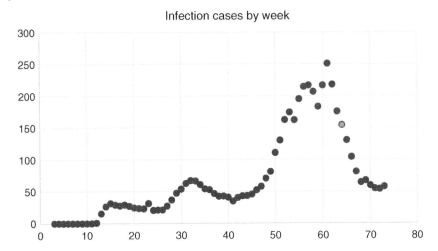

Figure 16.2 Pandemic health data plot.

one may take the ln of both sides of the equation

$$\ln y = (a_1 x)\ln(a_0/x) \tag{16.11}$$

This equation may be rewritten as

$$x \ln y = a_1 \ln(a_0/x) \tag{16.12}$$

$$= a_1 \ln a_0 + a_1 \ln(1/x) \tag{16.13}$$

Thus, a plot $x \ln y$ vs. $\ln 1/x$ produces a *linear* equation with intercept (b) and slope (m) so that

$$b = a_1 \ln a_0 \tag{16.14}$$

$$m = a_1 \tag{16.15}$$

Based on Figure 16.2, one may conclude that the *general* form of Equation (16.10), above, may be used to provide a very rough description to the pandemic health data presented in Table 16.1. More importantly, the results strongly suggest that this equation may be used as the *first* estimate to describe other pandemic health data. This equation, as well as five other models (Theodore and Theodore, personal notes 2021) that may also be employed for health data descriptive purposes, are listed below.

$$y = (a_0 x)^{a_1/x} \tag{16.10}$$

$$y = (a_0 x)^{a_1/x} \text{ to } y_{max}, x_{max}; y = (a_0 x)^{a_1/x} - ((x - x_{max})/x)^{a_2} \tag{16.16}$$

$$y = (a_0 x)^{[a_1/x-1]} \tag{16.17}$$

$$y = \frac{(a_0 x)^{a_1/x}}{x+1} \tag{16.18}$$

$$y = (a_0 x)^{(a_1/x)[1-(x/1+x)]} \tag{16.19}$$

$$y = (a_0)(x/a_1)^{a_2[1-(x/a_1)^{a_2}]} \tag{16.20}$$

$$y = (a_0) + (a_1)(x/a_2)^{a_3[1-(x/a_2)^{a_3}]} \tag{16.21}$$

Some comments on each of the above six proposed models are detailed below

- Equation (16.10). The simplest two coefficient equation that does and is amenable to analytical analyses and a reasonable job early, but it fails in the limit as $x \to \infty$.
- Equation (16.16). A two-range equation with *two*- and possibly *three*-coefficients that does a reasonable job.
- Equation (16.17). A two-coefficient equation that does a reasonable job but does not fail as $x \to \infty$.
- Equation (16.18). Another two-coefficient that does a reasonable job.
- Equation (16.19). Another two-coefficient model that does a good job.
- Equation (16.20). A three-coefficient model.
- Equation (16.21) A four-coefficient model.

The reader should note that model agreement with data increases with an increase in the number of coefficients (Shaefer and Theodore 2007).

As noted earlier, CDC COVID-19 data over the range of 1/28/20–3/24/21 have been plotted in Figure 16.2. Equation (16.21) has been superimposed on the graph

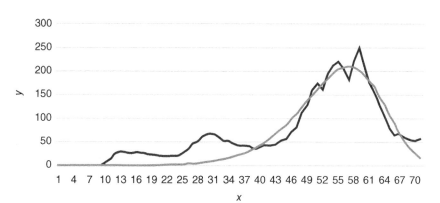

Figure 16.3 Pandemic health data plot comparison.

for comparative purposes and appears in Figure 16.3. Note once again that there is only one "peak", instead of the numerous "spikes" in health data rates of that period. In effect, these models cannot be used to model the multiple peaks and valleys that occur.

Seven general comments regarding the modeling of COVID-19 death/health data are presented below

1. One should realize that early medical data is often not available so that mathematical models describing the functional behavior of the disease requires assumptions concerning a host of different factors. However, despite the unavoidable errors that can arise, these equations can predict and provide important information – some of which may significantly reduce the impact of these diseases.

2. Lay personnel and government officials have come to distrust models because of their failure to accurately predict systems. These individuals usually cannot come to accept the fact that model results can also be adversely affected by a small number of observations.

3. Below are some additional problems confronting the modeler of pandemic public health data. These include people with co-morbid conditions and other factors such as:

   - The severity of the illness, which varies significantly between patients
   - The age variant
   - High blood pressure
   - Various heart diseases
   - Cancer
   - Diabetes
   - Obesity

   The reality is that, with COVID-19, individuals experience a range of symptoms ranging from asymptomatic to respiratory failure, i.e., death. Further details are provided in Part I, Chapter 3: Pandemics, Epidemics, and Outbreaks.

4. Model uncertainty, which is directly related to both risk and probability, often arise from the complexity of the values employed for the coefficients in the model. Since the ideal situation can rarely, if ever, be achieved, decisions generally have to be made with imperfect and/or incomplete data and information. In effect, there is often no satisfactory way of accurately relating their values with reality.

5. No discussion of the application of models would be complete without a comment on quality assurance/quality control (QA/QC). Much of this deals with validation studies. This can include not only updating earlier values and extrapolated values but also the effect of earlier time frames (i.e., are the values still representative?)

6. One of the authors is currently investigating the possibility of employing Monte Carlo methods for predicting purposes. The procedure involves repeating a calculation numerous times to determine validity estimates (L. Theodore, personal notes 2021).

The authors believe that the proposed health data models have successfully demonstrated the ability to reasonably predict outcomes. They should remain a valuable tool to engineers and public health officials for not only predictive purposes but also for policy planning and implementation.

As noted earlier in Section 16.2, the above material represents the authors' relatively simplistic approach to describing public health data modeling must remain a key tool in modeling policy decisions. There should also be a unified approach at the international level, which is an area where governments and leaders across the globe failed to accomplish with COVID-19 – especially early in the pandemic. The authors hope that the world will learn from past mistakes; in the meantime, efforts will continue to provide pandemic model solutions, particularly of an analytical nature (L. Theodore and M. Reynolds, personal notes 2022).

In conclusion, this section has discussed public health data mathematical models. It should be clear that they can be applied to guide public policy where any analysis should be considered along with several other factors such as health effects, economic impact(s), etc., when attempting to address the problem. Notwithstanding unavoidable uncertainties and shortcomings, mathematical models can highlight important principles and subsequently determine which actions are likely to reduce public health problems most effectively and protect humans from infectious diseases.

16.6 IN REVIEW

As mentioned earlier, the COVID-19 pandemic has affected virtually everyone around the world. It is crucial that governments seek to ameliorate the economic and social fallout and other consequences. Government assistance to their respective populations is necessary for workplaces, schools, etc., in their attempt to recover and return to (a new) normal. This assistance can vary from configuring various forms of solutions to preparing for crisis response and recovery. This would naturally be an integral and strategic part of any and all recovery efforts.

Pandemics in the past have repeatedly caught the technical community by surprise, resulting in massive health crises that could justifiably be described as chaotic. Many now believe that the next pandemic will overwhelm all aspects of present-day medical systems. There is also reason to believe that influenza may be the next airborne virus to emerge. The key is to successfully plan for worst-case scenarios so that for the next pandemic, governments and public health officials will be able to respond effectively. Focus should be on:

- Developing a vaccine
- Producing sufficient vaccines
- Correctly accumulating and storing any vaccines
- Proper and fair distribution of a vaccine
- Effectively storing any necessary medical equipment (i.e. ventilators, respirators, etc.)

Specific actions will also include nonpharmaceutical interventions (NPIs) to be implemented as necessary when populations face an imminent threat of infectious disease. According to the WHO, "Non-pharmaceutical interventions (NPIs) are the only set of pandemic countermeasures that are readily available at all times and in all countries." (WHO 2019) NPIs are shown to have a significant impact on local populations concerning infection rates. The goals are specifically to

"…delay the introduction of the pandemic virus into a population; delay the height and peak of the epidemic if the epidemic has started; reduce transmission by personal protective or environmental measures; and, reduce the total number of infections and hence the total number of severe cases." (WHO 2019)

As discussed in detail in Part II, Chapter 10: Ethical Considerations in Virology, examples of NPI include many that have been implemented during the COVID-19 pandemic, such as: (WHO 2019)

- Hygiene measures (hand washing, etc.,)
- Quarantining infected or exposed individuals
- Social distance measures to keep individuals physically separated from one another.
- Mask requirements to prevent the spread of the virus indoors or in close outdoor spaces.
- Limiting the number of people in a given physical space at one time, such as large public events (both indoors and outdoors), workplace and school closures, and restrictions placed on movement and transportation.

As noted, in both preparing for new waves of the pandemic as well as future crises, one potential approach is to develop simple pandemic health data models that incorporate some of the important pandemic parameters amendable to analytical analysis. This next-to-last chapter describes some of these models – including assumptions and numerical parameters – that may be employed for this purpose. Obviously, historical data must also be used in the hope that it is accurate and as complete as possible in order to ensure a properly calibrated model is employed. Mathematical models are important for guiding public health measures, and data from epidemiological modeling analyses should be considered as early as possible during a crisis. In general, modeling such real-world scenarios requires assumptions of different parameters. Therefore, models must be used in the context of other available information. While uncertainties exist, models can demonstrate important principles about outbreaks and determine which interventions are most likely to reduce case numbers effectively. However, models have demonstrated the ability to forecast outcomes and therefore must remain a key tool for pandemic policy (Thompson 2020).

As the world continues to change and knowledge of various health-related complexities expands, public health faces new and more complex challenges. In simpler times, the work of public health was limited to maintaining basic sanitary living conditions and preventing the spread of communicable diseases. Its primary work was to help coordinate the fight against tuberculosis in the late 1800s. Today, public health professionals must cope with an ever-expanding group of newly recognized diseases, bacterial, viral, etc., such as *E. coli*, Cyclospora, Cryptosporidium, Hantavirus, Ebola, Monkeypox, and of course, COVID-19. At the same time, illnesses once conquered through the use of antibiotics and other therapies are re-emerging as public health threats by becoming resistant to the very treatments once used to defeat them (CDC 2018a). Mathematical models will help meet this need.

In the final analysis, it is hoped that models presented in the earlier sections of this chapter will provide opportunities to save lives and reduce suffering during a pandemic. These models should allow the engineer and the applied scientist to provide the healthcare communities with the best judgments in making policy despite the uncertainties associated with the models. The practitioner must also be diligent and alert since the dynamics can quickly change with both existing pandemics as well as new, emerging diseases.

16.7 APPLICATIONS

Four illustrative examples complement this presentation on pandemic healthcare data modeling.

Illustrative Example 16.1 PANDEMIC PLANNING
What are the reasons to plan before a pandemic strikes?

Solution
The primary reasons for pandemic planning are as follows: (Theodore and McGuinn 1992)

- To minimize the effects of a disaster and the loss of life
- To respond immediately since it is sometimes difficult to think clearly during pandemic situation
- To reduce the chance of an improper response, which can make the situation worse
- To respond to the fact that pandemics and epidemics will always occur; they are essentially unavoidable and the best way to cope with them is to be prepared *prior* to the occurrence

Illustrative Example 16.2   PANDEMIC MODELING I

A new infectious disease – NAHUW4 (pronounced Nahoo) – is currently ravaging the nation of Pleh (pronounce plea) in the South Pacific. The Infectious Disease Council of Pleh (IDCP) has hired the Ryan Infectious Disease Experts (RIDE) to develop and provide the needed direction to assist in addressing this public health problem. The president accepted IDC's suggestion and RIDE was ultimately hired to provide answers/estimates to the following five questions.

1. *Develop a mathematical model that describes the weekly number of new NAHUW4 infections (NI) as a function of time (T).*
2. *Estimate approximately when the NI-T curve will "flatten".*
3. *Estimate the weekly NI per 100 citizens when the curve "flattens".*
4. *Estimate when the weekly NI will be reduced to half the maximum value.*
5. *Estimate when the weekly NI will be reduced to 10% of the maximum value.*

As a recently hired chemical engineer at RIDE from the prestigious Julian Institute of Technology (JIT), you have been assigned the above task. IDC has provided early NI-T data in Table 16.3. Provide a solution.

Table 16.3   New Infections (NI/100) vs Time in weeks (T)

T Weeks	NI/100	T Weeks	NI/100	T Weeks	NI/100
1	0.0	7	14.0	13	158
2	0.0	8	22.0	14	177
3	1.0	9	40.0	15	183
4	1.0	10	59.0	16	176
5	4.0	11	89.0	17	174
6	10.0	12	130	18	168

Solution

The young engineer first prepared a y-x plot (See Figure 16.4) of the data (see Table 16.3) and reviewed the various mathematical model presented earlier in Table 16.1 that can be deployed in describing functional relationships in equation form. Based on data from the recent COVID-19 pandemic (see also Figure 16.4), she concluded that the simplest model to select from those presented in Table 16.1 was

$$y = (a_0 x)^{a_1/x}$$

where y is the weekly number of infections (NI) and x is time in week (T).

1. Employing the method of least squares with the data provided in Table 16.3, EXCEL provide the following results.

    $$a_0 = 0.164$$

    $$a_1 = 85.6$$

    The describing equation is therefore

    $$y = (0.164x)^{(85.6/x)}$$

2. The plot in Figure 16.5 appears to flatten at approximately x = T = 17 weeks. A more exact answer could be obtained by taking the derivative dy/dx setting the result equal to zero and solving for x. (See also next illustrative example.)
3. Solving the above equation with T = 17 weeks gives

    $$y = NI = 178$$

4. The so-called half-life of NI is simply NI (50%). Solving the above equation with NI = 89 gives

    $$x = T = 31 \ weeks$$

5. Finally, employing the same procedure

    $$NI\,(10\%) = 17.8$$

    and

    $$x = T\,(10\%) = 74$$

The reader should note once again that it is not possible to calculate a valid correlation coefficient with non-liner regression. This concept is discussed in Chapter 14: Linear Regression.

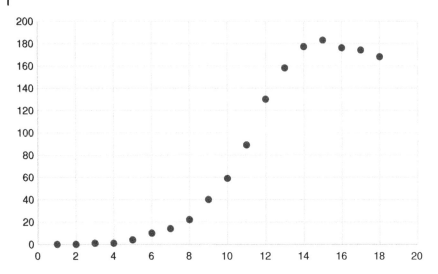

Figure 16.4 Infection data for *NAHUW4* virus.

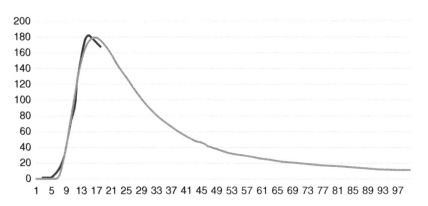

Figure 16.5 Plot of Health Data Model.

In addition, the solution to this Illustrative Example employing equation (16.20) is provided below.

1. See Problem #3.
2. 16
3. 190
4. 19.7
5. 22.7

Illustrative Example 16.3 MAXIMA AND MINIMA

Refer to question (2) in the previous illustrative example. Provide explanatory details regarding maxima and minima analysis.

Solution

Since the derivative dy/dx represents the rate of change of y with respect to the change in x, it is evident that if the function passes through a maximum or minimum, the derivative will be zero. If the occurrence of such a maximum or minimum is to be determined and located, the derivative is equated to zero, thus giving the condition for which the maximum and minimum exists. This procedure is of considerable value in numerous engineering calculations, particularly those involving optimization, since the location of a maximum or minimum is, frequently, of significant importance (Prochaska and Theodore 2018) and (Theodore and Behan 2018). This is particularly true in public health studies since the optimum condition often needs to be determined so that maximum prevention and/or minimum negative effects will result.

In Figure 16.6, the curve ABCDE represents the function $y = f(x)$. If values of y in the neighborhood of points A and C are all less than the values of y at A and C, the function is said to go through a maximum at these points. Similarly, because the values of y in the neighborhood of B are all greater than the value of B, the function is said to go through a minimum at B. This curve emphasizes that the terms "maximum" and "minimum" do not necessarily denote the greatest and least possible values a function may assume.

As noted above, it is evident from a geometrical perspective that the slope of the curve is zero at the maximum and minimum points, and since the slope is given by the derivative dy/dx, these points may be determined and located through

Figure 16.6 Maximum and minimum of a function with derived curve. Courtesy of Abhishek Garg, Adapted from Shaefer and Theodore 2007.

the solution of the equation $dy/dx = 0$. The roots of this equation merely locate the maximum and minimum points and cannot distinguish between them. Furthermore, the condition locates such points as E. One method of distinguishing between points A, C, and E is to calculate values of $f(x)$ in their immediate vicinity. These are known as points of inflection. A convenient rule of these relations is as follows:

$$\text{Maximum: } dy/dx = 0, \frac{d^2y}{dx^2} < 0$$

$$\text{Minimum: } dy/dx = 0, \frac{d^2y}{dx^2} > 0$$

On differentiating and equating to zero, one obtains the condition which determines the optimum value of the variable under consideration.

Note: Calculations to determine the optimum conditions from the point of view of costs and monetary return are termed economic balances. The basic criterion of the true optimum is maximum return on the investment, but the problem can frequently simplify to one of determining the minimum cost, the maximum production from a piece of equipment, the minimum power, etc.

Illustrative Example 16.4   PANDEMIC MODELING II

One of the major challenges in developing a reliable, predictive model of future infection rates is determining when a reliable predictor can be developed. Ideally a representative predictor equation would be developed as soon as possible after the start of an outbreak. The trends of the full data set may be difficult to predict based on a sparse initial data set, and this example explores how much data is enough from Illustrative Example 16.2 to generate a reliable predictor equation based on a more complete data set. Assuming a predictive model is developed during the course of the disease outbreak indicated in Table 16.3, answer the following questions.

1. *Develop a mathematical model that describes the weekly number of new NAHUW4 infections (NI) as a function of time (T) using data from Weeks 1 through 6.*
2. *Develop a mathematical model that describes the weekly number of new NAHUW4 infections (NI) as a function of time (T) using data from Weeks 1 through 11.*
3. *Develop a mathematical model that describes the weekly number of new NAHUW4 infections (NI) as a function of time (T) using data from Weeks 1 through 15.*
4. *Estimate approximately when the NI-T curve will reach the maximum rate based on the various modeling results in questions 1 through 3.*

5. *Estimate when the weekly NI will be reduced to 10% of the maximum value based the various modeling results in questions 1 through 3.*

6. *Discuss the implications of these findings in terms of reliable model development during a pandemic*

Solution

Using the same model as in Illustrative Example 16.2, the three models developed from a subset of the data in Table 16.3 are shown in Table 16.4, with the results of questions 1 through 5, along with the summary results from Illustrative Example 2 using the complete data set.

Figure 16.5 shows the results of the various models generated from each subset of complete data, along with the model generated from the complete data set from Illustrative Example 2. As indicated from Table 16.4 and Figure 16.7, the more

Table 16.4 Early New Infections (NI/100 People) Predictors versus Time Since Beginning of an Outbreak Based on Available Data Set

Available Data Set	a_0	a_1	Time to Peak Case Rate, Weeks Since Outbreak Started	Maximum Case Rate, NI/100	Time to 10% Peak Case Rate, Weeks Since Outbreak Started
Week 0 to Week 6	0.3929	40.02	7	325	27.0
Week 0 to Week 11	0.2743	45.01	10	94	54.3
Week 0 to Week 15	0.2241	62.89	12	178	54.4
Week 0 to Week 18	0.2227	63.42	12	180	55.0

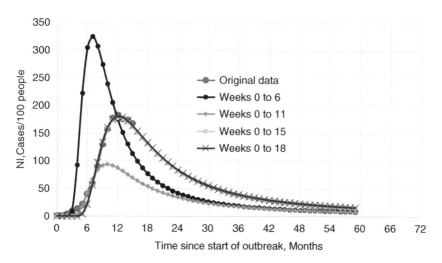

Figure 16.7 NAHUW4 infection rate data extrapolation based on data availability. Courtesy of Abhishek Garg, Adapted from Theodore and Dupont (2012).

complete a data set available, the more representative the model predictions. This is especially true the closer the data set is to including the first inflection point and the actual peak infection rate. With inclusion of the first inflection point (Week 0 to Week 11 data), the peak rate is significantly underestimated in this example, but the time for the epidemic to begin to subside is predicted relatively accurately. Once the peak infection rate is included in the model input (Week 0 to Week 15 data), the model prediction is quite representative of the more complete data set available 3 full weeks later. These results also suggest how important it is to make infection rate data quickly available and how critical it is for proper long-term responses to update these predictive models on essentially a continuous basis.

16.8 CHAPTER SUMMARY

- Life, as we knew it, changed forever with the arrival of COVID-19 (i.e. SARS-CoV-2 virus)
- More must be done in preparation for future pandemics. For decades, infectious disease specialists and scientists have warned that the influenza virus is capable of mutating into a more dangerous variation.
- Infectious diseases, once conquered through the use of antibiotics and other therapies, are reemerging as public health threats by becoming resistant to the very treatments once used to defeat them.
- The proper response requires an understanding of all the data that is available and a plan that can provide the practitioner with the information needed to respond quickly and properly during a health crisis.
- The proposed health data models have successfully demonstrated the ability to reasonably predict outcomes. They should remain a valuable tool to engineer and public health officials for not only predictive purposes but also for policy planning and implementation.
- Mathematical models are important for guiding public health measures, and data from epidemiological modeling analyses should be considered as early as possible during a crisis.
- Models can play an important role in the analysis of public health data.

16.9 PROBLEMS

1 Describe the correct way to write a pandemic response plan.

2 Comment on the plots of the equations in Table 16.3.

3 Refer to Illustrative Example 16.2. Employing the COVID-19 data provided, determine the coefficients for the equation

$$y = a_0(x_1/a_1)^{a_2[1-(x/a_1)^{a_2}]}$$

4 The temperature $(T,^\circ F)$ variation with time (t, hrs) in the emergency room of a hospital treating COVID-19 patients at 6:00 AM for a 6 hour period immediately after startup has been theoretically determined by RAT (Ricci and Theodore) Thermodynamic Consultants (a subsidiary of RAT Associates) to be

$$T = 2.2\,(1 + 1.7\sqrt[4]{t}); \qquad t = 0 - 6\,hr$$

During the last 6 years of operation, the room temperature has never exceeded $56^\circ F$, a temperature above which the COVID-19 patients would be adversely affected. As a recently hired chemical engineer at the hospital you have been assigned the task to determine whether the RAT model predicts an excursion will occur. You have also been asked to determine the maximum predicted temperature and its corresponding time during the *6:00AM* to noon period.

REFERENCES

Bertozzi, A.L., Franco, E., Mohler, G. et al. (2020). The challenges of modeling and forecasting the spread of covid-19. *Proceedings of the National Academy of Sciences* 117 (29): 16732–16738. https://doi.org/10.1073/pnas.2006520117.

Bhatty, V., Butron, S., and Theodore, L. (2017). *Introduction to Engineering: Fundamentals, Principles, and Calculations.* East Williston, NY: Theodore Tutorials and Butte, MT: Montana State University, Department of Environmental Engineering Graduate Division.

Centers for Disease Control and Prevention. (2022). CDC Covid Data tracker. Centers for Disease Control and Prevention. Retrieved March 25, 2022, from https://covid.cdc.gov/covid-data-tracker/

Centers for Disease Control and Prevention. (2018a, December 4). *Our history - our story. Centers for Disease Control and Prevention.* Retrieved November 28, 2021, from https://www.cdc.gov/about/history/index.html

Centers for Disease Control and Prevention (CDC). (2018b, March 21). *History of 1918 flu pandemic. Centers for Disease Control and Prevention.* Retrieved October 19, 2021, from https://www.cdc.gov/flu/pandemic-resources/1918-commemoration/1918-pandemic-history.htm.

Center on Budget and Policy Priorities. (2022, February 10). *Tracking the COVID-19 economy's effects on food, housing, and employment hardships.* Retrieved March 25, 2022, from https://www.cbpp.org/research/poverty-and-inequality/tracking-the-covid-19-economys-effects-on-food-housing-and

Davies, N.G., Kucharski A.J., Eggo R.M., Gimma A., (2016). CMMID COVID-19 working group, Edmunds W.J., *"The effect of non-pharmaceutical interventions on COVID-19 cases, deaths and demand for hospital service in the UK: a modelling study".* https://doi.org/10.1101/2016.04.01.20049908

Geng, X., Katul, G.G., Gerges, F. et al. (2021). A kernel-modulated SIR model for covid-19 contagious spread from county to continent. *Proceedings of the National Academy of Sciences* 118 (21): https://doi.org/10.1073/pnas.2023321118.

Krikorian, M. (1982). *Disaster and Emergency Planning*. Loganville, AL: Institute Press.

McGee, R.S., Homburger, J.R., Williams, H.E. et al. (2021). Model-driven mitigation measures for reopening schools during the COVID-19 pandemic. *PNAS* 118 (39): https://doi.org/10.1101/2021.01.22.21250282.

Moghadas, S.M., Shoukat, A., Fitzpatrick, M.C. et al. (2020). Projecting hospital utilization during the COVID-19 outbreaks in the United States. *Proceedings of the National Academy of Sciences* 117 (16): 9122–9126. https://doi.org/10.1073/pnas.2004064117.

Prochaska, C. and Theodore, L. (2018). *Introduction to mathematical methods for environmental engineers and scientists*. Beverly, MA: Scrivener Publishing.

Shaefer, S. and Theodore, L. (2007). *Probability and Statistics Applications in Environmental Science*. Boca Raton, FL: CRC Press/Taylor & Francis Group.

Theodore, L. (2009). *Air Pollution Control Equipment Calculations, Hoboken, NJ*. John Wiley & Sons.

Theodore, L. and Behan, K. (2018). *Introduction to Optimization for Environmental and Chemical Engineers*. Boca Raton: FL: CRC Press, Taylor & Francis Group.

Theodore, L. and Dupont, R.R. (2012). *Environmental Health and Hazard Risk Assessment: Principles and Calculations*. Boca Raton, FL: CRC Press/Taylor & Francis Group.

Theodore, L. and McGuinn, Y. (1992). *Heath, Safety and Accident Management; Industrial Application. A Theodore Tutorial*, East Williston, NY: Theodore Tutorials (originally published by the USEPA/APTI, RTP, NC).

Theodore, J. and Theodore, L. (2021). *personal notes*. New Haven, CT.

Thompson, R.N. (2020). Epidemiological Models are Important Tools for Guiding COVID-19 Interventions. *BMC Medicine* 18 (1): https://doi.org/10.1186/s12916-020-01628-4.

Wise, J. (2022). *Covid-19: Global Death Toll May be Three Times Higher than Official Records*. Study Suggests: *BMJ* https://doi.org/10.1136/bmj.o636.

World Health Organization (WHO) (2020, March 11). *WHO director-general's opening remarks at the media briefing on COVID-19 - 11 March 2020*. World Health Organization Retrieved March 18, 2022, from https://www.who.int/director-general/speeches/detail/who-director-general-s-opening-remarks-at-the-media-briefing-on-covid-19---11-march-2020.

World Health Organization (WHO) (2019). *Non-pharmaceutical public health measures for mitigating the risk and impact of epidemic and pandemic influenza*. World Health Organization Retrieved March 11, 2022, from https://www.who.int/publications/i/item/non-pharmaceutical-public-health-measuresfor-mitigating-the-risk-and-impact-of-epidemic-and-pandemic-influenza.

17

Optimization Procedures

As one might suspect, the term optimization has come to mean different things to different people. It has come to mean different things for different applications as well, i.e., it could involve a simple two-step calculation or one that requires the use of a detailed numerical method. To take this a step further, the authors were undecided as to how to present optimization in this chapter. After much deliberation and meditation, it was decided to present both qualitative and quantitative material. This decision was primarily influenced by the desire that this be a chapter that not only introduces optimization but also one that addresses the topic quantitatively.

Merriam-Webster defines optimization as "an act, process, or methodology of making something (such as a design, system, or decision) as fully perfect, functional, or effective as possible" (Merriam-Webster 2021). More succinctly, the optimization problem has been described by Aris as "getting the best you can out of a given situation" (Aris 1964).

Problems amenable to solution by mathematical optimization techniques generally have one or more independent variables whose values must be chosen to yield a viable solution and measure of "goodness" available to distinguish

A Guide to Virology for Engineers and Applied Scientists: Epidemiology, Emergency Management, and Optimization, First Edition. Megan M. Reynolds and Louis Theodore.
© 2023 John Wiley & Sons, Inc. Published 2023 by John Wiley & Sons, Inc.

between the many viable solutions generated by different choices of these variables. Mathematical optimization techniques are also used for guiding the problem solver to that choice of variables that maximizes the aforementioned "goodness" measure (e.g., profit) or that minimizes some "badness" measure (e.g., cost).

There are many important areas for which the application of mathematical optimization techniques can have a significant impact in the healthcare arena. Examples include:

- Generation of best functional representations (e.g. curve fitting).
- Design of optimal control systems for hospital operations.
- Determining the optimal length of a purifier in a pharmaceutical manufacturing line.
- Determining the optimal diameter of a magnetic resonance imaging (MRI) unit.
- Finding the best adjuvants for use in vaccine manufacturing.
- Designing equipment materials for construction.
- Generating surgical schedules in a hospital to minimize delays.
- Assessing organizational readiness.

In practice, this process or operation is required in the solution of many public health problems that can involve the maximization or minimization of a mathematical function. As one might suppose, many of those applications concern economic considerations (Theodore and Behan 2018).

17.1 THE HISTORY OF OPTIMIZATION

As discussed in Chapter 12, the subject of mathematics encompasses the study of relationships among quantities, magnitudes, and properties of logical operations by which unknown quantities, magnitudes, and properties may be deduced. *Primitive counting systems were almost certainly based on using the fingers on both hands, as evidenced by the predominance of the number ten as the base for many number systems employed today.* In the past, mathematics was regarded as a science involving geometry, numbers, and/or algebra. Toward the middle of the nineteenth century, however, mathematics came to be regarded increasingly as the science of relations, or as the science that draws necessary conclusions.

Interestingly, the mathematics of the late nineteenth and twentieth centuries is characterized by an interest in unifying elements across numerous fields of mathematical endeavor, especially in algebraic systems, linear algebra, modern algebra and vectors, matrices, logic, Venn and Euler diagrams, relations and functions, probability and statistics, linear programming, computer programming and of course, optimization.

As to the origin of optimization, it depends on who responds because there are so many aspects of optimization of interest to the practitioner. For example, some claim it was Thomas Edison when he developed a long-lasting, high-quality light bulb in the 1870s. His success was primarily the result of an extensive trial-and-error search for the optimum filament material. A few now refer to it as the Edisonian approach. One of the authors refers to it as the perturbation approach; on occasion, he has modestly termed it the *Theodore approach* (Theodore, personal notes 1964).

No discussion of optimization history would be complete without a discussion of the historic role of computers. In the early days of computing, they were viewed as fast calculators or special slide rulers, and their main role was to replace complicated mathematical operations which were, until then, done by hand or the calculator. Computers, at their most basic level are devices for manipulating symbolic information.

In the late 1960s, Famularo offered the following thoughts (Famularo, personal communication, 1967). "The digital computer can be viewed as a high-speed calculator, which, with the availability of subroutines and a compiling language, is able to perform many mathematical operations such as add, multiply, generate analytic functions, logic decisions, etc." Although details of coding are not in the scope of this chapter, there is an intermediate step between the equations and the coding program, and that is to arrange the computing procedure in block diagram or information flow form. The operation of each block is described inside the block, and the computer will perform the instructions around the loop until the condition in the decision block is satisfied. As the number of equations to be solved increases, the time required to obtain a solution also increases.

The rapid speed of development in computing has meant astonishing increases in processing ability and speed that could not have been imagined even a decade ago. This fact has had a tremendous impact on engineering design, scientific computation, and data processing. The ability of computers to handle large quantities of data and to perform the mathematical operations described above at tremendous speeds permits the analysis of many more applications and more engineering variables than could possibly be handled before. Calculations previously estimated in lifetimes of computation time are currently generated in microseconds, or even nanoseconds.

An example of the utility of optimization in public health is in analyzing the enormous and complex issue of healthcare availability in underserved populations in the United States such as rural areas and low-income neighborhoods. The topic of healthcare accessibility is multifaceted, with many variables which must be integrated for the purposes of gaining an accurate depiction of the scope of the problem. Optimization methods can be used to evaluate options to improve the distribution and relocation of healthcare resources in order to maximize physician coverage, maximize access quality, and minimize distances that patients must travel for proper medical care (Wang 2012).

17.2 THE SCOPE OF OPTIMIZATION

One can conclude that the theory and application of optimization is mathematical in nature, and it typically involves the maximization or minimization of a function (usually known), which represents the "performance" of some "system." This is carried out by determining the values for those variables which cause the function to yield an optimal value.

As previously mentioned, optimization is widely utilized in healthcare to reduce costs and streamline services. Hospitals can optimize metrics such as staffing allocation, capital investment, satellite clinics and offices, etc., to improve both efficiency and quality of care. To incentivize healthcare providers to maintain best practices at every level, the Centers for Medicare and Medicaid Services (CMS) developed the Hospital Quality Initiative (HQI)

> "… to improve the quality of care that hospitals provide and to distribute clearly defined and objective data about hospital performance. Among other benefits, this initiative,"…"encourages consumers to discuss and make more informed decisions on how to obtain the best hospital care, provides incentives for hospitals to improve care, and emphasizes public accountability. CMS uses a variety of tools to encourage improvements in the quality of care delivered by hospitals." (CMS 2021).

Perhaps the most important tool employed in optimization by public health professionals is linear programming. Linear programming consists of a linear multivariable function, which is to be optimized (maximized or minimized), subject to a particular number of constraints, to be discussed shortly. The constraints are normally expressed in *linear* form. Integer linear programming refers to optimization problems in which at least some of the variables must assume integer values. The reader should note that the terms linear programming and nonlinear programming are essentially similar from an applications perspective. There were problems around the middle of the last century because of some difficulties that arose in attempting to solve nonlinear programming problems. However, the arrival of the modern-day computer and sophisticated software (e.g. Excel) have removed those problems. Unless a solution is presented graphically in this chapter, it will be obtained directly from Excel so that the reader need not be concerned with whether a system's describing equations are linear or nonlinear. Thus, details regarding the solution methodology for both linear programming and nonlinear programming problems, and any corresponding illustrative examples to follow, will not be presented. They are beyond the scope of this chapter.

One of the major responsibilities in optimization is to construct a correct objective function to be maximized. In a simple environmental problem in healthcare, the objective function might be the cost associated with the disposal of waste for a hospital that must be separated into two contaminated categories as "medical" and "non-medical." One optimization problem could be to minimize costs from waste disposal by employing collection, separation, and storage systems in order to both maximize the hospital system's capacity and minimize the number of specialized medical waste, (which is much more costly than nonmedical waste), pickup each week. The problem might also be rephrased into the combination of the quantities of the two wastes streams that will cost the least while being subject to the constraints connected with each hospital's waste processing capabilities and different disposal requirements imposed on each class of waste.

The objective function may be quite simple and easy to calculate in some applications, or it may be complicated and difficult to not only calculate but also specify and/or describe. The objective function may also be very illusive due to the presence of conflicting or dimensionally incompatible sub-objectives; for example, one might be asked to minimize the overall costs for the aforementioned hospital system that not only maximizes waste storage space and minimizes disposal costs, but also prevents waste from building up at any one location so that everyday operations are not impacted. Thus, it may not be always possible to quantify an objective function and enable one to use any of the mathematical optimization procedures available.

There are a large number of mathematical optimization methods available in practice. Some of the simple ones are listed below

- A function of one variable with no constraints
- A function of two variables with no constraints
- A function of more than two variables with no constraints
- Simple perturbation schemes
- A function of one variable with constraints
- A function of two variables with constraints
- A function of more than two variables with constraints

Several of the above methods will be reviewed later in this chapter.

Some optimization problems can be divided into parts, for which each part is then optimized. And, in some instances, it is possible to attain the optimum for the original problem by simply realizing how to optimize these constituent parts. This process is very powerful, as it allows one to solve a series of smaller, easier problems rather than a large one. One of the best-known techniques to attack such problems is *dynamic programming*. This approach is characterized by a process which is performed in stages, such as in manufacturing processes. Rather than solving the problem as a whole, dynamic programming optimizes one stage at a time to produce an optimal set of decisions for the whole process. Although dynamic programming has applicability in some systems and processes, it is not reviewed in this chapter.

Finally, most engineering optimization applications involve economics – maximizing profit or minimizing cost, or both. Public health applications can include vaccine preparation and costs, virus side effects, minimizing health and safety risks, addressing hospital economic concerns, etc. The reader should also not lose sight that most real-world industry applications involving optimization usually require simple solutions.

One important use of optimization in healthcare is reflected in the management of patient flow during the early months of the COVID-19 epidemic. In 2020, when hospitals in New York City were first inundated and overwhelmed with the COVID-19 patients, medical providers were forced to delay or cease other procedures and elective surgeries. As the initial wave subsided, healthcare systems attempted to maximize patient flow in order to improve patient care. This ultimately improved survival in subsequent waves.

17.3 CONVENTIONAL OPTIMIZATION PROCEDURES

Pharmaceutical drug development and manufacturing naturally lend themselves to systematic optimization. To minimize costs while maximizing product quality and output, engineers and applied scientists employ these approaches in every stage of development from choosing the best drug candidates to manufacturing process design and production scale. For established optimization procedures, there are four main approaches, which can be described as direct methodologies (Theodore 1961):

- Perturbation methods
- Search methods
- Graphical approaches
- Analytical methods

Introductory details on these topics follow, noting that linear programming is addressed later in this chapter.

Perturbation methods involve a guessing game when attempting to solve an optimization problem. High-powered mathematics is usually not involved.

The same can be said for any trial-and-error method whether it be systematic or not. Numerous elementary search methods are available (Prochaska and Theodore 2018; Theodore and Behan 2018). Graphical approaches, as one might suppose, involve graphing available data and calculations in an attempt to arrive at an "optimization" solution. Analytical methods often employ ordinary and partial derivatives in obtaining a solution.

As noted, these topics have been defined as direct approaches, while linear programming is referred to as an indirect approach. As one might assume, indirect methods are generally preferred. There are, however, numerous other methods for solving optimization problems. These can involve the aforementioned procedures that simply require a yes or no answer, selecting the best option of the two alternatives, selecting the best of more than two options, and so on.

The selection of the independent variable(s) is often set in academic problems involving standard optimization calculations. The choice of these variables is usually based on past experience or sound engineering judgment. Reducing the number of variables to only what is critical to the solution reduces the problem to one that does not involve the need to resort to any sophisticated optimization mathematical methods. Irrelevant data is usually and understandably neglected. One should also note that these optimization problems, like many engineering problems, usually involve a simple two-step procedure:

- Developing a describing equation (or equations) for the system to be optimized.
- Solving the equation or equations.

The first step falls within the domain of the engineer. Step 2 traditionally fell in the domain of both the engineer and mathematician. However, the second step is essentially no longer a concern to many engineers with the advent of computers, particularly as it applies to contemporary optimization, a topic that is addressed in Section 17.5.

17.4 ANALYTICAL FOMULATION OF THE OPTIMUM

When more than one independent variable is involved in determining a function, e.g., profitability, a more elaborate treatment may be necessary other than what was employed in the past. For a system of $x_1, x_2, x_3, \ldots x_n$ independent variables, the function y (hereafter referred to as the objective function) will depend on these variables.

$$y \text{ (profit)} = y = f(x_1, x_2, x_3, \ldots x_n) \tag{17.1}$$

Instead of profit, y may be alternatively expressed in other convenient and equivalent forms. The problems are to specify $x_1, x_2, x_3, \ldots x_n$ so that y will be at a maximum, making y a profit function. In general, this will involve some type of trial-and-error calculation.

It is possible in some cases to set up an explicit function that will relate the objective function y to the independent variables. Then differentiation will result in a series of partial derivatives $\partial y/\partial x_1$, $\partial y/\partial x_2$, etc. Setting these partial derivatives equal to zero will result in the same number of simultaneous equations as the variables involved. These equations, often subject to "limitations" and "constraints," can be solved for the corresponding optimum values of these variables. The necessary conditions for a maximum at a point where the first derivatives become zero include restrictions on the second derivatives, continuity of the function involved throughout the range of independent variables, and absence of optimum conditions at limiting values of one or more of the variables. These conditions can restrict the applicability of this method of analysis (Happel 1958).

The reader was introduced to the concept of maximum and minimum earlier in Chapter 12. For the case of a two-variable function, i.e., $y = f(x_1, x_2)$, a point $x_1 = a, x_2 = b$ will have a maximum value $y = f(x_1, x_2)$ if the following conditions hold

$$\frac{\partial y}{\partial x_1} = \frac{\partial y}{\partial x_2} = 0 \tag{17.2}$$

with

$$\frac{\partial^2 y}{\partial x_1^2} < 0 \text{ and } \frac{\partial^2 y}{\partial x_2^2} < 0 \tag{17.3}$$

and

$$\left(\frac{\partial^2 y}{\partial x_1^2}\right)\left(\frac{\partial^2 y}{\partial x_2^2}\right) > \left(\frac{\partial^2 y}{\partial x_1 \partial x_2}\right)^2 \tag{17.4}$$

Similarly, if $\partial^2 y/\partial x_1^2, \partial^2 y/\partial x_2^2 \geq 0$ and Eq. (17.4) holds, there will be a minimum. But if

$$\left(\frac{\partial^2 y}{\partial x_1^2}\right)\left(\frac{\partial^2 y}{\partial x_2^2}\right) > \left(\frac{\partial^2 y}{\partial x_1 \partial x_2}\right)^2 \tag{17.5}$$

then the point will be a "saddle" point. The case where

$$\left(\frac{\partial^2 y}{\partial x_1^2}\right)\left(\frac{\partial^2 y}{\partial x_2^2}\right) = \left(\frac{\partial^2 y}{\partial x_1 \partial x_2}\right)^2 \tag{17.6}$$

is open and may be a maximum, a minimum, or neither (Theodore and Behan 2018).

Now consider an objective function specified as

$$f(x_1, x_2) = x_1^2 - 2x_1 + x_2^2 + x_1 x_2 + 1 \tag{17.7}$$

To determine if the function has a maximum or a minimum, proceed as follows:

$$\frac{\partial f}{\partial x_1} = 2x_1 - 2 + x_2 \tag{17.8}$$

$$\frac{\partial f}{\partial x_2} = 2x_2 + x_1 \tag{17.9}$$

When Eqs. (17.8) and (17.9) are set equal to zero and solved simultaneously, the following solution results.

$$x_1 = -\frac{4}{3} \tag{17.10}$$

$$x_2 = \frac{2}{3} \tag{17.11}$$

Since both second derivatives are positive, i.e.,

$$\frac{\partial^2 f}{\partial x_1^2} = 2 \tag{17.12}$$

$$\frac{\partial^2 f}{\partial x_2^2} = 2 \tag{17.13}$$

the above solutions for x_1 and x_2 represent a *minimum* with

$$y = f(x_1, x_2) = 5 \tag{17.14}$$

Consider now a traditional mathematician's exercise to determine the maximum or minimum of a function dependent on three variables (Bhatty et al. 2017)

$$P = 8x + 6y + 8z + xy + xz + yz - 2x^2 - 3y^2 - 4z^2 \tag{17.15}$$

Here

$$\frac{\partial P}{\partial x}\bigg|_{y,z} = 8 + y + z - 4x = 0 \tag{17.16}$$

$$\frac{\partial P}{\partial y}\bigg|_{x,z} = 6 + x + z - 6y = 0 \tag{17.17}$$

$$\frac{\partial P}{\partial z}\bigg|_{x,y} = 8 + y + x - 8z = 0 \tag{17.18}$$

The solution to the above three linear simultaneous equations can be shown to be

$$x = -4.62; y = 0.940; z = 0.54 \tag{17.19}$$

for which

$$P = -84.5 \tag{17.20}$$

To determine whether the above P is a maximum or minimum, calculate the three second order partial derivatives.

$$\frac{\partial^2 P}{\partial x^2} = -4 \tag{17.21}$$

$$\frac{\partial^2 P}{\partial y^2} = -6 \tag{17.22}$$

$$\frac{\partial^2 P}{\partial z^2} = -8 \tag{17.23}$$

Since the three derivatives are negative, the above value of P (-84.5) is a *maximum*, i.e., any other P for a different x, y, and z would produce a value of P less than (or more negative) than -84.5. The "proof" is left as an exercise for the reader. However, if one were to select values of x, y, and z straddling the supposed minimum set of values, P should be below -84.5. For example, the value of P for $x = -5$, $y = 1$, and $z = -0.5$ is -95.0.

17.5 CONTEMPORARY OPTIMIZATION: CONCEPTS IN LINEAR PROGRAMMING

This section introduces linear programming as a technique for optimization which has evolved into a useful procedure for the technical community. As noted earlier, the title results from the assumption inherent in the method that linear relationships describe the system under consideration. It deals with the determination of an optimum solution of a problem expressed in linear relationships where there are a large number of possible solutions. In general, the methods employed are trial-and-error, where the procedure systematically follows a mathematical treatment which minimizes the labor involved and ensures that a correct result has been obtained.

To apply linear programming to a problem, the relationships between variables must be expressed as a set of linear equations or inequalities. The variables must be otherwise independent and must exceed the number of equations. This may be mathematically expressed as,

$$\sum_i a_{ij} x_j = b_j; \ ^\circ (i = 1, 2, \ldots, m; \ j = 1, 2, \ldots, n; n > m) \tag{17.24}$$

where the sum of the coefficient a_{ij} times the value of each variable x_j equals the requirement b_j in the ith equation. As noted, the number of variables must exceed the number of equations so that an infinite number of solutions is possible. The best one is selected. The relationship in Eq. (17.24) is often expressed as inequalities. "Dummy" additional variables could be introduced to convert these inequalities into appropriate equations (Theodore and Behan 2018; Prochaska and Theodore 2018). Another requirement is that no quantity can be negative. Thus, mathematically

$$x_j \geq 0 \text{ and } b_j \geq 0 \tag{17.25}$$

This requirement will ensure that only useful answers are obtained.

The solution of a problem by linear programming also requires, for example, a profit or objective function to be maximized (or minimized). This requirement is stated as

$$P = \sum_i c_j x_i = \text{maximum (or minimum)} \tag{17.26}$$

where c_j is the profit (or cost) per unit x_i used. The solution is carried out by an iterative process, which at each stage of the calculation assigns either zero or positive values to all x_i variables. As the calculation proceeds, the values selected, while meeting the constraints imposed, will tend to increase the summation to be maximized at each stage. If a solution exists, a maximum (profit) case will be realized (Theodore and Behan 2018).

17.6 APPLIED CONCEPTS IN LINEAR PROGRAMMING

Consider now an example that involves minimizing an objective function, i.e., E (expenditure)

$$E = a_1 x_1 + a_2 x_2 \tag{17.27}$$

subject to two constraints

$$B = b_1 x_1 + b_2 x_2 \tag{17.28}$$

$$C = c_1 x_1 + c_2 x_2 \tag{17.29}$$

In addition

$$x_1 \geq 0 \tag{17.30}$$

$$x_2 \geq 0 \tag{17.31}$$

In terms of a solution, first examine the above system of equations graphically. The problem can be viewed in Figures 17.1 and 17.2. Note also that only the positive quadrant needs to be considered because Eqs. (17.30) and (17.31) must hold.

Equations (17.28) and (17.29) are now rewritten in the following form

$$x_2 = \frac{B}{b_2} - \left(\frac{b_1}{b_2}\right) x_1 \tag{17.32}$$

and

$$x_2 = \frac{C}{c_2} - \left(\frac{c_1}{c_2}\right) x_1 \tag{17.33}$$

Each appears in Figure 17.1a, b. It is apparent that the only values of x_1 and x_2 which will satisfy Eqs. (17.32) and (17.33) lie above the solid lines in Figure 17.1c

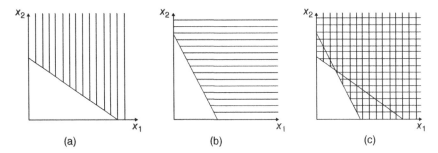

Figure 17.1 Two variable minimization problem. (a) Equation (17.28). (b) Equation (17.29). (c) Equations (17.28) and (17.29).

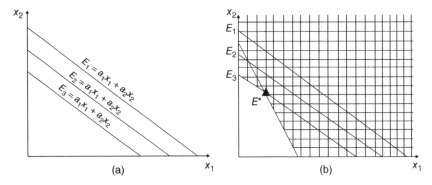

Figure 17.2 Solution to two-variable minimization problem. (a) Equation (17.27) with various values of E. (b) Figure 17.4a and Figure 17.1c superimposed.

(i.e. the shaded area). This shaded area is termed the constraint. Now place in Figure 17.2a a series of straight lines or contours corresponding to the expenditure Eq. (17.27) with a different value of E used for each line. Superimposing Figure 17.2a onto Figure 17.1c produces Figure 17.2b. Of all these lines, the smallest E for which x_1 and x_2 remain in the specified region in Figure 17.1c will be the minimum expenditure. This will yield that proportion of x_1 and x_2 which minimize the cost. Stated in other words, one desires the lowest-value contour having some point in common with the constraint set. For the particular values of a_1 and a_2 chosen for Eq. (17.27) the minimum point is seen to correspond to E where $E_3 = E^*$. Further if the constants c_1 and c_2 in Eq. (17.27) are changed, it is possible that the minimum point would occur elsewhere (Bhatty et al. 2017).

The point that is the minimum is determined by constants a_1 and a_2. It should be pointed out that if the expenditure equation was nonlinear the contour lines would be curved and if the constraint equations were nonlinear, the constraint lines would not be straight. It should be intuitively obvious that locating the

Figure 17.3 Possible solution to two variable maximization problem.

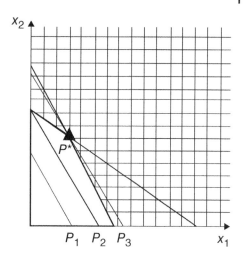

minimum point is more complicated in this case. The situation is essentially reversed for a maximization problem, i.e., for a profit P. This is demonstrated in Figure 17.3.

The above methodology can be extended to more than two variables and more than two constraints. The following would apply for three variables (x_1, x_2, x_3) and four constraints (b, c, d, e).

$$P = a_1x_1 + a_2x_2 + a_3x_3 \tag{17.34}$$

and

$$B = b_1x_1 + b_2x_2 + b_3x_3 \tag{17.35}$$

$$C = c_1x_1 + c_2x_2 + c_3x_3 \tag{17.36}$$

$$D = d_1x_1 + d_2x_2 + d_3x_3 \tag{17.37}$$

$$E = e_1x_1 + e_2x_2 + e_3x_3 \tag{17.38}$$

and

$$x_1 \geq 0 \tag{17.39}$$

$$x_2 \geq 0 \tag{17.40}$$

$$x_3 \geq 0 \tag{17.41}$$

Naturally, the above can be rewritten for n objective functions and m constraints.

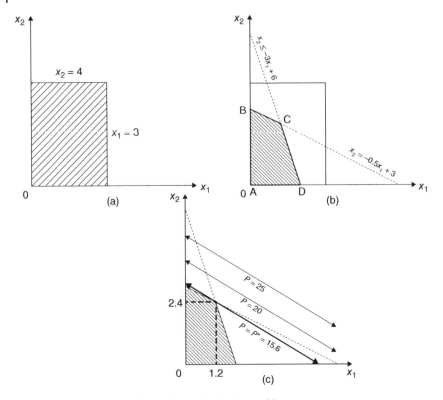

Figure 17.4 Graphical solution for maximization problem.

Suppose you have been requested to obtain the maximum value of P subject to the following conditions. The objective function is

$$P = 3x_1 + 5x_2 \tag{17.42}$$

and the constraints are given by

$$x_2 = -0.5x_1 + 3 \tag{17.43}$$

$$x_2 \le -3x_1 + 6 \tag{17.44}$$

$$x_1 \le 3 \tag{17.45}$$

$$x_2 \ge 4 \tag{17.46}$$

Once again, the solution is first provided in graphical form. Refer to Figure 17.4.

Conditions in Eqs. (17.45) and (17.46) require that the solutions, i.e., the values x_1 and x_2 that will maximize P, lie within the rectangle provided in Figure 17.4a. Conditions in Eqs. (17.43) and (17.44) require that the solution be within the area

ABCD in Figure 17.4b, Figure 17.4c results. When the objective function equation is superimposed on Figure 17.4b, Figure 17.4c results. The maximum P is located at $x_1 = 1.2$ and $x_2 = 2.4$ for which $P = 15.6$. Excel provides the same solution (Bhatty et al. 2017).

17.7 APPLICATIONS

The following four illustrative examples complement the presentation for this chapter on optimization.

Illustrative Example 17.1 OPTIMIZATION IN HEALTHCARE
Qualitatively discuss optimization and provide examples of how optimization could be used within healthcare-related fields.

Solution
As discussed, optimization is viewed by many as a tool in decision-making. It often aids the selection of values that allow practicing engineers and applied scientists to better solve a problem. In its elementary and basic form, optimization is concerned with the determination of the "best" solution to a given problem. Some examples for which this process is required is the maximization (or minimization) of a given function, the selection of an optimal vaccine dose, the scheduling of surgeries in a hospital, and the design of optimal layouts in a doctor's office.

One important area for the application of mathematical optimization techniques is in public health. In public health, one might be interested in vaccine potency, efficiency, cost, etc. Sophisticated optimization techniques are also routinely used in the design of pharmaceutical manufacturing plants and in hospitals to ensure ideal patient flow.

Illustrative Example 17.2 CONSTRAINED OBJECTIVE FUNCTION
The dependent variable (function) y is equal to five times x_1 minus four times x_2, where x_1 and x_2 are positive independent variables. The sum of x_1 and x_2 are both constrained by 0 and 50, and their cross product is limited to the range of 0 and 126. Define this constrained mathematical system.

Solution
The (objective) function is:

$$y = 5x_1 - 4x_2$$

The constraints are as follows:

$$0 \leq x_1 + x_2 \leq 50$$

$$0 \leq x_1 x_2 \leq 126$$

$$x_1 \geq 0 \text{ and } x_2 \geq 0$$

Illustrative Example 17.3   OPTIMIZING PROFITS AND MINIMIZING RISK

A pharmaceutical company manufactures two antiviral medications, A and B, with an accompanying unit profit of P_A and P_B, respectively. A minimum number of N medications can be produced on a daily basis. The fractional defective rate of producing A and B during the manufacturing process are a and b, respectively. The daily production rate is x_A and x_B, respectively. Any defect during production reduces the profits associated with A and B by c and d, per defect, respectively. Determine the equations that model the optimal production rate of A and B in order to maximize daily profits, P.

Solution

As noted, x_A and x_B are the daily production rates of A and B, respectively. Therefore,

$$P = x_A P_A + x_B P_B - x_A ac - x_B bd$$

with

$$x_A + x_B \geq N$$

and

$$x_A, x_B \geq 0$$

Illustrative Example 17.4   COST/BENEFIT ANALYSIS

A large hospital is considering expanding their intensive care unit (ICU) after subsequent overcapacity issues during the COVID-19 pandemic. The expansion would increase the number of ICU beds by 10 but decrease the number of (non-ICU) beds on the medical floor by 4. The alternative would be to open a new ICU in what is now a staff parking lot. This new ICU would increase bed capacity by 12 with no loss to the medicine department. The construction of an expansion would require less capital upfront, but the income would be higher with a new, larger ICU. As a member of the finance department, you have been asked for an optimization study to determine the best course of action. Based on the economic data in the following table, select the option that will yield the optimum annual profit. Calculations should be based on an interest rate of 12% and a lifetime of 12 years for both units (Theodore and Ricci 2010).

Costs/credits	Expanded ICU	New ICU
Capital investment	$1 875 000	$2 175 000
Overhead costs	$750 000	$800 000
Total capital	**$2 625 000**	**$2 975 000**
Installation	$1 575 000	$1 700 000
Operation (annual)	$400 000	$550 000
Maintenance (annual)	$650 000	$775 000
Income (annual)	$2 000 000	$2 500 000

Solution

This is a relatively simple economic optimization problem. Calculate the capital recovery factor, CRF (Shaefer and Theodore 2007):

$$CRF = \frac{(0.12)(1+0.12)^{12}}{(1+0.12)^{12}-1}$$
$$= 0.1614$$

Determine the annual capital and installation costs for the expanded ICU:

$$COST(EXP) = (2\,625\,000 + 1\,575\,000)(0.1614)$$
$$= \$677\,880/year$$

Determine the annual capital and installation costs for the new ICU:

$$COST(NEW) = (2\,975\,000 + 1\,700\,000)\,(0.1614)$$
$$= \$754\,545/year$$

See the table below for a comparison of costs and credits for both devices.

Costs/credits (annual)	Expanded ICU	New ICU
Total installed	$678 000	$755 000
Operation	$400 000	$550 000
Maintenance	$650 000	$775 000
Total annual cost	$1 728 000	$2 080 000
Income credit	$2 000 000	$2 500 000

Finally, calculate the profit for each on an annualized basis:

$$PROFIT(EXP) = 2\,000\,000 - 1\,728\,000 = +\$272\,000/year$$
$$PROFIT(NEW) = 2\,500\,000 - 2\,080\,000 = +\$420\,000/year$$

A new ICU should be constructed based on the above economic analysis.

17.8 CHAPTER SUMMARY

- The optimization problem has been described succinctly by Aris as "getting the best you can out of a given situation."
- Problems amenable to solution by mathematic optimization techniques generally have one or more independent variables whose values must be chosen to yield a viable solution and measure of "goodness" available to distinguish between the many viable solutions generated by different choices of these variables.
- The theory and application of optimization is mathematical in nature, and it typically involves the maximization or minimization of a function which represents the "performance" of some "system." This is carried out by the finding of values

for those variables which cause the function to yield an optimal value.

- Traditional optimization methods include perturbation methods, search methods, graphical approaches, and analytical methods.
- Contemporary optimization involving linear programming deals with the determination of an optimum solution of a problem expressed in linear relationships where there are a large number of possible solutions.

17.9 PROBLEMS

1 (a) Select a referred published article on optimization from the literature and provide a review.
(b) Provide some normal everyday domestic applications involving the general topic of optimization.
(c) Develop another (and hopefully improved) method of solving an optimization problem.

2 The dependent economic variable (function) y is equal to four times x_1 minus three times x_2, where x_1 and x_2 are positive independent variables. The sum of x_1 and x_2 are both constrained by 0 and 10, and their cross product is limited to the range of 0 and 36. Develop this constrained mathematical system.

3 A batch reactor at a pharmaceutical manufacturing site can produce either drug A, B, or C. The daily production levels of each are represented by x_A, x_B, and x_C, with an accompanying unit of profit of P_A, P_B, and P_C, respectively. Develop a model to optimize the profit P for this antiviral drug production process.

4 A profit equation for a vaccine is given by:

$$P = 162 - 2.33x_1 - 1.86x_2 - \frac{11{,}900}{x_1^2 x_2}$$

Calculate the values of x_1 and x_2 that would maximize the value of the function, P.

REFERENCES

Aris, R. (1964). *Discrete Dynamic Programming*. New York: Blaisdell.

Bhatty, V., Butron, S., and Theodore, L. (2017). *Introduction to Engineering: Fundamentals, Principles, and Calculations*. MT: Montana State University, Department of Environmental Engineering Graduate Division.

Centers for Medicare & Medicaid Services (CMS) (2021). Hospital quality initiative. https://www.cms.gov/Medicare/Quality-Initiatives-Patient-Assessment-Instruments/HospitalQualityInits (accessed 22 February 2022).

Happel, J. (1958). *Chemical Process Economics*. Hoboken, NJ: John Wiley & Sons.

Merriam-Webster (2021). Optimization. https://www.merriam-webster.com/dictionary/optimization (accessed 4 November 2021).

Prochaska, C. and Theodore, L. (2018). *Introduction to Mathematical Methods for Environmental Engineers and Scientists*. Beverly, MA: Scrivener-Wiley Publishing.

Shaefer, S. and Theodore, L. (2007). *Probability and Statistics Applications for Environmental Science*. Boca Raton, FL: CRC Press/Taylor & Francis Group.

Theodore, L. (1961). *Personal notes*. NY: East Williston.

Theodore, L. and Behan, K. (2018). *Introduction to Optimization for Chemical and Environmental Engineers*. Boca Raton, FL: CRC/Press/Taylor & Francis Group.

Theodore, L. and Ricci, F. (2010). *Mass Transfer Operations for the Practicing Engineer*. Hoboken, NJ: John Wiley & Sons.

Wang, F. (2012). Measurement, optimization, and impact of health care accessibility: a methodological review. In: *Annals of the Association of American Geographers*. Association of American Geographers. https://doi.org/10.1080/00045608.2012.657146.

Index

A Guide to Virology for Engineers and Applied Scientists: Epidemiology, Emergency Management, and Optimization, First Edition. Megan M. Reynolds and Louis Theodore.
© 2023 John Wiley & Sons, Inc. Published 2023 by John Wiley & Sons, Inc.

Printed and bound by CPI Group (UK) Ltd, Croydon, CR0 4YY

16/04/2025